Neutrino Oscillations and their Origin

The Third International Workshop on

Neutrino Oscillations and their Origin

University of Tokyo, Japan   5 – 8 December 2001

Edited by

Y. Suzuki, M. Nakahata, Y. Fukuda,
Y. Takeuchi, T. Mori & T. Yoshida

University of Tokyo, Japan

World Scientific

New Jersey • London • Singapore • Hong Kong

Published by

World Scientific Publishing Co. Pte. Ltd.

P O Box 128, Farrer Road, Singapore 912805

USA office: Suite 202, 1060 Main Street, River Edge, NJ 07661

UK office: 57 Shelton Street, Covent Garden, London WC2H 9HE

British Library Cataloguing-in-Publication Data
A catalogue record for this book is available from the British Library.

Proceedings of the Third International Workshop on
NEUTRINO OSCILLATIONS AND THEIR ORIGIN (NOON 2001)

ISBN 981-238-177-5

This book is printed on acid-free paper.

Printed in Singapore by Uto-Print

PREFACE

The aim of this workshop is to understand the neutrino oscillations more deeply and discuss the future direction of the oscillation study. In 1998 the SK collaboration has shown the evidence of neutrino oscillations in the study of the atmospheric neutrinos, and in June 2001, the result of the solar neutrino flux measurement in SK together with the SNO charged current result has also shown a strong evidence for the neutrino oscillation. Therefore, one of the stress of this workshop was made on the active discussion of the recently established solar neutrino oscillation not only by the experimental point of view but also by the phenomenological and theoretical point of views. We also covered the discussion on the recent progress on the atmospheric neutrinos, the absolute neutrino mass measurements, the on-going and future accelerator neutrino oscillations, the various particle models behind of the neutrino oscillations and the lepton flavor violations

This is the 3rd workshop in this series and the first one was held in February, 2000 at Fuji-Yoshida. This workshop was supported by the grant-in-aid of the specific fields on NEUTRINO, which started in 1999 for the five year grant.

The workshop was held from 5th of December to 8th of December at Kashiwa Campus of the University of Tokyo and we had 91 participants from 23 institutions of 9 different countries. We thank all the participants and all the speakers who made this workshop very successful.

The program was arranged by the program committee, the members of which were M. Bando, T. Kajino, T. Kajita, H. Minakata, T. Mori, M. Nakahata, H. Shibahashi, Y. Suzuki, M. Yamaguchi, O. Yasuda, and T. Yoshida.

Yoichiro Suzuki
Chairman of the Workshop

In Kamioka
September, 2002.

CONTENTS

Session X. Supernova Neutrinos

The latest solar neutrino results in Super-Kamiokande

Y.Koshio
Higashi-Mozumi, Kamioka-town, Gifu-pre.,506-1205,Japan
E-mail: koshio@icrr.u-tokyo.ac.jp

The results of 1500 days of solar neutrino measurement in Super-Kamiokande are presented, those are absolute solar neutrino flux, flux time variation, energy spectrum, hep flux limit. The precise measurement of the solar neutrino energy spectrum and day-night flux differences are the most important issue to understand neutrino oscillations. So it is very important to precisely determine the energy scale of the detector. In Super-Kamiokande, we have used an electron LINAC and ^{16}N calibration, and it can reduce the energy scale uncertainty 0.64%.

1 Solar neutrinos

The origin of the energy in the sun is the following nuclear fusion reaction which generates two neutrinos,

$$4p \rightarrow \alpha + 2e^+ + 2\nu_e \tag{1}$$

The neutrino fluxes and spectra in these reactions are predicted by so called Standard Solar Models (SSMs). And the results of several solar neutrino experiments up to now are reported[2]. Compared to the prediction of SSMs, the flux observed by all of the independent solar neutrino experiments are significantly small as shown in Table 1. Here, we use the model called "BP2000" as SSM[1].

Table 1: Summary of the several solar neutrino flux measurement.

Experiment	data/SSM(BP2000)
Cl (Homestake)	0.33 ±0.03
Ga (GNO+GALLEX)	0.57 ±0.05
Ga (SAGE)	0.58 ±0.06
Water (Super-K)	0.465 ±0.016
heavy water (SNO)	0.347 ±0.029

Various explanations of the solar neutrino problem have been proposed, and the most possible explanation is assuming neutrino oscillations[3]. Combine all the results of several experiments, the possible solutions of neutrino oscillation are shown in Fig 1, those are called Large Mixing Angle solution, Small Mixing Angle solution, LOW solution, and Vacuum solution. The most effective method to solve the problem is an independent measurement on solar models, because SSMs have been improved but the flux calculation from SSMs has remained ambiguous. Super-Kamiokande can measure the energy spectrum and the time dependence of the solar neutrino flux, (day/night or seasonal differences), and these are solar model independent measurements. Therefore, Super-Kamiokande is expected to provide definitive evidence of any possible new neutrino physics independent of the uncertainties in the solar models.

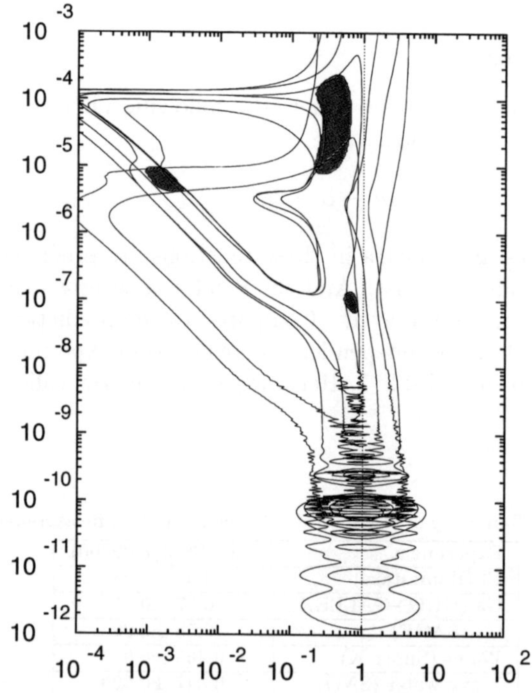

Figure 1: Allowed parameter region of neutrino oscillation from flux measurements in several experiments.

2 Super-Kamiokande

2.1 General properties

Super-Kamiokande is a 50000 tons of ring imaging water Cherenkov detector, and it is located 1000m underground (2700m of water equivalent), to shield against cosmic ray muons, in the Kamioka mine in Gifu Prefecture, Japan. Cherenkov lights generated by charged particles scattered by neutrinos in water are detected by 11146 20-inch photomultiplier tubes. The experiment started normal data taking from April 1st in 1996. After reduction of several background event, the analysis threshold have been achieved to 5.0MeV.

2.2 Detection method

For solar neutrino observation in Super-Kamiokande, recoil electrons through the following reaction are measured;

$$\nu + e^- \rightarrow \nu + e^- \tag{2}$$

The direction of recoil electrons has strong peak to the direction of original neutrinos, so we can identify solar neutrino signal. And it is real time measurement, so we can measure day-night and seasonal effect of the solar neutrino flux. The other advantage of this method is also measurement of the recoil electron energy, it can get the solar neutrino spectrum.

2.3 Detector calibration

As the detector calibration, we use mono-chromatic electron sources from a linear accelerator (LINAC calibration). The incident electron energy from LINAC is determined by germanium detector with less than 20keV[4]. For the solar neutrino analysis, it is especially important that the uncertainty of absolute energy scale is reduced. We have done the energy calibration at several energy points (from 4MeV to 17MeV which fit the energy range of solar neutrinos) and several positions. From this calibration, we can achieve the total systematic uncertainty of the absolute energy scale 0.64% .

3 Results

3.1 Absolute flux

We have obtained 1496 days of data between 31 May 1996 and 13 July 2001. The observed solar neutrino flux in Super-Kamiokande is

$$2.35 \quad \pm 0.02(stat.) \quad \pm 0.08(sys.) \quad [\times 10^6/cm^2/sec] \tag{3}$$

Comparing the result to SSM (BP2000), $5.05[\times 10^6/cm^2/sec]$,

$$\frac{Data}{SSM} = 0.465 \quad \pm 0.005(stat.) \ {}^{+0.016}_{-0.015}(sys.) \tag{4}$$

the observed flux is significantly smaller than predicted flux.

3.2 Day/Night flux differences

The day-night flux differences are also observed, and the result is,

$$\frac{\phi_{night} - \phi_{day}}{(\phi_{night} + \phi_{day})/2} = 0.021 \pm 0.020(stat.)^{+0.013}_{-0.012}(sys.), \tag{5}$$

And Fig 2 shows shows the solar neutrino flux in the daytime and the five night-time bins. The MSW effect through the earth, which is ν_e regeneration, causes flux difference between daytime and night-time, and the flux ratio depends on distances through the earth and the different electron density. Fig 2 also shows the expected solar neutrino flux in each bin assuming the three typical MSW parameters, corresponding to large mixing angle solutions and LOW solution.

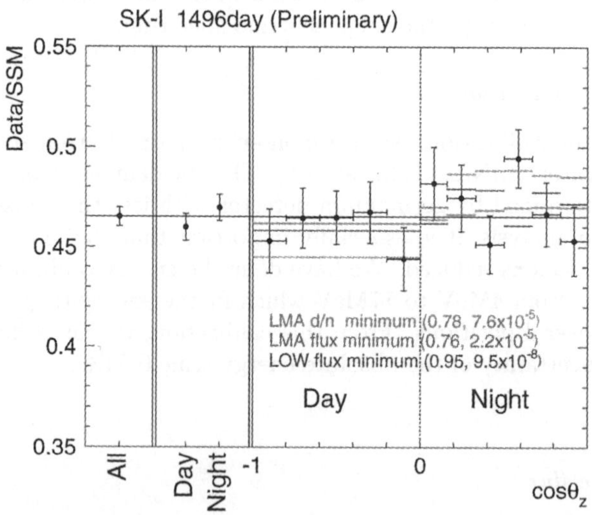

Figure 2: ***Flux diveded by daytime and five night time bins, and the expected flux differences assuming several neutrino oscillation solutions.***

3.3 Seasonal flux variation

The seasonal flux variation in Super-Kamiokande is measured as shown in Fig 3. The result is consistent with the expectation with eccentricity correction. The χ^2 for the expected after eccentricity correction is 4.7/7 (69%C.L.), and the χ^2 for flat is 10.3/7 (17%C.L.). The seasonal variation is especially sensitive to vacuum solution among several allowed neutrino oscillation parameter region, however, the statistics are not enough for oscillation analysis at present.

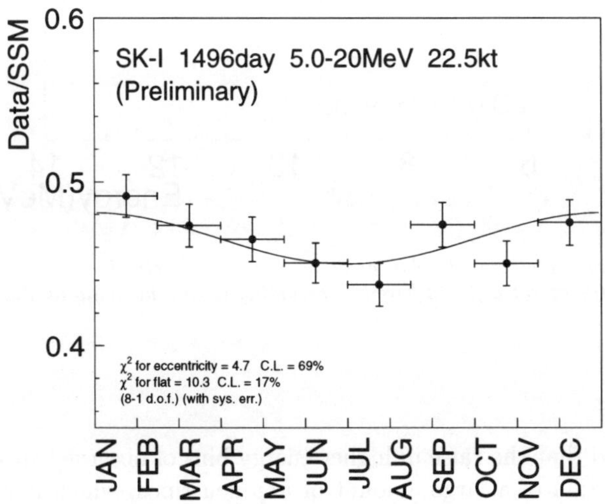

Figure 3: *Seasonal solar neutrino flux variation.*

3.4 Energy spectrum

Fig.4 shows the observed spectrum of solar neutrino events normalized by the predicted energy spectrum. The chi-square for flat distribution is $\chi^2/\nu = 17.4/18$. (49.6% C.L.) In this figure, the spectrum assuming the several couples of neutrino oscillation parameters are also shown. We cannot see any distortion of the solar neutrino spectrum. This result strongly constrains the parameter region of solar neutrino oscillation. It will be discussed later in several papers in this proceeding.

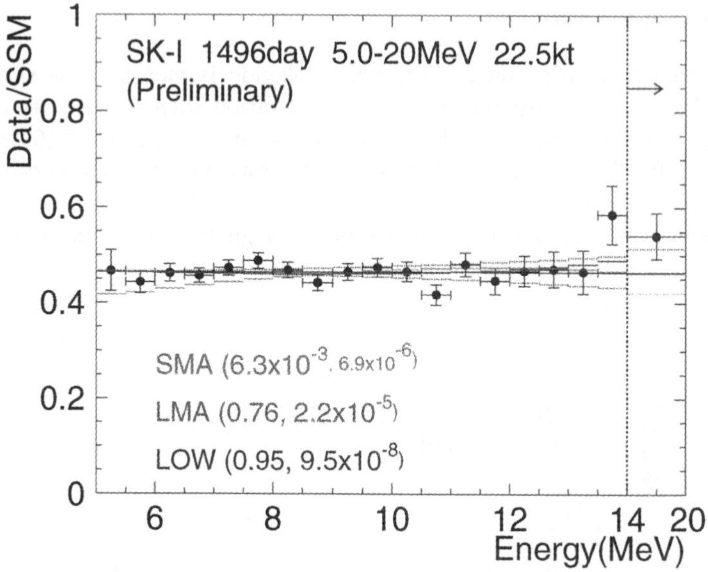

Figure 4: *Energy spectrum of solar neutrino events normalized by the predicted energy spectrum, and the expected flux differences assuming several neutrino oscillation solutions.*

3.5 Hep flux

It is suggested that the raise in higher energy bins of observed spectrum could be a contribution of a large amount of *hep* neutrinos, which is generated by $^3\text{He} + p \rightarrow ^4\text{He} + e^+ + \nu_e$ reaction in the center of the sun. The contribution of the *hep* neutrino flux at the higher energy region is only 1~2 %. However, if we fit the observed energy spectrum with free *hep* neutrino flux, the *hep* flux limit would be 7.9 times as much as the expectation from the SSM (BP2000) at 90%C.L.. Such large flux of *hep* neutrinos should be investigated in the nuclear physics point of view. Whereas the statistics of solar neutrino events in higher region are still poor, so the *hep* flux will be able to extract by the spectral shape with enough statistics in future.

4 Conclusions

First of all, we achieved that the total systematic error coming from the energy scale has been less than 0.64% using LINAC calibration. The solar neutrino spectrum is observed in 1496 days of data. The measured absolute flux comparing the expectation is 0.465, and the flux differences between daytime and night-time is 0.9σ level. The comparison of the spectral shape between observation and expectation gives a chi-square of 17.4 with 18 d.o.f. (49.6 % C.L.).

Acknowledgments

I gratefully acknowledge the cooperation of Kamioka Mining and Smelting Company. My participation of this useful meeting was financially supported by the Japanese Ministry of Education, Science and Culture. Finally, I appreciate all the members who organized this meeting.

References

1. J. Bahcall *et al*, Phys. Lett. **B433**, 1 (1998).
2. B.T. Cleveland *et al*, Nucl. Phys. B(Proc. Suppl.), **38**, 47 (1995), P. Anselmann *et al*, Phys. Lett. **B342**, 440 (1995), J.N. Abdurashitov *et al*, Phys. Lett. **B328**, 234 (1994), B.K. Kim *et al*, Proc. 26th ICRC (Salt Lake City, 1999).
3. S.P. Mikheyev & A.Y. Smirnov, Sov. Jour. Nucl. Phys. **42**, 913 (1985), L. Wolfenstein Phys. Rev. **D17**, 2369 (1978).
4. M. Nakahata *et al*, Nucl. Instr. Meth. **A421**, 113 (1999) (1995).

NEUTRINO OSCILLATION RESULTS FROM THE SUDBURY NEUTRINO OBSERVATORY

SCOTT M. OSER, FOR THE SNO COLLABORATION

Department of Physics and Astronomy, University of Pennsylvania, Philadelphia, PA 19104-6396

We describe here the measurement of ^{8}B solar neutrinos by the Sudbury Neutrino Observatory (SNO). The neutrinos were detected via the charged current (CC) reaction on deuterium and by the elastic scattering (ES) of electrons. The CC reaction is sensitive exclusively to ν_e's, while the ES reaction also has a small sensitivity to ν_μ's and ν_τ's. The flux of ν_e's from ^{8}B decay measured by the CC reaction rate is $\phi^{\rm CC}(\nu_e) = 1.75 \pm 0.07\,({\rm stat.})^{+0.12}_{-0.11}\,({\rm sys.}) \pm 0.05\,({\rm theor.}) \times 10^6\ {\rm cm}^{-2}{\rm s}^{-1}$. Assuming no flavor transformation, the flux inferred from the ES reaction rate is $\phi^{\rm ES}(\nu_x) = 2.39 \pm 0.34\,({\rm stat.})^{+0.16}_{-0.14}\,({\rm sys.}) \times 10^6\ {\rm cm}^{-2}{\rm s}^{-1}$. Comparison of $\phi^{\rm CC}(\nu_e)$ to the Super-Kamiokande Collaboration's precision value of $\phi^{\rm ES}(\nu_x)$ yields a 3.3σ difference, assuming the systematic uncertainties are normally distributed, providing evidence that there is an active flavor neutrino component which is not ν_e in the solar flux. The total flux of active ^{8}B neutrinos is thus determined to be $5.44 \pm 0.99 \times 10^6\ {\rm cm}^{-2}{\rm s}^{-1}$, in close agreement with the predictions of solar models.

1 The Sudbury Neutrino Observatory

The Sudbury Neutrino Observatory (SNO) is a water Cherenkov detector for ^{8}B solar neutrinos located 2 km underground in INCO Ltd's Creighton nickel mine near Sudbury, Ontario.[1] SNO's central target volume consists of 1000 tons of ultra-pure heavy water (D_2O), contained in a 12 m diameter acrylic vessel. The heavy water is viewed by an array of $\sim$ 9500 inward-looking photomultiplier tubes arranged on a 17.8 m diameter geodesic sphere. The region between the photomultipliers and the acrylic vessel contains ultra-pure light water for shielding.

The heavy water target allows SNO to detect solar neutrinos through three separate interactions:

$$\begin{array}{ll} \nu_e + d \rightarrow p + p + e^- & \text{(CC)} \\ \nu_x + e^- \rightarrow \nu_x + e^- & \text{(ES)} \\ \nu_x + d \rightarrow p + n + \nu_x & \text{(NC)} \end{array}$$

The charged current (CC) reaction is sensitive only to ν_e's, and so measures the total flux of electron neutrinos. The kinetic energy of the electron is highly correlated with the initiating neutrino's energy, allowing one to measure the energy spectrum of the ν_e's as well. The Q value for the CC reaction is -1.4 MeV. The direction of the relativistic electron produced in the CC reaction is weakly anticorrelated with the incoming neutrino's direction. For solar neutrinos, the angular distribution of the CC reaction is approximately $1 - (1/3)\cos\theta_{\odot}$.[2]

The elastic scattering (ES) reaction occurs for all active neutrino flavors. The cross-section for scattering by ν_e's is 6.5 times larger than the cross-section for ν_μ/ν_τ. The rate of elastic scattering interactions is thus proportional to $\phi(\nu_e) + 0.154(\phi(\nu_\mu) + \phi(\nu_\tau))$. The direction of the scattered electron is very strongly correlated with the initial neutrino's direction. This strong directionality makes it easy to distinguish the ES signal from other neutrino signals or backgrounds. The energy of the scattered electron is weakly correlated with the initial neutrino's energy.

The neutral current (NC) reaction occurs for all active neutrino flavors with equal probability. SNO measures NC interactions by observing the neutrons liberated by this reaction. In pure D_2O, approximately 30% of these neutrons capture on a deuteron, emitting a 6.25 MeV γ-ray. This γ-ray Compton scatters, and SNO then detects the resulting Compton electron. The neutral current reaction provides no directional information about the interacting neutrinos, and simply measures the total neutrino rate above its threshold energy of 2.2 MeV.

By measuring solar neutrinos with these interactions, SNO can directly probe for neutrino flavor conversion. If ν_e's are being converted into ν_μ or ν_τ, the ^{8}B flux inferred from the NC or ES interactions, which each have sensitivity to all neutrino flavors, will exceed the rate inferred from the CC interaction, which measures only ν_e's.

2 Data Analysis

SNO published first results for the CC and ES fluxes in June 2001.[3] This data set consisted of 240.95 live-days, and spanned the period from 2 Nov 1999 to 15 Jan 2001.

Figure 1 schematically illustrates the analysis flow. The engine of the analysis is signal extraction, which fits probability distribution functions (PDFs) for the 3 signal reactions to the data to determine the rates of each. Each neutrino signal (CC, ES, NC) has its own characteristic energy, radial, and angular distribution (see Figure 2). The inputs to the signal extraction procedure are a clean, reconstructed neutrino data set, and well-characterized signal PDFs.

SNO uses a series of instrumental background cuts to remove non-Cherenkov events from the data stream. These events include electrical pickup

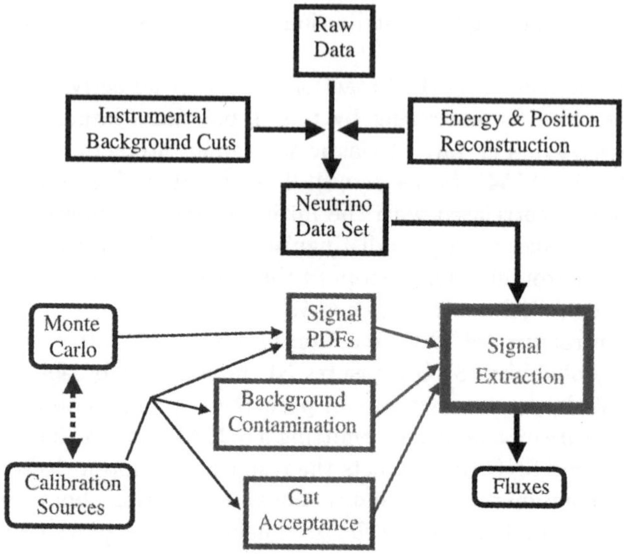

Figure 1. Analysis Flow Chart.

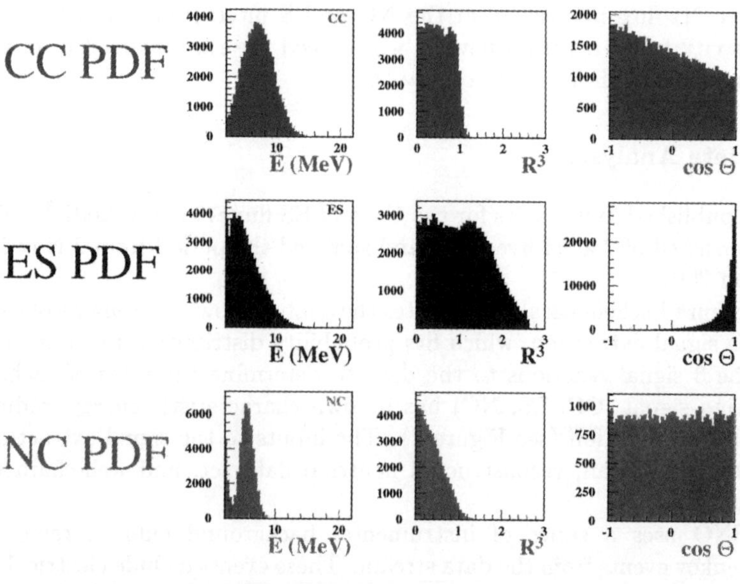

Figure 2. Energy, radial, and angular PDFs for the 3 signal reactions

and light from electrical discharges in the detector. Cuts on low level detector properties, such as charge and timing distributions, together with a cut on the number of tubes hit, remove $\sim$ 99% of data triggers (see Table 1), while removing 0.33% of neutrino events. A muon follower cut removes any events (e.g. spallation events or neutrons) occurring within 20 seconds of an incident muon.

The remaining events are reconstructed in time, position, and direction by a maximum likelihood fitter which fits a timing and angular PDF to the hit PMTs. The reconstructed vertex is used to define two "Cherenkov box cuts" designed to remove any residual instrumental contamination. One parameter measures the fraction of hit tubes which were hit in a narrow time coincidence with the reconstructed vertex (accounting for time of flight)— for Cherenkov events most hits will be in time coincidence with a single vertex. A second parameter is the average angle between all pairs of hit PMTs, relative to the vertex—this will have a characteristic distribution of values for single ring Cherenkov events. Cutting on these two parameters defines a "Cherenkov box" which characterizes Cherenkov events.

The number of phototubes hit within a prompt time window around the reconstructed vertex (again correcting for time of flight) is used to estimate the event's energy. The number of prompt phototube hits is corrected for position and directional dependencies in the detector's optical response, and an event energy is assigned. Fiducial volume and energy cuts of $R < 550$ cm and $T > 6.75$ MeV define the final signal region. This restricted fiducial volume and high energy threshold are designed to reduce contamination by low energy radioactivity to negligible levels. The final result is a data set with 1169 events consisting almost entirely of neutrinos.

Combinations of calibration data and Monte Carlo are used to construct signal PDFs and to measure background contamination, cut acceptance, and reconstruction properties. A tagged ^{16}N source of 6.1 MeV γ-rays deployed on a grid inside the AV and H_2O provides the primary energy and reconstruction calibration. A ^{8}Li β source and a proton-tritium source of 19.8 MeV γ-rays provide additional calibration points at higher energies.[4] These calibration sources are used to measure the cut acceptance for real Cherenkov events of all cuts. The signal loss from the instrumental and Cherenkov box cuts is $\sim$ 1.3%. The calibration sources also measure detector response parameters such as energy, vertex, and angular resolution.

Optical calibrations from a pulsed laser source are used as inputs to the Monte Carlo, which is tuned to reproduce source data. The performance of the Monte Carlo is verified by comparison to calibration data. The Monte Carlo is then used to model the PDF shapes for the 3 neutrino signals, and to calculate the predicted rate for each signal given a solar model flux. The shapes of the signal PDFs are verified by comparison to an analytic calculation which uses detector response parameters measured directly from calibration data without Monte Carlo.

The signal extraction fits linear combinations of signal PDFs to the three dimensional distribution of the cleaned neutrino data in T, R^3, and $\cos\theta_{sun}$. For a *constrained* signal extraction, the data is fit assuming the CC PDF follows an undistorted ^{8}B spectrum.[5] This assumption is appropriate for testing the null hypothesis of Standard Model neutrino physics. An *unconstrained* fit is also done in which the amplitude of the CC PDF is allowed to vary in different energy bins. This allows one to measure the CC energy spectrum. The constrained signal extraction provides the best statistical uncertainty on the CC flux, since it uses spectral information to statistically separate CC, ES, and NC events.

3 Results and Implications

Fitting the signal PDFs to the 1169 neutrino candidates yields 975.4 ± 39.7 CC events, 106.1 ± 15.2 ES events, and 87.5 ± 24.7 neutron capture events (errors are statistical only). Systematic errors are estimated by modifying the PDFs by varying low level detector response parameters (e.g. energy scale, vertex resolution) within their uncertainties, and then repeating the signal extraction. Correcting for cut acceptances, these event totals translate into CC and ES fluxes of:

$$\phi^{\rm CC}(\nu_e) = 1.75 \pm 0.07\ (\text{stat.})^{+0.12}_{-0.11}\ (\text{sys.}) \pm 0.05\ (\text{theor.}) \times 10^6\ \text{cm}^{-2}\text{s}^{-1}$$
$$\phi^{\rm ES}(\nu_x) = 2.39 \pm 0.34\ (\text{stat.})^{+0.16}_{-0.14}\ (\text{sys.}) \times 10^6\ \text{cm}^{-2}\text{s}^{-1}$$

Figure 3 shows the radial and angular distribution of the data sample. Table 2 summarizes the various sources of systematic error. Uncertainty on the absolute energy scale ($\pm 1.4\%$) and fitter radial bias ($\pm 1\%$) are the major systematics.

SNO's measured ES rate exceeds its measured CC rate by 1.6σ. However, the published high statistics ES result from Super-Kamiokande of 2.32 ± 0.03 (stat.)$^{+0.08}_{-0.07}$ (sys.) exceeds SNO's CC rate by 3.35σ.[6] Standard electroweak theory predicts that the ^{8}B fluxes inferred from the CC and ES reaction rates should be equal.

The high value of the ES flux compared to the CC flux implies non-standard neutrino physics. Either the flux measured by the ES reaction contains active neutrino flavors besides ν_e which account for the additional rate, or else a differential spectral distortion between SNO and Super-K's energy

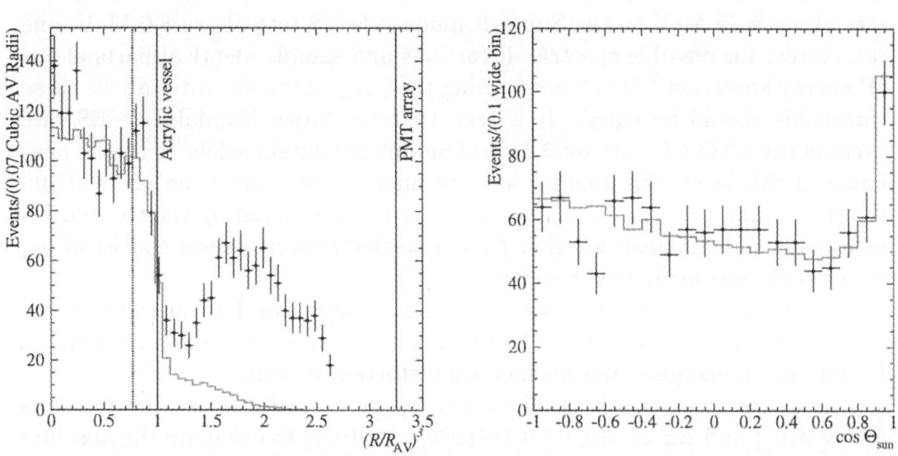

Figure 3. Radial and angular distributions of candidate neutrino events. The superimposed curves are scaled Monte Carlo predictions for ^{8}B neutrinos. The fiducial volume cut at $R = 550$ cm (dotted line) removes events from low energy radioactivity at high radii.

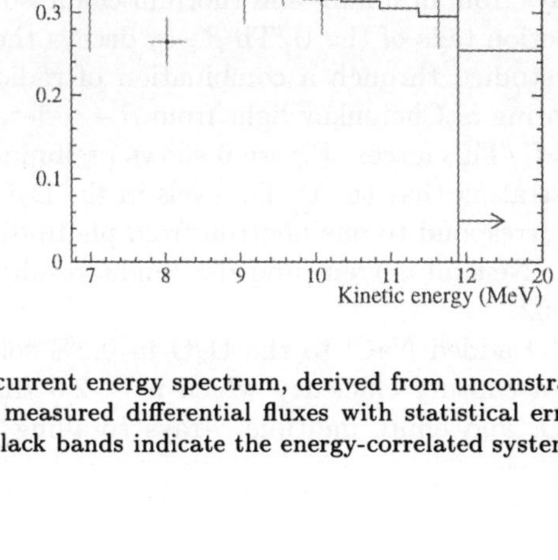

Figure 4. Charged current energy spectrum, derived from unconstrained signal extraction. The points are the measured differential fluxes with statistical errors, normalized to the BPB01 SSM. The black bands indicate the energy-correlated systematic errors.

thresholds accounts for the difference in rate (possibly allowing for sterile neutrino oscillations). Fogli et al. have shown that by comparing the SNO CC rate above 6.75 MeV to the Super-Kamiokande ES rate above 8.6 MeV, one can correct for possible spectral distortions and sample identical parts of the ^{8}B energy spectrum.[7] If ν_e's are turning to $\nu_{sterile}$, then the rates above these thresholds should be equal. It is seen that the Super-Kamiokande ES rate exceeds the SNO CC rate by 3.1σ, ruling out oscillation solely to sterile neutrinos at this level. Put another way, we infer the presence of active neutrino flavors besides ν_e in the ^{8}B flux at the $\sim 3\sigma$ level. Comparing the CC and ES rates cannot distinguish whether these non-electrons neutrinos consist of ν_μ, ν_τ, or their anti-neutrinos, however.

Figure 4 shows the measured CC energy spectrum derived from an unconstrained signal extraction. The CC flux is suppressed relative to Super-K's ES flux at all energies, and no spectral distortion is seen.

Assuming that this is oscillation to ν_μ or ν_τ, we can use the relations $CC \propto \phi(\nu_e)$ and $ES \propto \phi(\nu_e) + 0.154(\phi(\nu_\mu) + \phi(\nu_\tau))$ to calculate the absolute fluxes of ν_e and ν_μ/ν_τ. The derived flux of $\phi(\nu_{\mu\tau})$ is $3.69 \pm 1.13 \times 10^6$ cm^{-2}s^{-1}. The total flux of ^{8}B neutrinos is derived to be $5.44 \pm 0.99 \times 10^6$ cm^{-2}s^{-1}. This value is in excellent agreement with predictions of standard solar models.[8,9] Figure 5 shows graphically the derived fluxes of ν_e and $\nu_{\mu\tau}$.

4 Future Prospects

In the near future, SNO will produce neutral current and day/night results from its initial D_2O data. The NC measurement requires lowering the energy threshold of the analysis so as to cover a substantial fraction of the NC PDF. This also requires a careful accounting of possible backgrounds from low energy radioactivity. These sources include neutrons from deuteron photodisintegration by γ-rays from uranium- and thorium-chain isotopes, as well as energy and reconstruction tails of the U/Th $\beta - \gamma$ decays themselves. These backgrounds can be studied through a combination of radiochemical water assays, *in situ* monitoring of Cherenkov light from $\beta - \gamma$ decays, and deployment of encapsulated U/Th sources. Figure 6 shows preliminary results from water assays, demonstrating that the U/Th levels in the D_2O are well below target values which correspond to one neutron from photo-disintegration per day for each species. Neutral current and day/night results from SNO are expected in spring 2002.

In June 2001 SNO added NaCl to the D_2O in 0.2% solution. Neutron capture on ^{35}Cl has a capture efficiency which is ~ 2.6 times higher than capture on pure D_2O, and emits multiple γ-rays totalling 8.6 MeV. Thus

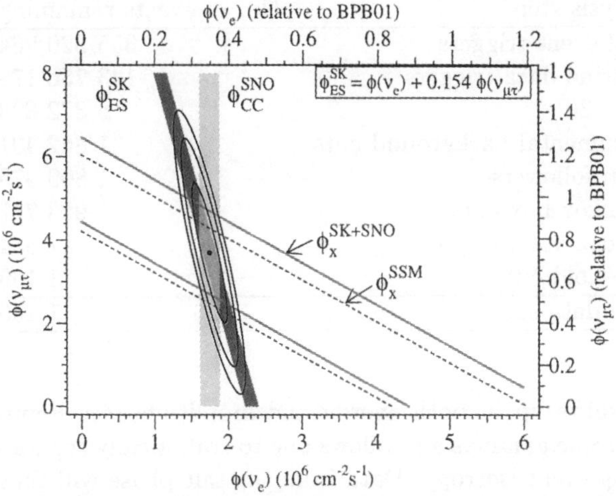

Figure 5. Flux of ^{8}B neutrinos which are $\nu_{\mu\tau}$ vs. flux of ν_e, as deduced from the SNO and Super-K data. The intersection of the diagonal bands with the axes show the $\pm 1\sigma$ limits on the total derived ^{8}B flux, along with the BPB01 SSM prediction. The ellipses are 68%, 95%, and 99% contours.

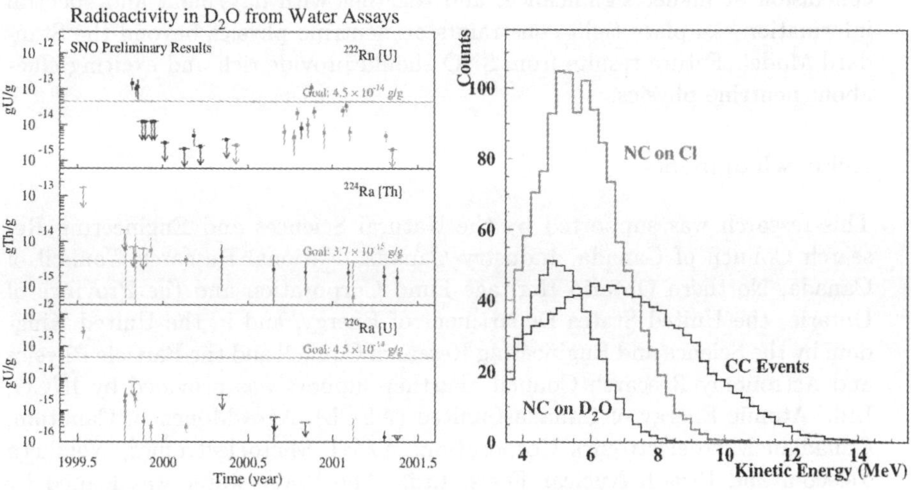

Figure 6. D_2O purity, and salt PDFs. Left: Preliminary limits on U-chain and Th-chain isotope concentrations in the D_2O from radiochemical assays. Right: CC and NC energy PDFs for pure D_2O and for salt-phase data, from Monte Carlo.

Table 1. Data processing steps.

Analysis step	No. of events remaining
Total event triggers	355 320 964
Neutrino data triggers	143 756 178
$N_{\text{hit}} \geq 30$	6 372 899
Instrumental background cuts	1 842 491
Muon followers	1 809 979
Cerenkov box cuts	923 717
Fiducial volume cut	17 884
Threshold cut	1 169
Final data set	1 169

the NC PDF shifts up in both energy and amplitude (see Figure 6). The presence of multiple gammas also allows one to statistically separate NC from CC events using event isotropy. Data from the salt phase will thus provide a second ^{8}B flux measurement from the NC interaction with high statistics and improved CC/NC separation.

SNO's first results, when combined with Super-Kamiokande's ES measurements, provide direct evidence for neutrino flavor conversion of solar neutrinos. NC results from SNO's pure D_2O and salt phase can confirm this conclusion at higher significance, and together with day/night and spectral information can place tight constraints on neutrino physics beyond the Standard Model. Future results from SNO should provide rich and exciting clues about neutrino physics.

Acknowledgments

This research was supported by the Natural Sciences and Engineering Research Council of Canada, Industry Canada, National Research Council of Canada, Northern Ontario Heritage Fund Corporation and the Province of Ontario, the United States Department of Energy, and in the United Kingdom by the Science and Engineering Research Council and the Particle Physics and Astronomy Research Council. Further support was provided by INCO, Ltd., Atomic Energy of Canada Limited (AECL), Agra-Monenco, Canatom, Canadian Microelectronics Corporation, AT&T Microelectronics, Northern Telecom and British Nuclear Fuels, Ltd. The heavy water was loaned by AECL with the cooperation of Ontario Power Generation. SNO gratefully acknowledges the efforts of its technical and support staff.

Table 2. Systematic uncertainties on fluxes.

Error source	CC error (percent)	ES error (per cent)
Energy scale	-5.2, +6.1	-3.5 ,+5.4
Energy resolution	±0.5	±0.3
Energy non-linearity	±0.5	±0.4
Vertex accuracy	±3.1	±3.3
Vertex resolution	±0.7	±0.4
Angular resolution	±0.5	±2.2
High energy γ background	-0.8, +0.0	-1.9, +0.0
Low energy backgrounds	-0.2, +0.0	-0.2, +0.0
Instrumental backgrounds	-0.2, +0.0	-0.6, +0.0
Trigger efficiency	0.0	0.0
Live time	±0.1	±0.1
Cut acceptance	-0.6, +0.7	-0.6, +0.7
Earth orbit eccentricity	±0.1	±0.1
CC interactions on ^{17}O, ^{18}O	0.0	0.0
Experimental uncertainty	-6.2, +7.0	-5.7, +6.8
Cross section	3.0	0.5
Solar Model	-16, +20	-16, +20

References

1. The SNO Collaboration, Nucl. Instr. and Meth. **A449**, 172 (2000).
2. J.F. Beacom and P. Vogel, hep-ph/9903554, Phys. Rev. Lett. **83**, 5222 (1999).
3. Q.R. Ahmad *et al.*, Phys. Rev. Lett. **87**, 071301 (2001)
4. A.W.P. Poon *et al.*, Nucl. Instr. and Meth. **A452**, 115, (2000).
5. C.E. Ortiz *et al.*, Phys. Rev. Lett. **85**, 2909 (2000).
6. S. Fukuda *et al.*, Phys. Rev. Lett. **86**, 5651 (2001).
7. G. L. Fogli, E. Lisi, A. Palazzo, and F.L. Villante Phys. Rev. D **63**, 113016 (2001); F.L. Villante, G. Fiorentini and E. Lisi Phys. Rev. D **59** 013006 (1999).
8. J.N. Bahcall, M. H. Pinsonneault, and S. Basu, astro-ph/0010346 v2. The reference ^{8}B neutrino flux is 5.05×10^6 cm^{-2}s^{-1}.
9. A.S. Brun, S. Turck-Chièze, and J.P. Zahn, Astrophys. J. **525**, 1032 (1999); S. Turck-Chièze *et al.*, Ap. J. Lett., v. **555** July 1, 2001.

RECENT STATUS OF ASTROPHYSICAL S_{17}

T. MOTOBAYASHI

*Department of Physics, Rikkyo University, 3-34-1 Nishi-Ikebukuro, Toshima, Tokyo 171-8501, Japan**

E-mail: motobaya@rarfaxp.riken.go.jp

The present status of the astrophysical S factor for the ^{7}Be(p,γ)^{8}B reaction is reviewed. Because of its importance for the solar neutrino problem, the reaction has been extensively studied. Three independent methods, the direct capture, the Coulomb dissociation and the ANC method, give almost consistent results within 10-20% accuracy.

1 Introduction

The main nuclear burning process in the sun is called the pp chain, which converts four protons to an alpha particle. Recently one of the reactions involved in the chain, ^{3}He(^{3}He,2p)^{4}He, could be measured at around 20 keV, the effective energy (Gamow energy) of the solar burning for the first time [1]. The result is consistent with the earlier estimate from higher energy data. The recent compilation on the solar fusion reactions [2] evaluated the errors for the reaction rates to be within 5% except for ^{7}Be(p,γ)^{8}B. The ^{7}Be(p,γ)^{8}B reaction at low energies is of crucial importance, because it is the source of high-energy solar neutrinos. The measurement at Homestake mine[3,4] is sensitive mainly to the neutrinos from ^{8}B through its β^+ decay ^{8}B$\to^8$Be(2^+)+e$^+$+ν with a high end-point energy of about 14 MeV (Note that the 2^+ state of ^{8}B is a broad resonance with Γ=1.5 MeV.). The solar neutrino measurements in Kamioka mine[5] are sensitive almost only to the ^{8}B neutrino. Since the flux of the solar neutrino originating from ^{8}B depends directly on the ^{7}Be(p,γ)^{8}B cross section at around 20 keV, the Gamow energy or the effective burning energy in the sun, its experimental information is of crucial importance in predicting the flux corresponding to the Homestake and Kamioka experiments.

Much effort has been devoted to the determination of the astrophysical S factor of the ^{7}Be(p,γ)^{8}B reaction, S_{17}. The S factor is related to the energy-dependent cross section $\sigma(E) = SE\exp[-2\pi\eta]$. The term $E\exp[-2\pi\eta]$ accounts for the steep energy dependence of $\sigma(E)$ due to the Coulomb penetration in S-wave, where η denotes the Sommerfeld parameter $e^2Z_1Z_2/\hbar v$. Since ^{7}Be is an unstable isotope with 53 day half life, direct (p,γ) experiments

*Present address: RIKEN, 2-1 Hirosawa, Wako, Saitama 351-0198, Japan

should employ a radioactive target. This requires special attention in determination of the beam-target luminosity. Since the first experimental study by Kavanagh was reported in 1960[6], many experiments have performed for accurate measurements of direct capture [7–16]. The most recent recommendation for the S factor at zero energy is given as $S_{17}(0)=19^{+4}_{-2}$ eV-b [2], which is obtained by evaluating the direct capture data up to 1998.

Attempts to employ indirect methods have also been made for determination of S_{17} hoping that the measurements are independent of the difficulties associated with the direct measurements. So far two methods have been investigated, Coulomb dissociation with intermediate-energy ^{8}B beams [17–22] and low-energy transfer reactions with ^{7}Be beams to extract the asymptotic normalization coefficient (ANC) for the ^{8}B ground state [23,24]. In the following sections the recent status of these investigations to determine S_{17} will be reported.

2 Recent studies on the ^{7}Be(p,γ)^{8}B reaction

2.1 Direct capture measurements

After the evaluation on $S_{17}(0)$ was made in 1998, four direct-capture experiments were performed [13,14,15,16]. The results of these measurements are consistent with the recommendation $S_{17}(0)=19^{+4}_{-2}$ eV-b [2]. The most recent study at Seattle [16] gives a large value of 22.3 eV-b with a very small uncertainty of 4%[a]. The experiment introduced a new scheme of normalization. The incident proton beam was controlled so as to achieve a uniform intensity distribution on the target in order to be free from possible non-uniformity in the thickness of the radioactive ^{7}Be target. For the precise argument on the neutrino oscillation, an accuracy of the order of 5% is required [2]. Therefore, the quoted small error in ref. [16] might have to be examined, for example, by independent experiments.

2.2 Coulomb dissociation of ^{8}B

For studying astrophysical (p,γ) processes involving unstable nuclei, the Coulomb dissociation method is sometimes very useful. It uses intermediate-energy RI beams, which are now available at many laboratories with variety of ions. Low-energy cross sections can be extracted indirectly from the data of Coulomb dissociation, for example, at several tens MeV per nucleon. For

[a]Recently they revised the S_{17} values to be slightly lower.

the case of the ^{7}Be(p,γ)^{8}B, the residual nucleus ^{8}B bombards a high-Z target and is Coulomb-excited to an unbound state that decays to the ^{7}Be+p channel. Since the process is regarded as absorption of a virtual photon, *i.e.* ^{8}B(γ,p)^{7}Be, the radiative capture (the inverse of the photoabsorption) cross section can be extracted from the dissociation yield. This idea was first proposed by Baur, Bertulani and Rebel[25] based on the virtual photon theory. Topical reviews were given by Baur and Rebel[26].

A series of experiments has been performed for the Coulomb dissociation of ^{8}B at RIKEN. It should be noted that the astrophysical process is dominated by the E1 transition to the unbound continuum state of ^{8}B. Therefore the Coulomb dissociation to continuum states should be studied.

The first measurement is described in two articles[17,18], and the results of the second experiment are reported in another two articles[19,20]. Beams of ^{8}B were produced by the ^{12}C+^{9}Be interaction at 92 MeV/u for the first two experiments and 135 MeV for the third one. The ^{8}B energies in the center of the target, 50 mg/cm^2 ^{208}Pb, were approximately 50 MeV/u. A p-^{7}Be relative energy spectrum could be constructed from the measurement and it was converted to the ^{7}Be(p,γ)^{8}B cross section with the help of a Monte-Carlo simulation calculation on detection efficiency and theoretical calculation for the Coulomb dissociation mechanism.

In the second and third experiments, the DALI setup[27] was also used to measure the deexcitation γ rays from the first excited state of ^{7}Be at 429 keV populated in the dissociation process. The contribution from this process was measured to be about 5% of the Coulomb dissociation yield.

In Fig. 1 the astrophysical S-factors obtained in the first and second experiments are shown together with the ones determined in direct (p,γ) measurements. Our Coulomb dissociation data are consistent within errors with the results by Filippone *et al.*[11] and Vaughn *et al.*[9].

The post Coulomb-acceleration effect is expected to be small owing to the special situation for ^{8}B where the proton binding energy is only 137 keV. Trajectories very far from the target are enough to excite the continuum state of ^{8}B, because a low energy virtual photon is responsible for the excitation. Recent theoretical calculations supports this picture[28,29]

In general, higher order processes including the post acceleration is less important at higher incident energy. We also performed a Coulomb dissociation experiment at GSI using ^{8}B beams of 254 MeV/u and a ^{208}Pb target[21]. The resultant S_{17} factors are quite similar to the Coulomb dissociation results obtained in the second RIKEN experiment, and therefore to the direct (p,γ) results by Filippone *et al.*[11] and Vaughn *et al.*[9]. The agreement between the two results obtained at the different energies (50 MeV/u at RIKEN and

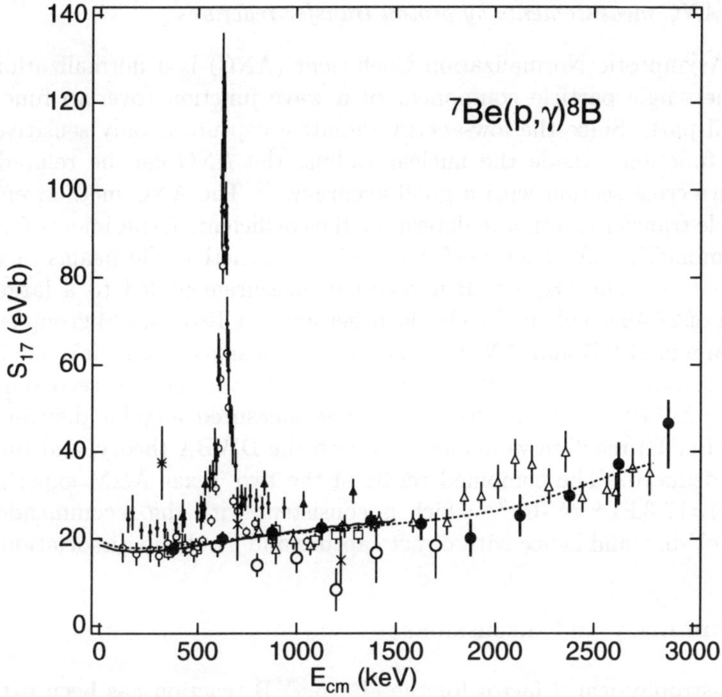

Figure 1. Astrophysical S_{17}-factors for the ^{7}Be(p,γ)^{8}B reaction extracted from the first (thick crosses) and second (large solid dots) experiments. Existing direct (p,γ) data are also shown.

250 MeV/u at GSI) might indicate smallness of the higher-order contribution.

Recently another p-^{7}Be coincidence experiment was performed at MSU with 83 MeV/nucleon ^{8}B beams[22]. The reaction products were momentum-analyzed by a dipole magnet, and detected by multiwire drift chambers for tracking their trajectories. Though the experimental results are essentially consistent with those of the previous studies at RIKEN and GSI, the authors propose a slightly lower $S_{17}(0)$ value of $17.8^{+1.4}_{-1.2}$ eV-b by subtracting the E2 component evaluated from their parallel-momentum distribution measurements[30].

2.3 ANC measurements by proton transfer reactions

The Asymptotic Normalization Coefficient (ANC) is a normalization factor for the single particle component of a wave function (overlap function) in its tail part. Since the low-energy radiative capture is only sensitive to the wave function outside the nuclear radius, the ANC can be related to the capture cross section with a good accuracy. [31] The ANC method employs a particle transfer reaction to determine the coefficient. Experiments for the S_{17} determination have been performed with radioactive ^{7}Be beams in reversed kinematics. The ^{2}H(^{7}Be,^{8}B)n reaction measurement led to a large $S_{17}(0)$ factor of 27.4±4.4 eV-b. [23]. On the other hand, a Texas A&M group employed the targets of ^{10}B and ^{14}N. Radioactive ^{7}Be beams of 12.1 MeV/nucleon was produced by the ^{1}H(^{7}Li,^{7}Be)n reaction, and delivered to the secondary target by the MARS recoil spectrometer. The measured angular distributions of the (^{7}Be,^{8}B) reaction were analyzed with the DWBA theory, and the ANC's were deduced. The combined result of the two Texas A&M experiments is $S_{17}(0)$=17.3±1.8 eV-b, [24] which is consistent with the recommended value 19^{+4}_{-2} eV-b, [2] and hence with direct capture and Coulomb dissociation results.

3 Summary and discussions

The astrophysical S factor for the ^{7}Be(p,γ)^{8}B reaction has been extensively studied experimentally. Recent progress of solar neutrino measurements stimulates further effort to improve the accuracy and reliability. Three methods have been employed so far: the direct capture measurement, the Coulomb dissociation method and the ANC determination by transfer reaction measurements.

Figure 2 summarizes the $S_{17}(0)$ values reported so far with the three methods. The results are generally in good agreement with the recent evaluation [2] This suggests that there is little possibility that the high-energy neutrino flux takes a very different value from the presently expected one.

For more accurate determination, however, further studies are necessary to evaluate possible corrections due to the mixture of E2 and nuclear components and higher-order processes for the Coulomb dissociation method. More detailed confirmation of the ANC method should also be made by careful investigations of multi-step processes, choice of optical potentials, and so on. Further confirmation of the new accurate determination by the direct capture experiment of Seattle group [16] is desirable. Finally it should be pointed out that either the direct or Coulomb dissociation measurements could not reach the Gamow energy of the solar burning. Extrapolation is necessary to esti-

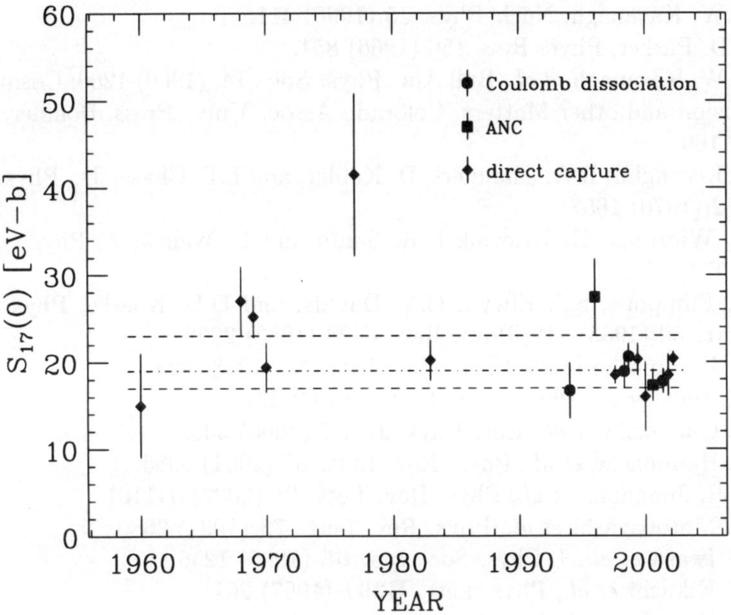

Figure 2. The $S_{17}(0)$ values extracted by the Coulomb dissociation method (solid circles) and the ANC method (solid squares). The ones obtained by direct capture measurements are also shown by solid diamonds. The latest recommendation $S_{17}(0)=19^{+4}_{-2}$ eV-b is indicated by the dashed and dot-dashed lines. Note that the most recent direct capture result is renormalized to the recent revised value.

mate S_{17} at lower energies with a theoretical energy-dependence. Effort to obtain lower energy cross section is desirable to reduce the ambiguity of the $^{7}Be(p,\gamma)^{8}B$ reaction rate.

References

1. R. Bonetti *et al.*, Phys. Rev. Lett. 82 (1999) 5205.
2. E.G. Adelberger *et al.*, Rev. Mod. Phys. 70 (1998) 1265.
3. R. Davis Jr., D.S. Harmer and K.C. Hoffman, Phys. Rev. Lett. 20 (1968) 1205.
4. K. Lande *et al.*, Nucl. Phys. B91 (2001) 50.
5. Y. Suzuki *et al.* (Super-Kamiokande Collaboration), Nucl. Phys. B91 (2001) 29.

6. R.W. Kavanagh, Nucl. Phys. 15 (1960) 411.
7. P.D. Parker, Phys. Rev. 150 (1966) 851.
8. R.W. Kavanagh *et al.*, Bull Am. Phys. Soc., 14, (1969) 1209; Cosmology, Fusion and other Matters, Colorado Assoc. Univ. Press, Boulder, 1972, P. 169.
9. F.J. Vaughn, R.A. Chalmers, D. Kohler, and L.F. Chase, Jr., Phys. Rev. C 2 (1970) 1657.
10. C. Wiezorek, H. Krawinkel, R. Santo and L. Wallek, Z. Phys. A 282 (1977) 121.
11. B. Filippone, S.J. Elwyn, C.N. Davids, and D.D. Koetke, Phys. Rev. Lett. 50 (1983) 412; Phys. Rev. C 28 (1983) 2222.
12. F. Hammache *et al.*, Phys. Rev. Lett. 80 (1998) 928.
13. M. Hass *et al.*, Phys. Lett. B 462 (1999) 237.
14. L. Gialanella *et al.*, Eur. Phys. J. A 7 (2000) 303.
15. F. Hammache *et al.*, Phys. Rev. Lett. 86 (2001) 3985.
16. A.R. Junghans *et al.*, Phys. Rev. Lett. 88 (2002) 041101.
17. T. Motobayashi *et al.*, Phys. Rev. Lett. **73** (1994) 2680.
18. N. Iwasa *et al.*, J. Phys. Soc. Jpn. **65** (1996) 1256.
19. T. Kikuchi *et al.*, Phys. Lett. **B391** (1997) 261.
20. T. Kikuchi *et al.*, Eur. Phys. J. **A3** (1998) 209.
21. N. Iwasa *et al.*, Phys. Rev. Lett. **83** (1999) 2910.
22. B. Davids *et al.*, Phys. Rev. Lett. 86 (2001) 2750.
23. W. Liu *et al.*, Nucl. Phys. A 616 (1997) 131c.
24. C.A. Gagliardi, Nucl. Phys. A 682 (2001) 369c.
25. G. Baur, C.A. Bertulani, and H. Rebel, Nucl. Phys. **A458** (1986) 188.
26. G. Baur and H. Rebel, J. Phys. **G20** (1994) 1; Ann. Rev. Nucl. and Part. Sci. **46** (1996) 321.
27. T. Nishio *et al.*, RIKEN Accel. Prog. Rep. **29** (1996) 184.
28. C.A. Bertulani, Phys. Rev. **C49** (1994) 2688.
29. S. Typel and G. Baur, Phys. Rev. **C50** (1994) 2104.
30. B. Davids *et al.*, Phys. Rev. Lett. 81 (1998) 2209.
31. H.M. Xu *et al.*, Phys. Rev. Lett 73 (1994) 2027; A.M. Mukhamedzhanov *et al.*, Phys. Rev. C 56 (1995) 1302.

Weak Current in Deuterium

T. Sato
Department of Physics, Graduate School of Science
Osaka University
Toyonaka, Osaka 56-0043, JAPAN
E-mail: tsato@phys.sci.osaka-u.ac.jp

Accurate estimates of the νd reaction cross sections are of great importance in interpreting the SNO experiments. We improve our previous estimates by updating inputs and by taking into account the results of a recent effective-field-theoretical calculation. The new cross sections are slightly (~1 %) larger than the previously reported values with smaller uncertainty.

1 Introduction

The establishment of the Sudbury Neutrino Observatory (SNO) [1,2] has motivated intensive theoretical effort to make reliable estimates of the neutrino-deuteron reaction cross sections [3,4,6]. The first report from SNO [2] has shown a strong evidence for ν_e oscillations by combining the SNO data on the charged-current(CC) reaction with the Super-Kamiokande data on neutrino-electron elastic scattering(ES) [7]. The sharpness of this important conclusion depends on the precision of theoretical estimates for the νd-reaction cross sections.

Theoretical estimations of νd reactions are to consider one-body and two-body nuclear weak currents acting on the nuclear wave function. In the standard nuclear physics approach(SNPA), the two-body exchange current is derived from meson exchange model. The most recent and detailed study of νd reaction in SNPA is done by Nakamura et al.(NSGK)[4]. A new approach based on the effective field theory(EFT)[5] has been applied to νd reactions by Butler et al.(BCK)[6]. BCK adjusted unknown low energy constant associated with the two-body axial vector current by fitting the cross section of NSGK. With this low energy constant, results of NSGK and BCK agree with each other within 1% level over the entire solar neutrino energy region.

In this note we report our improved results of our previous work(NSGK) on the νd reaction cross sections based on Ref. 8. At first we updated the value of the axial coupling constant g_A pointed out by Beacom and Parke [9]. Secondly, we examine the consequences of using the full set of exchange currents [12], which include some sub-dominant diagrams left out in NSGK. Furthermore, we adopt the strength of axial vector exchange current $\boldsymbol{A}_{\rm EXC}$ controlled by Γ_t^β ($\boldsymbol{A}_{\rm EXC}(\beta)$). NSGK adopted the strength of $\boldsymbol{A}_{\rm EXC}$ adjusted to reproduce the measured $np \to \gamma d$ rate ($\boldsymbol{A}_{\rm EXC}(\gamma)$). However, $np \to d\gamma$ reaction governed by the vector current cannot be considered as a better constraint than Γ_t^β for monitoring the strength of $\boldsymbol{A}_{\rm EXC}$. Thirdly, at the level of precision in question, radiative corrections become relevant[9,10,11]. Here we will not address radiative corrections but study a related problem on the weak coupling constant. NSGK used the standard Fermi constant, G_F, which has been derived from μ-decay. Here we employ an effective coupling constant (denoted by $G_F^{\prime}$) that includes the so-called inner radiative corrections for nuclear β-decay.

In addition to these improvements within the framework of SNPA, we present a comparison with the work by Ando et al.[13] based on an EFT approach developped by Park *et al.* [14,15,16]. Here the electroweak transition operators are derived with a cut-off scheme EFT (à la Weinberg [17]) and the initial and final wave functions are obtained from the high-quality phenomenological nuclear interactions(EFT*). EFT* applied to the Gamow-Teller transitions contains one unknown LEC determined *directly* from Γ_t^β [16], which is a parameter-free EFT calculation of $\sigma_{\nu d}$.

In the next section we briefly discuss our new results of νd reactions [8].

2 Results

In this report we concentrate on the quantities directly relevant to the SNO solar neutrino experiments and limit ourselves to the neutrino reactions for the charge-current(CC) and neutral current(NC) with $E_\nu \leq 20$ MeV. The calculational framework is exactly the same as NSGK. Here we improve the following points on the nuclear weak current. For the weak coupling constant, instead of $G_F = 1.16637 \times 10^{-5}$ GeV^{-2} employed in NSGK, we adopt $G_F^{\prime} = 1.1803 \times 10^{-5}$ GeV^{-2} obtained from $0^+ \rightarrow 0^+$ nuclear β-decays [18] with $\cos\theta_C = 0.9740$. $G_F^{\prime}$ subsumes the bulk of the *inner* radiative corrections. For g_A, we adopt the current standard value, g_A=1.267, instead of g_A=1.254 used in NSGK. We employ here the meson exchange axial vector current of Schiavilla *et al.* [12], which reproduces the experimental value of Γ_t^β. It consists of the π-pair, ρ-pair, π and ρ exchange Δ-excitation currents and $\pi\rho$-exchange current. The treatment of the exchange vector currents are the same those of as NSGK.

Using the improved nuclear current and nuclear wave function obtained from ANLV18 potential[19], which we refer as *standard case*, the calculated cross sections $\sigma_{\nu d}(stnd)$ are about 1% larger than those of NSGK $\sigma_{\nu d}(NSGK)$ shown in Table 1. The origin of this difference is a ~4% enhancement of $\sigma_{\nu d}$ due to the changes in the Fermi constant and g_A and a ~3% reduction due to the use of the $\boldsymbol{A}_{\rm EXC}(\beta)$ instead of using $\boldsymbol{A}_{\rm EXC}(\gamma)$. These changes cancel each other to some extent giving the net results shown in Table 1.

In NSGK, estimation of theoretical error of $\sigma_{\nu d}$ was essentially based upon the difference between the cross sections obtained by using $\boldsymbol{A}_{\mathrm{EXC}}(\beta)$ and $\boldsymbol{A}_{\mathrm{EXC}}(\gamma)$. As we have discussed, $\boldsymbol{A}_{\mathrm{EXC}}(\gamma)$ fails to explain Γ_t^β and should not given the same status as $\boldsymbol{A}_{\mathrm{EXC}}(\beta)$. The exchange current tested for Γ_t^β may not be necessary valid for the kinematics of the solar neutrino energy region. Our analysis[8] shows that when we normalize the strength of $\boldsymbol{A}_{\mathrm{EXC}}$ at threshold then the difference between two models of exchange current is only up to 0.2% over the entire solar neutrino energy region. We propose this difference as a measure of the model dependence of exchange current.

Table 1: Comparison of the present results with those of NSGK. The ratio $\sigma_{\nu d}(stnd)/\sigma_{\nu d}(NSGK)$ is given for representative value of E_ν

E_ν (MeV)	5	10	15	20
$\nu_e d \to e^- pp$(CC)	1.013	1.008	1.006	1.004
$\nu d \to \nu pn$(NC)	1.011	1.006	1.003	1.001

An additional important measure of reliability of our SNPA calculation is obtained by comparing with the results of EFT* calculation. Ando *et al.* [13] have carried out a parameter-free EFT-motivated calculation of $\sigma_{\nu d}$by fixing the low-energy constant to reproduce the experimental value of Γ_t^β [16]. The uncertainty in $\sigma_{\nu d}$ in this study is dominated by the 0.5% error resulting from the uncertainty in the experimental value of Γ_t^β. We compare the cross section of s-wave final NN state in SNPA $\sigma_{\nu d}$(s-wave) with $\sigma_{\nu d(\mathrm{EFT}^*)}$, since only s-wave is included in Ref.[13]. The ratio $\eta \equiv \sigma_{\nu d(\mathrm{EFT}^*)}/\sigma_{\nu d}(s\text{-wave})$ is shown in Table 2 from which we can conclude that SNPA and EFT give identical results at the 1% level.

Table 2: Comparison of SNPA and EFT* calculations.

E_ν (MeV)	5	10	15	20
$\nu_e d \to e^- pp$	1.003	1.001	0.999	0.998
$\nu d \to \nu pn$	1.004	1.003	1.002	1.001

The ratio $R \equiv \sigma_{\nu d}^{NC}/\sigma_{\nu d}^{CC}$ is also an important quantity for SNO experiments, which will place constraints on various neutrino oscillation scenarios as emphasized by Bahcall and Lisi [20]. The calculated R in the *standard case* is shown as R_{stnd} in table 3 together with the deviations of R in various models from R_{stnd}. The table indicates that the model dependence of R is smaller than that of $\sigma_{\nu d}^{CC}$ and $\sigma_{\nu d}^{NC}$ themselves. The simple IA calculation gives a value of R that agrees with R_{stnd} within 0.2%. The variance between the *standard* *case* and the $R_{stnd-\gamma}$ obtained using $\boldsymbol{A}_{\rm EXC}(\gamma)$, is less than 0.5%. It is to be recalled that results using $\boldsymbol{A}_{\rm EXC}(\gamma)$ is a rather extreme. Furthermore, the difference between the *standard case* and the EFT* calculation does not exceed 0.4%. We therefore consider it reasonable to assign 0.5% accuracy to R. This is an improvement by a factor of ~2 over the precision reported in [3,4]. This stability of R can be understood by noting the dominance of the iso-vector axial-vector current on νd reactions in the solar neutrino energy region.

Table 3: The ratio $R \equiv \sigma_{\nu d}^{NC}/\sigma_{\nu d}^{CC}$.

E_ν (MeV)	R_{stnd}	$R_{\rm IA}/R_{stnd}$	$R_{stnd-\gamma}/R_{stnd}$	$R_{\rm EFT*}/R_{stnd}$
5	0.2764	1.000	1.004	1.001
10	0.4087	1.001	1.004	1.002
15	0.4378	1.002	1.004	1.004
20	0.4464	1.002	1.005	1.004

3 Summary

In summary, we have improved NSGK's calculation [4] for the νd reactions by updating its inputs and with the use of the axial-vector exchange current whose sthrength is controlled by Γ_t^β. We have also taken into account the results of a recent parameter-free EFT* calculation [13]. The new value of $\sigma_{\nu d}$is slightly larger than $\sigma_{\nu d}(NSGK)$ reported in [4]. Based on the arguments presented above, we consider it reasonable to assign 1% uncertainty to $\sigma_{\nu d}(stnd)$, and ~0.5% uncertainty to R. The results, however, do not include radiative corrections except for those already incorporated into the empirical value of $G_F^{\prime}$, which subsumes the bulk of the inner radiative corrections for nuclear β-decay. With the inclusion of the remaining radiative corrections, $\sigma_{\nu d}^{CC}$ is likely to become larger than $\sigma_{\nu d}^{CC}(stnd)$ by up to ~2 %, while $\sigma_{\nu d}^{NC}$ is expected to lie within the quoted ~1 % error of $\sigma_{\nu d}^{NC}(stnd)$.

1. The SNO Collaboration, Phys. Lett. B **194**, 321 (1987), and references therein.
2. Q. Ahmad *et al.*, Phys. Rev. Lett. **87**, 071301 (2001).
3. N. Tatara, Y. Kohyama and K. Kubodera, Phys. Rev. C **42**, 1694 (1990); S. Ying, W. C. Haxton and E. M. Henley, Phys. Rev. C **45**, 1982 (1992); for a review, see K. Kubodera and S. Nozawa, Int. J. Mod. Phys. E **3**, 101 (1994).

4. S. Nakamura, T. Sato, V. Gudkov and K. Kubodera, Phys. Rev. C **63**, 034617 (2001).
5. For a review, see U. van Kolck, Prog. Part. Nucl. **43**, 337 (1999); S.R. Beane *et al.*, in "At the Frontiers of Particle Physics - Handbook of QCD", ed. M. Shifman, vol. 1 (World Scientific,Singapore, 2001).
6. M. Butler and J.-W. Chen, Nucl. Phys. **A675**, 575 (2000); M. Butler, J.-W. Chen and X. Kong, Phys. Rev. C **63**, 035501 (2001).
7. S. Fukuda *et al.*, Phys. Rev. Lett. **86**, 5651 (2001).
8. S. Nakamura, T. Sato, S. Ando, T.-S. Part, F. Myhrer, V. Gudkov and K. Kubodera, to be appear in Nucl. Phys.
9. J. F. Beacom and S. J. Parke, Phys. Rev. D **64**, 091302 (2001); J. F. Beacom, private communication.
10. I. S. Towner, Phys. Rev. C **58**, 1288 (1998).
11. A. Kurylov, M. J. Ramsey-Musolf and P. Vogel, nucl-th/0110051.
12. J. Carlson, D. O. Riska, R. Schiavilla and R. B. Wiringa, Phys. Rev. C **44**, 619 (1991). R. Schiavilla, V. G. J. Stoks, W. Glöckle, H. Kamada, A. Nogga, J. Carlson, R. Machleidt, V. R. Pandharipande, R. B. Wiringa, A. Kievsky, S. Rosati and M. Viviani, Phys. Rev. C **58**, 1263 (1998).
13. S. Ando *et al.*, in preparation.
14. T.-S. Park, D.-P. Min and M. Rho, Phys. Rev. Lett. **74**, 4153 (1995).
15. T.-S. Park, K. Kubodera, D.-P. Min and M. Rho, Nucl. Phys. **A684**, 101 (2001), and references therein.
16. T.-S. Park, L. E. Marcucci, R. Schiavilla, M. Viviani, A. Kievsky, S. Rosati, K. Kubodera, D.-P. Min and M. Rho, nucl-th/0106025; nucl-th/0107012.
17. S. Weinberg, Phys. Lett. B **251**, 288 (1990); Nucl. Phys. **B363**, 3 (1991).
18. I. S. Towner and J. C. Hardy, in Physics beyond the standard model, eds. P. Herczeg, C.M. Hoffman and H.V. Klapdor-Kleingrothaus, (World Scientific, Singapore, June 1999) p. 338.
19. R. B. Wiringa, V. G. J. Stoks and R. Schiavilla, Phys. Rev. C **51**, 38 (1995).
20. J. N. Bahcall and E. Lisi, Phys. Rev. D **54** 5417 (1996).

SOLAR MODEL BASED ON HELIOSEISMOLOGY

SATORU WATANABE AND HIROMOTO SHIBAHASHI

Department of Astronomy, School of Science, The University of Tokyo, Bunkyo-ku, Tokyo 113-0033, Japan

E-mail: watanabe@astron.s.u-tokyo.ac.jp

Imposing the constraints of both of the sound-speed and the density profiles determined from helioseismology, we construct a solar model which is faithfully consistent with observations of helioseismology. We discuss the theoretically expected neutrino fluxes and their uncertainties based on this faithful solar model.

1 Introduction

Helioseismology is a powerful tool to study the internal structure of the Sun. From inversion of a large number of very accurately measured eigenfrequencies of the Sun, we now know the sound-speed profile in the Sun, $c(r)$, within the errors of a few tenths of a percent. With the imposition of a constraint of the sound-speed profile thus obtained, the profiles of temperature, density, pressure, luminosity, and chemical composition can be deduced (the seismic solar model; Shibahashi [22]; Watanabe and Shibahashi [26]).

Although the conventional discussion about the solar neutrino problem has been based on evolutionary solar models, the seismic solar model has the advantages over them as follows:

- We can construct a model of the present-day Sun without assumptions about the evolutionary history of the Sun, which cannot be directly justified by observations.

- The location of the base of the convection zone, $r_{\rm conv}$, is accurately known by helioseismology. Knowing $r_{\rm conv}$ allows us to treat the radiative core only without worrying about the treatment of convection, which is not well-described theoretically (Takata and Shibahashi [23]).

- The model is naturally consistent with the helioseismically determined sound-speed profile, while the standard evolutionary solar models are not necessarily so.

Therefore we can discuss about the solar neutrino problem from the new point of view through the seismic solar model.

2 Equations of the Seismic Solar Model

We solve the basic equations governing the radiative core of the Sun with the imposition of information from helioseismology. They are the continuity equation,

$$\frac{dM_r}{dr} = 4\pi r^2 \rho, \tag{1}$$

the hydrostatic equation,

$$\frac{dP}{dr} = -\frac{GM_r\rho}{r^2}, \tag{2}$$

the energy equation,

$$\frac{dL_r}{dr} = 4\pi r^2 \rho\varepsilon, \tag{3}$$

and the energy transfer equation,

$$\frac{dT}{dr} = -\frac{3}{4ac}\frac{\kappa\rho}{T^3}\frac{L_r}{4\pi r^2}. \tag{4}$$

In addition to these differential equations, we need auxiliary equations —that is, the equation of state, the equations for the opacity and the nuclear reaction rates:

$$\rho = \rho(P, T, X_i), \quad \kappa = \kappa(P, T, X_i), \quad \text{and} \quad \varepsilon = \varepsilon(P, T, X_i), \tag{5}$$

which link the thermal quantities and the chemical abundances. In making standard evolutionary models, the chemical abundance distribution is obtained by following the time evolution:

$$\frac{\partial X_i}{\partial t} = \left(\frac{\partial X_i}{\partial t}\right)_{\text{nuclear}} + \left(\frac{\partial X_i}{\partial t}\right)_{\text{diffusion}}, \tag{6}$$

where the nuclear reactions and the diffusion processes are taken into account. In the standard scenario, the chemical composition is assumed to be uniform at zero-age; $\partial X_i(t=0)/\partial r = 0$.

On the other hand, in constructing seismic solar models, we do not follow the time evolution. Instead, we impose the helioseismically determined sound-speed and density profiles as constraints on the model. Note that, if we distinguish only the hydrogen (^{1}H) and helium (^{4}He) separately as X and Y, respectively, and treat all the other elements collectively as heavy elements, Z, then the sound speed and the density can be regarded as functions of the chemical composition, X and Z, and any two other thermodynamical quantities such as P and T;

$$c(P, T, X, Z) = c_{\rm obs}(r) \tag{7}$$

and

$$\rho(P, T, X, Z) = \rho_{\rm obs}(r). \tag{8}$$

These inversely relate the hydrogen abundance, X, and the heavy elements abundance, Z, at a given r with the pressure, temperature, the sound speed, and the density;

$$X = X(P, T, c_{\rm obs}, \rho_{\rm obs}) \quad \text{and} \quad Z = Z(P, T, c_{\rm obs}, \rho_{\rm obs}). \tag{9}$$

Then both the opacity and the nuclear reaction rate are, in turn, given in terms of $(P, T, c_{\rm obs}, \rho_{\rm obs})$ by Eq. (5). Thus all the variables appearing in the right-hand side of equations (1)–(4) can be expressed in terms of the dependent variables in the left-hand side, and hence these equations are solvable. Note that in this way we obtain directly a model of the present day's Sun without following the evolution of the Sun. Note also that we do not need to make assumptions about the chemical composition profiles in the Sun, but obtain the X-, Y-, and Z-profiles as a part of the solutions of the fourth-order differential equations of stellar structure (1)–(4).

The boundary conditions at the center are trivial:

$$M_r = 0 \quad \text{and} \quad L_r = 0 \quad \text{at} \quad r = 0. \tag{10}$$

The location of the base of the convection zone is estimated from the inverted sound-speed profile. We adopt the helioseismically determined depth of the convection zone, $r = r_{\rm conv}$, and set the outer boundary conditions at the base of the convection zone (Takata and Shibahashi [23]);

$$\nabla_{\rm ad} = \nabla_{\rm rad} \equiv \frac{3\kappa L_r P}{16\pi acGM_rT^4} \quad \text{and} \quad L_r = L_\odot \quad \text{at} \quad r = r_{\rm conv} \tag{11}$$

This means that we do not need to care about the convective heat transport, which has theoretical uncertainties. Moreover, chemical homogeneity in the convection zone requires Z/X at the base of the convection zone should be identical with the value at the photosphere, $(Z/X)_{\rm surf}$, which is determined spectroscopically.

Table 1. The Microphysics and the physical parameters of seismic solar models.

Physics/Parameters		Reference
Equation of state	OPAL	Rogers *et al.* [18]
Opacity	OPAL	Iglesias and Rogers [15]
Nuclear cross-section		Adelberger *et al.* [1]; Marcucci *et al.* [17]
Screening effect	Weak screening	Salpeter [20]; Gruzinov and Bahcall [14]
$(Z/X)_{\rm surf}$	0.023	Grevesse and Sauval [13]
Luminosity $L_\odot$		Bahcall and Pinsonneault [4]
Sound-speed profile	SOHO MDI	Basu [6]
Density profile	SOHO MDI	Basu [6]
$r_{\rm conv}$	$0.713R_\odot$	Basu and Antia [7]

3 The Practical Method for Constructing the Seismic Solar Model

Table 1 summarizes the microphysics and the physical parameters adopted in this paper to construct seismic solar models. They are the same as those adopted by Bahcall et al. [5] (hereafter BP00). As for the helioseismically determined sound-speed and density profiles, we adopt Basu's [6] results obtained by inversion of the MDI velocity data of the first 144 days. The formal error of the sound-speed inversion is smaller than 0.05 % in the region $0.2 \le r/R_\odot \le 0.8$ and does not exceed 0.3 % in the entire region. The density profile has the error of the order of 1% in the entire region. In the central core ($r/R_\odot \le 0.05$), we extrapolate the inverted profiles and their error levels to the center.

The mathematical procedure has been described in the previous section. It is, however, difficult in practice to determine the Z-profile directly in that way, since the dependence of the equation of state upon Z is very weak. In this section, instead, we construct seismic solar models with a constraint of the helioseismically determined sound-speed profile by assuming various Z-profiles, and we search among them for the model of which the density profile fits best with the helioseismically determined density profile. If the sound-speed profile and the Z-profile are given, X is given in terms of, $(P, T, c_{\rm obs}, Z)$ by Eq. (7). Then the density, the opacity and the nuclear reaction rate are given in terms of $(P, T, c_{\rm obs}, Z)$ by Eq. (5). Therefore the basic equations (1)–(4) can be solved with an assumed Z-profile.

Our goal is to find among various seismic solar models the model which minimizes the following integral, which is called the fitness indicator:

$$F = \int_0^{r_{\rm conv}} \left(\frac{\rho({\rm obs}) - \rho({\rm model})}{\sigma_\rho({\rm obs})} \right)^2 \frac{dr}{R_\odot}, \tag{12}$$

where $\sigma_\rho(\text{obs})$ is the error of the helioseismically determined density.

We assume that the Z-profile is a monotonically decreasing or increasing function of radius. Based on numerical experiments, we found that seismic solar models are not so sensitive to the detail of Z-profiles. The properties of the model with Z-profile to the fourth decimal place and those of the model with Z-profile to the third decimal place (cf. figure 1) are almost same. Therefore we calculate Z-profile to the third decimal place and treat it as a step function like the dashed line in figure 1. We try to find the best-fit Z-profile by changing the number of steps and the width and the height of each step systematically. In the actual calculations, the resolution of the width is chosen as $0.1R_\odot$, and that of the height is 0.001. The resolution of the width is exceptionally $0.2R_\odot$ in the core, $r/R_\odot \leq 0.2$, because we found that the properties of the model, particularly the model's density, is not so sensitive to the resolution of the width in the solar core. We try all the possible Z-profiles under the above assumptions and evaluate the fitness indicator, F, given in Eq. (12). The results are shown in figure 2, in which the ^{8}B-neutrino flux is plotted against the fitness indicator, F, for each seismic solar model differing in Z-profile. There are two tendencies:

- A higher Z (but lower than a critical value) in the core leads to a smaller F and a larger ^{8}B-neutrino flux.

- On the other hand, a higher Z than a critical value in the core inversely leads to a larger F and a much larger ^{8}B-neutrino flux.

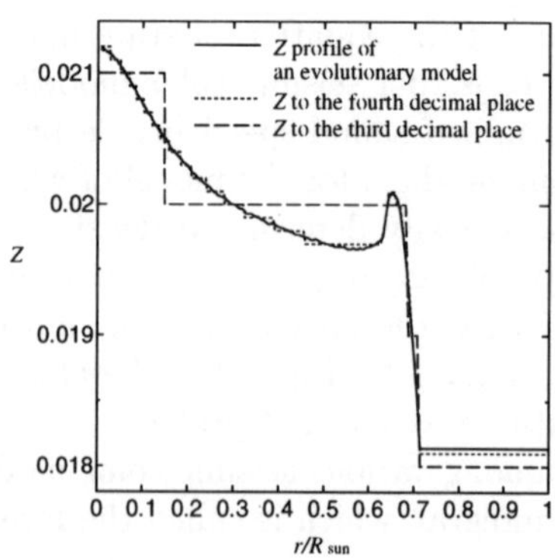

Figure 1. Example of approximated Z-profiles. The dotted and dashed lines are the approximations of the solid line to the fourth and the third decimal place, respectively.

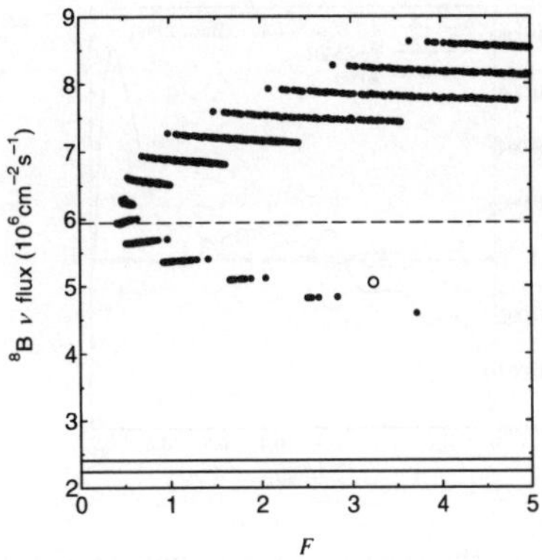

Figure 2. The ^{8}B-neutrino flux is plotted against the fitness indicator, F, for each seismic solar model differing in Z-profile. The open circle at (F, ^{8}B ν flux) = (3.24, 5.05) is BP00. The ^{8}B-neutrino flux for the seismic solar model with the best-fit Z-profile corresponding to the minimum value of F is shown by the horizontal dashed line. For a comparison, ± 1 σ errors of the ^{8}B-neutrino flux detected by the Super-Kamiokande (Fukuda *et al.* [10]) are shown by the horizontal solid lines.

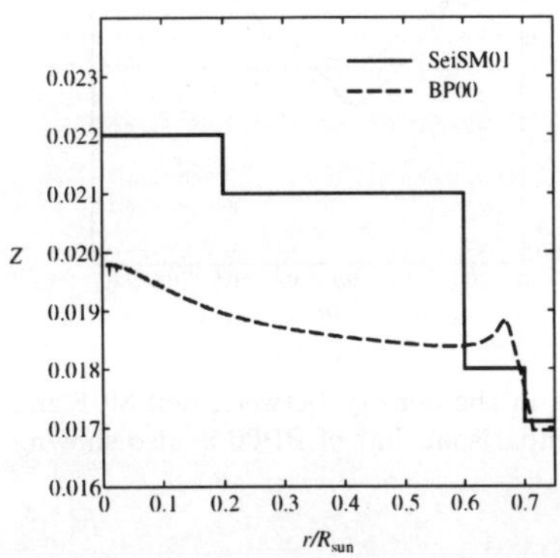

Figure 3. The best-fit Z-profile. For a comparison, the Z-profile of BP00 is also shown.

We call the Z-profile minimizing the fitness indicator, $F_{\text{min}} \approx 0.42$, as the best-fit Z-profile (figure 3) and the corresponding seismic solar model as the seismic solar model 2001 (hereafter SeiSM01; Watanabe and Shibahashi 2001b [27]).

Comparison of the sound-speed and the density profiles between the helioseismically determined profiles and SeiSM01 are shown in figure 4 and 5.

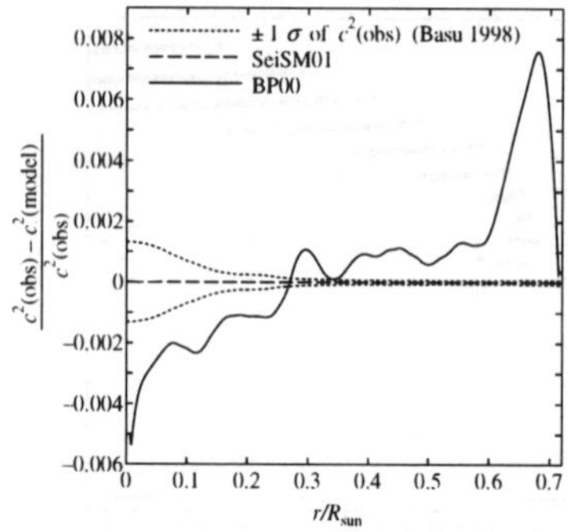

Figure 4. Relative difference in the square of the sound-speed between SeiSM01 and the helioseismically determined profile [6]. For a comparison, that of BP00 is also shown.

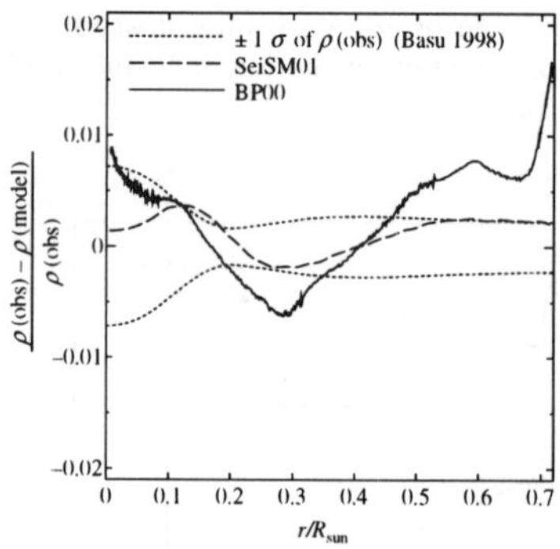

Figure 5. Relative difference in the density between SeiSM01 and the helioseismically determined profile [6]. For a comparison, that of BP00 is also shown.

4 Theoretical Uncertainties

In order to estimate the uncertainty of the model, we first evaluate the sensitivity of various physical quantities of the model, such as the central temperature and the theoretically expected neutrino fluxes, to each of the significant input parameters such as the nuclear cross-section S-factor. Let y_i and x_j be a certain physical quantity of the output model and a certain input parameter, respectively. We evaluate numerically the logarithmic partial derivative of y_i with respect to x_j,

$$\alpha_{i,j}|_{\mathrm{SeiSM01}} = \left(\frac{\partial \ln y_i}{\partial \ln x_j}\right)_{x_k,\ L_\odot,\ r_{\mathrm{conv}},\ (Z/X)_{\mathrm{surf}},\ c(r),\ \rho(r)} , \tag{13}$$

by adopting slightly different ($\pm 1\ \sigma \sim \pm 2\ \sigma$ level of the uncertainty) values from the most likely value of the parameter x_j and then calculating SeiSM01 with each value of x_j. In this procedure, we tune up the Z-profile so that the fitness indicator, F, always turns to be a minimum: SeiSM01 always reproduce the helioseismically determined density profile. In Eq. (13), $\{x_k\}$ are input parameters except for x_j. The net uncertainty, δy_i, in the physical quantity, y_i, of the output model is then obtained by

$$\frac{\delta y_i}{y_i} = \left[\sum_j (\alpha_{i,j}\delta x_j / x_j)^2\right]^{1/2} , \tag{14}$$

where $\{\delta x_j\}$ are the uncertainty in the input parameters, x_j.

Comparison of observations and solar models are summarized in table 2. The theoretical uncertainties in the ^{8}B-neutrino flux and the neutrino capture rates for the chlorine and gallium experiments of SeiSM01 are larger than those of BP00. This is caused by two reasons.

First, there are still considerable unknowns in many of the physical assumptions in the standard evolutionary solar models as follows (e.g. Gough

Table 2. Comparison of observations and solar models. T_c are the central temperature. Note that the concept of uncertainty of SeiSM01 is different from that of BP00. Therefore it is meaningless to compare the values of them literally and we propose the careful uncertainty of SeiSM01 (see text).

	Observations	SeiSM01	BP00
ν for Cl(SNU)	2.56 ± 0.23 [8]	$8.7^{+2.4}_{-2.8}$	$7.6^{+1.3}_{-1.1}$
ν for Ga(SNU)	$75.4^{+7.8}_{-7.4}$ [11] $74.1^{+6.7}_{-6.8}$ [3]	132^{+12}_{-13}	128^{+9}_{-7}
^{8}B ν flux (10^6 cm^{-2} s^{-1})	$2.32^{+0.09}_{-0.08}$ (ES) [10] $1.75^{+0.15}_{-0.14}$ (CC), $2.39^{+0.38}_{-0.37}$ (ES) [2] 5.44 ± 0.99 ($\nu_{e,\ \mu,\ \tau}$) [2]	$5.94^{+1.90}_{-2.17}$	$5.05^{+1.01}_{-0.81}$
^{7}Be ν flux (10^9 cm^{-2} s^{-1})	—	$5.18^{+0.62}_{-0.81}$	4.77 ± 0.48
$T_c(10^7$K)	—	$1.579^{+0.026}_{-0.032}$	~1.570

& Toomre [12]; Roxburgh [19]; Elliott [9]):

- The initial chemical composition is assumed to be uniform for simplicity, but this assumption cannot be justified directly by observations (Larson [16]).
- Mass loss and accretion, which are known to be important processes in stellar evolution, are ignored.
- The model adopts the mixing length theory, which is only a rough description of the convective transport.
- ^{3}He-induced g-mode instability, which may affect the evolution of the Sun, is ignored.

While the systematic errors coming from these ignored factors should be included in the error estimation of the standard evolutionary models, SeiSM01 does not have to worry about them.

Second, the procedure of estimating the uncertainty of models are different between SeiSM01 (Eq. (14)) and BP00;

$$\alpha_{i,j}|_{\mathrm{BP00}} = \left(\frac{\partial \ln y_i}{\partial \ln x_j}\right)_{x_k,\ L_\odot,\ R_\odot,\ (Z/X)_{\mathrm{surf}},\ (\text{evolutionary assumptions})} \tag{15}$$

BP00 evaluate derivatives by using solar models, which are not necessarily reproduce all observations. In our opinion, the standard evolutionary solar models gets apparent small uncertainties simply because they ignore the poor fitness of the sound-speed and the density profiles. For example, we can see how much the sound-speed and the density profiles of the standard evolutionary solar models are influenced by a slight change of microphysics in Schlattl *et al.* [21] and Turck-Chièze et al.[24,25]. Therefore we stress again that, in evaluating the uncertainties of the theoretically expected neutrino fluxes, we should use only solar models that reproduce all observations including the helioseismically determined sound-speed and density profiles.

5 Conclusion

Imposing the constraints of both of the sound-speed and the density profiles determined from helioseismology, we have succeeded in constructing the seismic solar model 2001 (SeiSM01) that is faithfully consistent with almost all observations (sound-speed profile, density profile, r_{conv}, $(Z/X)_{\mathrm{surf}}$, $L_\odot$) except for the neutrino fluxes. We evaluate the theoretically expected neutrino fluxes and their uncertainties based on SeiSM01. The theoretically expected neutrino fluxes of SeiSM01 are still significantly more than the observations and somewhat more than the latest standard evolutionary solar model. The theoretically expected ^{8}B-neutrino flux of SeiSM01 is, however, in agreement with the total flux of the active neutrino flux, $\phi(\nu_{e,\ \mu,\ \tau})$, which is determined by comparing the measured neutrino flux of SNO and that of the Super-Kamiokande. Therefore the neutrino oscillation turns out to be the solution of the solar neutrino problem. SeiSM01 will remain important in this field as a tool for constraining the parameters of the neutrino oscillation.

Acknowledgments

We would like to thank M. Takata for many useful discussions. This research was supported in part by the Grant-in-Aid for Scientific Research on Priority Areas by the Ministry of Education, Science, Sports and Culture of Japan (12047208).

References

1. E.G. Adelberger *et al.*, Rev. Mod. Phys. **70**, 1265 (1998).
2. Q.R. Ahmad, *et al.*, *Phys. Rev. Lett.* **87**, 071301 (2001).
3. M. Altmann, *et al.*, *Phys. Lett.* B **490**, 16 (2000).
4. J.N. Bahcall and M.H. Pinsonneault, Rev. Mod. Phys. **67**, 781 (1995).
5. J.N. Bahcall, M.H. Pinsonneault, and S. Basu, ApJ **555**, 990 (2001).
6. S. Basu, MNRAS **289**, 719 (1998).
7. S. Basu and H.M. Antia, MNRAS **287**, 189 (1997).
8. B.T. Cleveland *et al.*, ApJ **496**, 505 (1998).
9. J.R. Elliott, A&A **334**, 703 (1998).
10. Y. Fukuda *et al.*, *Phys. Rev. Lett.* **86**, 5651 (2001).
11. V.N. Gavrin, Nuclear. Phys. B Proc. Suppl. **91**, 36 (2000).
12. D. Gough and J. Toomre, ARA&A **29**, 627 (1991).
13. N. Grevesse and A.J. Sauval, Space Sci. Rev. **85**, 161 (1998).
14. A.V. Gruzinov and J.N. Bahcall, ApJ **504**, 996 (1998).
15. C.A. Iglesias and F.J. Rogers, ApJ **464**, 943 (1996).
16. R.B. Larson, MNRAS **145**, 271 (1969).
17. L.E. Marcucci *et al.*, Phys. Rev. C **63**, 015801 (2001).
18. F.J. Rogers, F.J. Swenson, and C.A. Iglesias, ApJ **456**, 902 (1996).
19. I.W. Roxburgh, Bull. Astron. Soc. India **24**, 89 (1996).
20. E.E. Salpeter, Australian J. Phys. **7**, 373 (1954).
21. H. Schlattl *et al.*, *Phys. Rev.* D **60**, 113002 (1999).
22. H. Shibahashi, in *Frontiers of Neutrino Astrophysics*, ed. Y. Suzuki and K. Nakamura (Universal Academy Press, Tokyo 1993) p93.
23. M. Takata and H. Shibahashi, ApJ **504**, 1035 (1998).
24. Turck-Chièze *et al.*, ApJL **555**, 69 (2001a).
25. Turck-Chièze *et al.*, Sol. Phys. **200**, 323 (2001b).
26. S. Watanabe and H. Shibahashi, PASJ **53**, 565 (2001a).
27. S. Watanabe and H. Shibahashi, 2001b, submitted to PASJ.

A UNIQUE OSCILLATION SOLUTION TO THE SOLAR NEUTRINO PROBLEM?

M. B. SMY

Department of Physics and Astronomy

4182 Frederick Reines Hall, University of California, Irvine, CA 92697-4575

A global two-neutrino oscillation fit combining Super-Kamiokande solar neutrino data with the solar neutrino rates measured by Homestake, Gallex/GNO, SAGE and SNO prefers a single allowed area, the Large Mixing Angle solution, at about 95% confidence level. The mass^2 difference between the two mass eigenstates is $\Delta m^2 \approx$3–25$\times 10^{-5}$eV^2, the mixing angle θ is $\tan^2\theta \approx$0.21–0.67.

1 Introduction

Neutrino flavor oscillations are currently the most favored way to understand the larger solar ^{8}B neutrino interaction rate measured by Super-Kamiokande[1] (SK) with respect to the ^{8}B rate reported by the SNO-collaboration[2]. The SNO rate includes only e-type ^{8}B neutrinos while SK measures solar neutrinos via neutrino-electron elastic scattering which has a small sensitivity to other (active) flavors as well. The SK excess is then interpreted as a hint for appearance of other active flavors in the ^{8}B neutrino flux (born in the sun as purely e-type). Neutrino oscillations also explain the deficit of both rates as well as the deficits of the "Chlorine rate" (Homestake[3]) and the "Gallium rates" (Gallex/GNO[4], SAGE[5]) with respect to the Standard Solar Model[6] (SSM). Chlorine and Gallium rates include only e-type solar neutrinos.

The large neutrino flavor mixing between the second and third generation inferred from atmospheric neutrino data[7] and the absence of an oscillation signal in the CHOOZ reactor neutrino experiment[8] requires the neutrino flavor mixing between the first and third generation to be small. Solar neutrino oscillations can therefore be approximated by a two-neutrino description with the parameters θ (mixing angle) and Δm^2 (difference in mass^2). The mixing angle ranges between 0 and $\pi/2$, since Δm^2 is defined to be positive. For Δm^2 between $\approx 10^{-8}$eV^2 and $\approx 10^{-3}$eV^2, the matter density in the sun and earth can strongly affect the oscillation probability. For $\theta < \pi/4$, resonant enhancement of the oscillations (MSW effect[9]) can occur in the sun. On the "dark side" of the parameter space ($\theta > \pi/4$), an anti-resonance can suppress the oscillations. Below $\approx 10^{-9}$eV^2 (quasi-vacuum/vacuum region), the oscillation probability is more affected by the oscillation phase than by matter effects.

2 Parameter Estimation for Two-Neutrino Oscillations

The simplest way to constrain the oscillation parameters is a χ^2 fit to all e-type neutrino rates, that is Gallex/GNO and SAGE (combined into a single "Gallium" number), Homestake and SNO. The dotted lines of figure 1a) show the allowed regions at 95% C.L. (which can be understood as an "overlay" of the allowed regions[10] from the three experimental rates mentioned above and their SSM-based predictions). The area above $\Delta m^2 \approx 10^{-5}\text{eV}^2$ near maximal mixing is called the Large Mixing Angle (LMA) solution. The Small Mixing Angle Solution (SMA) is located between $\Delta m^2 \approx 10^{-6}\text{eV}^2$ and $\approx 10^{-5}\text{eV}^2$ at $\tan^2\theta \approx 10^{-3}$. The LOW solution is the large Δm^2 part of the extensive region(s) between $\approx 10^{-7}\text{eV}^2$ and $\approx 10^{-9}\text{eV}^2$. The lower part of this region is called the quasi-vacuum (quasi-VAC) solution. The vacuum solutions (VAC) are below $\approx 10^{-10}\text{eV}^2$. All regions have similar χ^2 (LOW fits slightly worse).

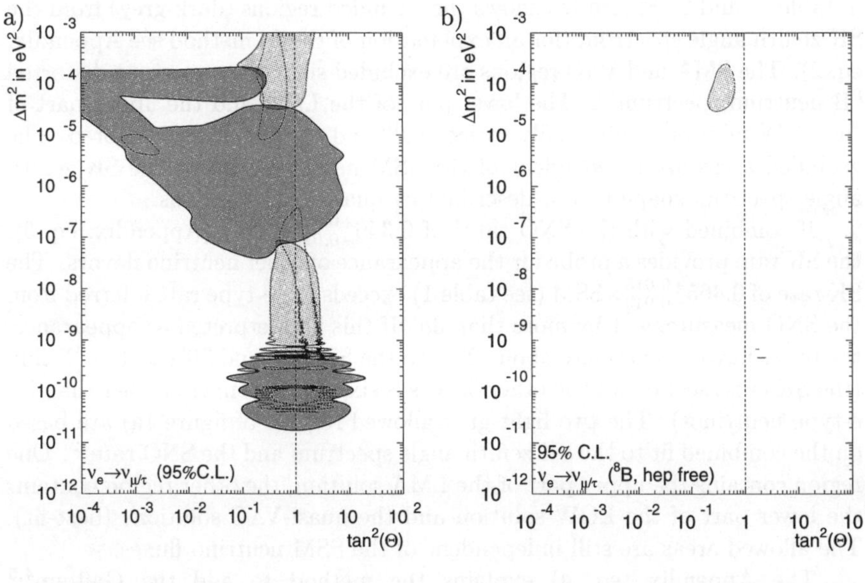

Figure 1. SK Zenith Spectrum/SNO (a) and global fit (b) contours at 95% C.L. The dark-grey area in a) is excluded by the shape of the SK Zenith Spectrum. Adding the SNO rate measurement results in the (light-grey) allowed areas. Both regions do not depend on neutrino flux predictions by the SSM. Overlaid are the allowed regions (inside dotted lines) based on the Gallex/GNO, SAGE, Homestake and SNO rates and the SSM. In b) the allowed areas from a global fit (adding the rates of Gallex/GNO, SAGE and Homestake to the SK Zenith Spectrum–SNO fit) are predominantly LMA solutions.

To break the degeneracy in χ^2 of these regions, more information is needed than provided by the suppression of the $e-$type rates. Indeed, neutrino oscillations not only cause a 'disappearance' of the $e-$type fluxes but also produce other neutrino flavors (appearance). Furthermore, they can induce distortions of the neutrino spectrum and time variations of the solar neutrino fluxes. In the MSW region, the time variations arise from matter effects inside the earth (daily variations). In the vacuum regions, the time variations are the consequence of the change of the oscillation phase due to the yearly variation of the oscillation baseline (distance between sun and earth).

SK can search for spectral distortions through analysis of the spectrum of the recoiling electron as well as daily or yearly time variations. The analysis of the shape of the SK "zenith angle spectrum"[10,11] combines spectrum and daily variation analyses. Neither spectral distortion nor daily variation was found. The centerpiece of this work is the preliminary SK zenith angle spectrum based on the entire SK data set (1496 days). The numerical results are listed in tables 1 and 2. Figure 1a) shows the excluded regions (dark-grey) from the SK zenith angle spectrum (for an explanation of the fit method see Appendix, eq. 2). The SMA and VAC regions are excluded since they require a distorted ^{8}B neutrino spectrum[10]. The lower part of the LMA and the upper part of the LOW solution predict daily variations[10] and are therefore disfavored. The excluded areas are independent of the SSM neutrino fluxes. The SK zenith angle spectrum shape is best described by quasi-VAC solutions.

If combined with the SNO rate[2] of $0.346^{+0.029}_{-0.028}\times$SSM (Appendix, eq. 3), the SK rate provides a probe for the appearance of other neutrino flavors. The SK rate of $0.465^{+0.015}_{-0.013}\times$SSM (see table 1) exceeds the e-type rate inferred from the SNO measurement by more than 3σ. If this is interpreted as appearance, the other flavors contribute about 25% to the SK rate and 70% to the ^{8}B flux (the SK cross section for the other flavors is six to seven times smaller than for e-type neutrinos). The two light-grey allowed regions of figure 1a) are based on the combined fit to the SK zenith angle spectrum and the SNO rate[12]. One region contains the upper part of the LMA solution, the other region contains the lower part of the LOW solution and the quasi-VAC solutions (best-fit). The allowed areas are still independent of the SSM neutrino fluxes.

The Appendix (eq. 4) explains the method to add the Gallium[4,5] ($74.8^{+5.1}_{-5.0}$SNU or $0.584^{+0.040}_{-0.039}\times$SSM) and Chlorine rates[3] (2.56 ± 0.23SNU or $0.337\pm0.030\times$SSM) to the fit. The allowed areas of this global fit (shown in figure 1b) look quite different when compared with the $e-$type rate fit: Only the upper part of the LMA survives (and two tiny quasi-VAC solutions). The disappearance of the LOW solution is due to predicted daily variations. The SK rate requires a larger ^{8}B flux than the Chlorine rate allows, so almost all

Table 1. SK Rate and Uncertainty for eight energy bins. The rates, statistical and systematic uncertainties (of the spectrum shape) in the third column are presented in units of SSM expectation. These systematic uncertainties are assumed to be uncorrelated in energy. The uncertainties in the fourth (uncertainty of the ^{8}B neutrino spectrum), fifth (uncertainty of the energy scale of 0.64%) and sixth column (uncertainty of the energy resolution of 2.5%) are fully correlated in energy (but uncorrelated with each other). The combined uncertainty (last row) is based on the sum of all bins of the zenith angle spectrum. The combined rate has an additional systematic uncertainty of $\sigma_{SK} =^{+2.9}_{-2.6}$ % (excluding ^{8}B shape, energy scale and resolution), which was added to the uncorrelated systematic uncertainties of the spectrum shape.

Bin	Range [MeV]	Rate±stat±syst [SSM]	^{8}B Spectrum	E-Scale	E-Resol.
1	5.0-5.5	$0.4671 \pm 0.0404\,^{+0.0165}_{-0.0138}$	$^{+0.04}_{-0.02}$ %	$^{+0.09}_{-0.01}$ %	$^{+0.23}_{-0.21}$ %
2	5.5-6.5	$0.4580 \pm 0.0141\,^{+0.0066}_{-0.0065}$	$^{+0.13}_{-0.09}$ %	$^{+0.20}_{-0.16}$ %	$^{+0.19}_{-0.17}$ %
3	6.5-8.0	0.4729±0.0084±0.0065	$^{+0.41}_{-0.38}$ %	$^{+0.63}_{-0.62}$ %	$^{+0.17}_{-0.16}$ %
4	8.0-9.5	0.4599±0.0093±0.0063	$^{+0.89}_{-0.85}$ %	$^{+1.3}_{-1.3}$ %	$^{+0.12}_{-0.12}$ %
5	9.5-11.5	0.4627±0.0103±0.0063	$^{+1.7}_{-1.6}$ %	$^{+2.5}_{-2.4}$ %	$^{+0.16}_{-0.18}$ %
6	11.5-13.5	0.4621±0.0168±0.0063	$^{+3.1}_{-2.7}$ %	$^{+4.4}_{-4.1}$ %	$^{+1.1}_{-1.1}$ %
7	13.5-16.0	0.5666±0.0390±0.0078	$^{+5.1}_{-4.2}$ %	$^{+7.0}_{-6.4}$ %	$^{+3.2}_{-3.2}$ %
8	16.0-20.0	0.5554±0.1458±0.0076	$^{+7.7}_{-5.6}$ %	$^{+10.6}_{-9.6}$ %	$^{+8.4}_{-7.9}$ %
Comb.	5.0-20.0	$0.4653 \pm 0.0047\,^{+0.0138}_{-0.0122}$	$^{+1.15}_{-1.04}$ %	$^{+1.66}_{-1.58}$ %	$^{+0.33}_{-0.34}$ %

Table 2. Subdivision of bins 2–7 according to the solar zenith angle θ_z. The range of $\cos\theta_z$ is given for each bin: $\cos\theta_z < 0$ is 'Day' and $\cos\theta_z > 0$ is 'Night' ('Mantle' and 'Core'). The rates are given in units of 0.001×SSM. Only statistical uncertainties are quoted. All systematic uncertainties (see table 1) are assumed to be fully correlated in zenith angle.

	Day			Mantle			Core
Bin	-1–0	0.00–0.16	0.16–0.33	0.33–0.50	0.50–0.67	0.67–0.84	0.84–1
2	453±20	442±53	379±49	472±45	522±45	503±49	426±52
3	474±12	530±34	506±30	438±26	478±26	451±28	439±31
4	448±13	463±36	470±33	462±29	509±29	461±32	451±35
5	453±15	449±40	502±38	451±32	473±32	477±35	483±40
6	477±25	509±67	351±55	391±49	498±53	434±56	521±64
7	511±54	570±150	831±167	694±131	665±127	441±118	469±131

the quasi-VAC regions disappear. The fit does not depend on the ^{8}B and *hep* neutrino flux predictions by the SSM, which suffer from the largest uncertainties. However, it depends on the other SSM neutrino fluxes, in particular the ^{7}Be flux (10% uncertainty) and the neutrino fluxes of the CNO cycle (≈ 20% uncertainty). Those fluxes contribute[6] about 15%(^{7}Be) and 6%(CNO) to the Chlorine rate and 27%(^{7}Be) and 7%(CNO) to the Gallium rate in the SSM.

Table 3. Parameters for the best-fit points. The probabilities given in the fourth row are based on the difference of χ^2 (with respect to the minimum). The five rows below show the five independent parts of the fit: the $\Delta\chi^2$ from a fit to the shape of the SK zenith spectrum (Appendix, eq. 2) and four interaction rates (deviation probabilities are given in units of Gaussian standard deviation σ). The last four rows show the values of the minimized fit parameters. The ^{8}B and the *hep* fluxes are free, the ^{8}B neutrino spectrum shift as well as the SK energy scale and resolution shifts are constrained within the systematic uncertainties.

Solution	Large Mixing Angle (LMA)	Quasi-Vacuum (Quasi-VAC)	Low Δm^2 (LOW)	Small Mixing Angle (SMA)
Δm^2	6.0×10^{-5}	4.57×10^{-10}	5.0×10^{-8}	4.8×10^{-6}
$\tan^2\theta$	0.35	2.1	0.83	0.00044
χ^2 (45 dof; p_{χ^2} [%])	43.4 (53.9)	48.5 (33.4)	51.2 (24.3)	54.2 (16.2)
$\Delta\chi^2$(2 dof;$p_{\Delta\chi^2}$[%])	0.0 (100.0)	5.1 (7.9)	7.8 (2.0)	10.8 (0.5)
$\Delta\chi^2_{SK}$ ($p_{\Delta\chi^2}[\sigma]$)	3.4 (1.3σ)	3.1 (1.3σ)	3.9 (1.5σ)	5.0 (1.7σ)
Ga Rate [SNU]	73.2 (-0.3σ)	69.6 (-1.0σ)	68.2 (-1.3σ)	75.1 ($+0.1\sigma$)
Cl Rate [SNU]	2.97 ($+1.8\sigma$)	3.18 ($+2.8\sigma$)	3.13 ($+2.5\sigma$)	2.67 ($+0.5\sigma$)
SK Rate [%SSM]	46.4 (-0.1σ)	44.7 (-1.4σ)	44.9 (-1.2σ)	44.1 (-1.9σ)
SNO Rate [%SSM]	32.8 (-0.7σ)	37.1 ($+0.8\sigma$)	38.5 ($+1.3\sigma$)	43.8 ($+3.1\sigma$)
$\phi_{^8B}$ [$10^6/(\mathrm{cm}^2\mathrm{s})$]	5.62 ($+0.6\sigma$)	3.71 (-1.7σ)	4.04 (-1.2σ)	2.71 (-2.9σ)
ϕ_{hep} [$10^3/(\mathrm{cm}^2\mathrm{s})$]	40	0	21	8
^{8}B Spectrum Shape	-0.3σ	-0.7σ	-0.1σ	$+0.1\sigma$
SK E-scale/resol.	$-0.4\sigma/-0.1\sigma$	$-1.0\sigma/\ 0.0\sigma$	$-0.1\sigma/-0.2\sigma$	$+0.1\sigma/-0.4\sigma$

3 Results of the global fit

Table 3 compares the four smallest local minima of the χ^2 describing the fit. The best-fit is located in the upper LMA area. The ^{8}B flux resulting from this fit is somewhat higher than expected by the SSM ($5.05^{+1.01}_{-0.81}\times10^6/(\mathrm{cm}^2\mathrm{s})$) but well within the uncertainty. The *hep* flux is considerably higher than expected by the SSM ($9.3\times10^3/(\mathrm{cm}^2\mathrm{s})$); however, the uncertainty of this prediction is thought to be very large. The fit agrees with the SK zenith angle spectrum moderately well. Figure 2a) shows the SK zenith angle spectrum and this best fit. Figure 2b) gives a magnified view of the LMA region (in a linear scale of $\tan^2\theta$). The 1σ (black) and 3σ (light grey) contours are also shown. The fit accommodates well the Gallium, SK and SNO rates, however the predicted Chlorine rate is about 2σ too high. This worsens the otherwise very good best-fit χ^2. As a consequence, quasi-VAC solutions, which don't fit well the interaction rates in general, appear at 95% C.L.

The quasi-VAC solution describes the SK zenith angle spectrum slightly better than the LMA. The fit is helped by a statistical up-fluctuation of the last two bins (see figure 3). Together with a down-shift of energy scale and ^{8}B spectrum, the required oscillation minimum can be 'generated' in the data.

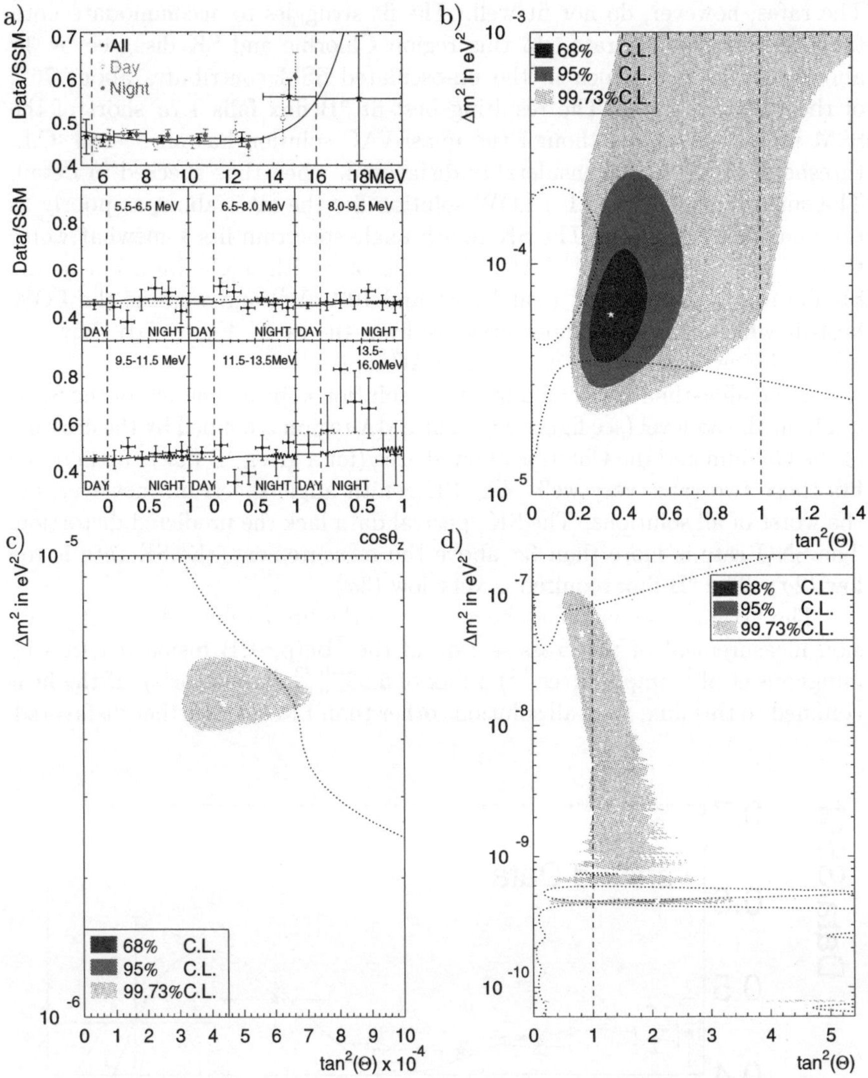

Figure 2. SK Zenith Spectrum (a) and LMA (b), SMA (c) and LOW/Quasi-VAC (d) contours resulting from the global fit. The top panel of a) shows the SK energy spectrum during the day (open circles), the night (filled circles) as well as day and night combined (stars). Between 5.5 and 16 MeV, zenith-angle distributions are shown (lower 6 panels of a). The dotted line in each of the lower panels is the combined rate in that bin. Error bars reflect the statistical uncertainty only. Superimposed is an oscillation prediction ($\tan^2\theta = 0.34$ and Δm^2=6.0$\cdot10^{-5}$eV^2) near the best-fit point. In b)-d) the one (black), two (dark-grey) and three σ (light-grey) contours of the LMA, SMA and LOW/Quasi-VAC solutions are shown in a linear scale in $\tan^2\theta$. The best fits are indicated by the white stars. Superimposed is the 95%C.L. excluded area from the SK Zenith Spectrum (inside dotted lines). SMA, LOW and Quasi-VAC are disfavored.

The rates, however, do not fit well. The fit struggles to accommodate both the Chlorine and SK rate. In this region Chlorine and SK disagree by 3σ about the ^{8}B flux (which in the un-oscillated SSM contributes about 76% of the Chlorine rate). The resulting best-fit ^{8}B flux falls 1.7σ short of the SSM prediction. Even though the quasi-VAC solution has an overall 'C.L. threshold' of 8%, it is considerably disfavored, when thus checked in detail. The surviving (at 98% C.L.) LOW solution fits the rates about as poorly as the quasi-VAC solution. The SK zenith angle spectrum fits somewhat worse than either LMA or quasi-VAC. The lack of zenith angle variation in the SK data reduces the Δm^2 (usually around 10^{-7}eV^2) and worsens the LOW best-fit which is already under pressure from the rates. Figure 2d) shows an enlarged view of the LOW and quasi-VAC region.

A "smaller-than-small mixing angle" solution appears just left of the SMA at about the 3σ level (see figure 2c). The SMA region is defined by the crossing of the Gallium and the Chlorine allowed area (for a given ^{8}B flux). It therefore fits those two rates very well. The SK zenith angle spectrum, however, fits the worst of all solutions: The SK spectral data lack the predicted distortion. The SNO rate is more than 3σ above the measurement, the SK rate is too low (2σ). The ^{8}B flux required is very low (3σ).

The ^{8}B flux comparisons above refer to the SSM value. A recent precision measurement of the cross section of the ^{7}Be(p,γ)^{8}B fusion reaction by Junghans et al.[13] implies (see [14]) a flux of $5.93^{+0.83}_{-0.89} \times 10^6/(\text{cm}^2\text{s})$. If the fit is confined to this flux, then all solutions other than LMA are further disfavored.

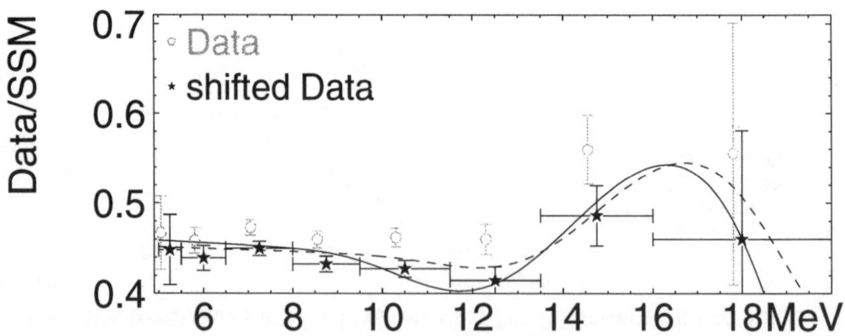

Figure 3. SK Spectral distortion for two quasi-VAC solutions (solid line: best fit at $4.57 \cdot 10^{-10}$eV^2, dashed line: $6.68 \cdot 10^{-10}$eV^2) and data (open circles). Although the data do not support spectral distortions, a shift (stars) in energy scale (-1.0σ) and ^{8}B shape (-0.7σ) and systematic uncertainty of the combined rate (-1.4σ) can accommodate the prediction.

4 Conclusion

A global fit to all solar neutrino data was performed using a two-neutrino oscillation model. At 95% C.L. only the upper LMA solution and two tiny quasi-VAC regions are still allowed. The quasi-VAC regions are disfavored at more than 92% C.L. caused by a disagreement about the ^{8}B flux between SK and the Chlorine rate of 3σ; the resulting quasi-VAC ^{8}B flux fit is 1.7σ smaller than the Standard Solar Model. The LMA solutions above $\Delta m^2 = 3\cdot 10^{-5}\text{eV}^2$ are therefore the only viable solution at 95% C.L.

Acknowledgments

This analysis relies crucially on Super-Kamiokande data. The author acknowledges the extensive help and cooperation of the Super-Kamiokande collaboration as well as the Kamioka Mining and Smelting Company. The Super-Kamiokande detector has been built and operated from funding by the Japanese Ministry of Education, Culture, Sports, Science and Technology, the U.S. Department of Energy, and the U.S. National Science Foundation.

Appendix: Fitting Method

To estimate and limit oscillation parameters, a χ^2 describing the shape of the SK zenith spectrum is extended to take into account the measured neutrino interaction rates by various experiments. Using oscillation probabilities (obtained as in [11]) and SSM neutrino fluxes, the expected interaction rates $B^{\text{osc}}_{i,z}$ (due to ^{8}B neutrinos) and $H^{\text{osc}}_{i,z}$ (due to *hep* neutrinos) in energy bin i and zenith-angle bin z are calculated. These rates as well as the SK measurements $D_{i,z}$ are then normalized by the SSM expectations without oscillations:

$$b_{i,z} = \frac{B^{\text{osc}}_{i,z}}{B^{\text{SSM}}_{i,z} + H^{\text{SSM}}_{i,z}}, \; h_{i,z} = \frac{H^{\text{osc}}_{i,z}}{B^{\text{SSM}}_{i,z} + H^{\text{SSM}}_{i,z}}, \; d_{i,z} = \frac{D_{i,z}}{B^{\text{SSM}}_{i,z} + H^{\text{SSM}}_{i,z}}. \qquad (1)$$

The zenith components of the flux difference vector $\overrightarrow{\Delta_i}$

$$\Delta_{i,z}(\beta,\eta) = (\beta \cdot b_{i,z} + \eta \cdot h_{i,z}) \times f(E_i, \delta_B, \delta_S, \delta_R) - d_{i,z}$$

allow for arbitrary neutrino fluxes (through the free parameters β and η). The combined rate prediction is modified by the energy-shape factors

$$f(E_i, \delta_B, \delta_S, \delta_R) = f_B(E_i, \delta_B) \times f_S(E_i, \delta_S) \times f_R(E_i, \delta_R)$$

with δ_B describing the ^{8}B neutrino spectrum shape uncertainty, δ_S describing the uncertainty of the SK energy scale (0.64%) and δ_R describing the uncertainty of the SK energy resolution (2.5%) (The shapes f_B, f_S, f_R are given in table 1). All three uncertainties affect the bins of the SK zenith angle spectrum in a correlated way. The 7×7 matrices V_i describe statistical and energy-uncorrelated uncertainties; the latter are assumed to be fully correlated in zenith angle. For any given parameters δ_k, the χ^2

$$\chi_0^2 = \sum_{i=1}^{8} \vec{\Delta}_i \cdot V_i^{-1} \cdot \vec{\Delta}_i = \chi_{0,m}^2 + \left(\vec{\phi} - \vec{\phi}_{0,m}\right) C_0 \left(\vec{\phi} - \vec{\phi}_{0,m}\right)$$

$$\text{with} \quad C_0 = \sum_{i=1}^{8} \begin{pmatrix} \vec{b}_i \cdot V_i^{-1} \cdot \vec{b}_i & \vec{h}_i \cdot V_i^{-1} \cdot \vec{b}_i \\ \vec{h}_i \cdot V_i^{-1} \cdot \vec{b}_i & \vec{h}_i \cdot V_i^{-1} \cdot \vec{h}_i \end{pmatrix} \quad \text{and} \quad \vec{\phi} = \begin{pmatrix} \beta \\ \eta \end{pmatrix}$$

can be written as a quadratic form of $\vec{\phi}$ and the curvature matrix C_0. The minimum is

$$\chi_{0,m}^2 = \sum_{i=1}^{8} \vec{d}_i \cdot V_i^{-1} \cdot \vec{d}_i - C_{0,m} \quad \text{with} \quad C_{0,m} = \vec{\phi}_{0,m} C_0 \vec{\phi}_{0,m}.$$

If the minimum flux vector is scaled by α ($\vec{\phi} = \alpha \times \vec{\phi}_{0,m}$) then χ_0^2 constrains α to be $\alpha = 1 \pm \sqrt{1/C_{0,m}}$. To take into account the systematic uncertainty of the SK combined rate $\sigma_{SK} =^{+2.9}_{-2.6}$ % (which is completely correlated in zenith-angle and energy), χ_0^2 is modified to

$$\chi_1^2 = \chi_{0,m}^2 + \left(\vec{\phi} - \vec{\phi}_{0,m}\right) C_1 \left(\vec{\phi} - \vec{\phi}_{0,m}\right) \quad \text{with} \quad C_1 = \frac{1/\sigma_{SK}^2}{C_{0,m} + 1/\sigma_{SK}^2} \times C_0.$$

χ_1^2 constrains α to be $\alpha = 1 \pm \sqrt{1/C_{0,m} + \sigma_{SK}^2}$, that is, the minimum is unchanged, but the allowed parameter range for β and η is larger. The total χ^2 for the SK zenith spectrum shape is then

$$\chi_{\text{SK}}^2 = \text{Min}\left(\chi_1^2(\beta, \eta, \delta_B, \delta_S, \delta_R) + \left(\frac{\delta_B}{\sigma_B}\right)^2 + \left(\frac{\delta_S}{\sigma_S}\right)^2 + \left(\frac{\delta_R}{\sigma_R}\right)^2\right) \tag{2}$$

where all δ_k as well as β, η are minimized.

A combined fit with the rate measured by the SNO collaboration[2] can be done by defining $b_{\text{SNO}}, h_{\text{SNO}}$ and $\Delta_{\text{SNO}}(\beta, \eta)$ in a similar way as in (1) and form

$$\chi_{\text{SK-SNO}}^2 = \text{Min}\left(\chi_1^2 + \left(\frac{\delta_B}{\sigma_B}\right)^2 + \left(\frac{\delta_S}{\sigma_S}\right)^2 + \left(\frac{\delta_R}{\sigma_R}\right)^2 + \left(\frac{\Delta_{\text{SNO}}(\beta, \eta)}{\sigma_{\text{SNO}}}\right)^2\right) \tag{3}$$

To add the radio-chemical rate measurements of Homestake[3], Gallex/GNO[4], and SAGE[5] (all "Gallium" rates are combined into a single rate), the χ^2

$$\chi^2_{\rm RC}(\beta, \eta)$$

takes into account the correlations between Gallium and Chlorine measurements. The ^{8}B and *hep* fluxes are constrained by the minimization of $\chi^2_{\text{SK-SNO}}$, not by the Standard Solar Model. The total χ^2 is then a simple addition

$$\chi^2 = \chi^2_{\text{SK-SNO}} + \text{Min}\left(\chi^2_{\rm RC}(\alpha\beta_{\rm min}, \alpha\eta_{\rm min}) + \left(\frac{\alpha - 1}{\sigma_\alpha}\right)^2\right) \tag{4}$$

where $1/\sigma_\alpha^2 = C_{\text{SK-SNO,m}} = \vec{\phi}_{\text{SK-SNO,m}} \cdot C_{\text{SK-SNO}} \cdot \vec{\phi}_{\text{SK-SNO,m}}$. Numerically, σ_α is found to be about 2.7% (The accuracy of SK and SNO combined)

References

1. S. Fukuda et al., *Phys. Rev. Lett.* **86**, 5651 (2001).
2. Q.R. Ahmad et al. *Phys. Rev. Lett.* **87**, 71301 (2001).
3. B.T. Cleveland et al., *Astrophys. J.* **496**, 505 (1998).
4. E. Bellotti, *Nucl. Phys. B(Proc. Suppl.)* **91**, 44 (2001);
   W. Hampel et al., *Phys. Lett.* B **388**, 364 (1996);
   P. Anselmann et al., *Phys. Lett.* B **342**, 440 (1995).
5. V. Gavrin, *Nucl. Phys. B(Proc. Suppl.)* **91**, 36 (2001);
   J.N. Abdurashitov et al., *Phys. Lett.* B **328**, 234 (1994).
6. J.N. Bahcall, M.H. Pinsonneault, S. Basu, *Astrophys. J.* **555**, 990 (2001).
7. Y. Fukuda et al., *Phys. Rev. Lett.* **81**, 1562 (1998).
8. M. Apollonio, *Phys. Lett.* B **466**, 415 (1999).
9. S.P. Mikheyev and A.Y. Smirnov, *Sov. Jour. Nucl. Phys.* **42**, 913 (1985); L. Wolfenstein, *Phys. Rev.* D **17**, 2369 (1978).
10. For individual allowed areas, predictions of the spectrum and daily variations of oscillation solutions as well as a discussion of zenith angle spectrum vs. day/night spectrum or spectrum see M. Smy, *hep-ex/0106064*, to be published in proceedings of XXXVIth Rencontres de Moriond on Electroweak Interactions and Unified Theories.
11. S. Fukuda et al., *Phys. Rev. Lett.* **86**, 5656 (2001).
12. M. Smy in *Neutrino Oscillations in Venice*, ed. M. Baldo Ceolin, (Venezia, 2001) 35.
13. A.R. Junghans et al., *Phys. Rev. Lett.* **88**, 041101 (2002).
14. J.N. Bahcall, M.C. Gonzales-Garcia and Carlos Peña Garay, *hep-ph/0111150* (2001).

Solar Neutrino Oscillations

M. C. Gonzalez-Garcia
Theory Division, CERN, CH-1211 Geneva 23, Switzerland,
Instituto de Física Corpuscular, Universitat de València - C.S.I.C.
Edificio Institutos de Paterna, Apt 22085, 46071 València, Spain,
and C.N. Yang Institute for Theoretical Physics
State University of New York at Stony Brook Stony Brook,NY 11794-3840, USA
email:concepcion.gonzalez-garcia@cern.ch

1 Introduction: Fluxes and Data

The sun is a source of $\nu_e^{\prime}s$ which are produced in the different nuclear reactions taking place in its interior. Along this talk we will use the ν_e fluxes from Bahcall–Pinsonneault calculations [1], but we include the impact of the new measurement of $^7\mathrm{Be}(\mathrm{p},\gamma)^8\mathrm{B}$ [2]

$$S_{17}(0) = (22.3 \pm 0.7\,(\mathrm{expt}) \pm 0.5\,(\mathrm{theor}))\,\mathrm{eVb}.$$

instead of the previous(1998) standard value of $S_{17}(0) = 19^{+4}_{-2}\mathrm{eVb}$. This induces a change in the predicted ^{8}B flux

$$\begin{array}{ll} \mathrm{BP00} & 5.05 \times 10^{-6}\,(1.00^{+0.20}_{-0.16})\,\mathrm{cm}^{-2}\mathrm{s}^{-1} \\ \mathrm{NEW} & 5.93 \times 10^{-6}\,(1.00^{+0.14}_{-0.15})\,\mathrm{cm}^{-2}\mathrm{s}^{-1} \end{array}$$

Solar neutrinos have been detected at the Earth by seven experiments which use different detection techniques: The chlorine experiment at Homestake[3], the water cerencov experiments Kamiokande and Super–Kamiokande (SK) [5] and the radiochemical Gallex, GNO and Sage experiments[4] and the SNO[6]. Due to the different energy threshold for the detection reactions, these experiments are sensitive to different parts of the solar neutrino spectrum. They all observe a deficit between 30 and 60 % (see table 1) which seems to be energy dependent. SK has also presented their results on the time and energy variation of their observed events. In this talk we will use their measured energy spectrum during the day and during the night corresponding to 1258 days of data [5]. This will be referred in the following as the day–night spectra data which contains 2×19 data bins.

In this talk we will present a summary the results of Ref. [7] on our analysis of the solar neutrino data and we will refer to these publications for references and further details.

Source	Flux $\left(10^{10}\ \mathrm{cm}^{-2}\mathrm{s}^{-1}\right)$	Cl SNU	Ga SNU	SK $10^6\ \mathrm{cm}^{-2}\mathrm{s}^{-1}$	SNO(CC) $10^6\ \mathrm{cm}^{-2}\mathrm{s}^{-1}$
pp	$5.95\left(1.00^{+0.01}_{-0.01}\right)$	0.0	69.7	0.0	0.0
pep	$1.40\times10^{-2}\left(1.00^{+0.015}_{-0.015}\right)$	0.22	2.8	0.0	0.0
hep	9.3×10^{-7}	0.04	0.1	0.0093	0.0093
^{7}Be	$4.77\times10^{-1}\left(1.00^{+0.10}_{-0.10}\right)$	1.15	34.2	0.0	0.0
^{8}B	$5.93\times10^{-4}\left(1.00^{+0.14}_{-0.15}\right)$	6.76	14.2	5.93	5.93
^{13}N	$5.48\times10^{-2}\left(1.00^{+0.21}_{-0.17}\right)$	0.09	3.4	0.0	0.0
^{15}O	$4.80\times10^{-2}\left(1.00^{+0.25}_{-0.19}\right)$	0.33	5.5	0.0	0.0
^{17}F	$5.63\times10^{-4}\left(1.00^{+0.25}_{-0.25}\right)$	0.0	0.1	0.0	0.0
Total		$8.6^{+1.1}_{-1.2}$	130^{+9}_{-7}	$5.93^{+0.89}_{-0.83}$	$5.93^{+0.89}_{-0.83}$
Measured		2.56 ± 0.226	75.6 ± 4.8	2.32 ± 0.085	1.75 ± 0.148
$\frac{\text{Measured}}{\text{SSM}}$		0.298 ± 0.049	0.581 ± 0.055	0.391 ± 0.060	0.295 ± 0.051

Table 1: SSM predictions: solar neutrino fluxes and neutrino capture rates for the different experiments, with 1σ uncertainties.

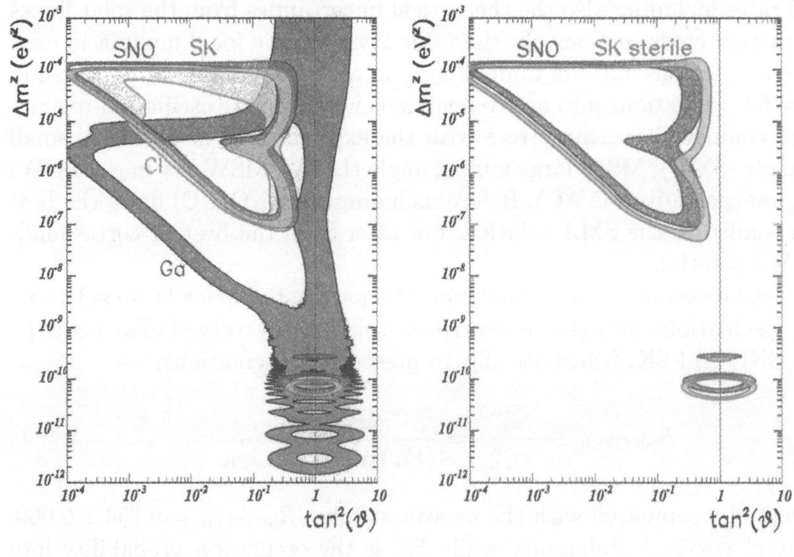

Figure 1: (Left) Oscillation parameters to describe the measured deficit with 3σ (only experimental errors) for SNO, Gallium, Chlorine and SK experiments for oscillations into active neutrinos. (Right) Oscillation parameters to describe the measured deficit with 3σ (only experimental errors) for SNO and SK experiments for oscillations into sterile neutrinos.

2 Total Rates

The most generic and popular explanation of the solar neutrino anomaly is in terms of neutrino masses and mixing leading to oscillations of ν_e into an active (ν_μ and/or ν_τ) or sterile neutrino, ν_s. We first present the results from the analysis of the total event rates using the CC event rate measured by SNO [6], the Chlorine [3] and Gallium event rates [4] (we use here the weighted averaged of the GALLEX/GNO and SAGE rates). In Fig. 1(left) we plot the oscillation parameters to describe the measured deficit with 3σ (only experimental errors) for each of the four experiments for oscillations into active neutrinos. From the figure we already see that there are several possible region of parameters compatible with all the observations in this case. For the case of oscillations into sterile neutrinos the allowed region for SK is different as a consequence of the absence of the neutral current contribution to the event rates. In Fig. 1(right) we plot the corresponding allowed region for SK in this case together with the one for SNO. Already at this level we see the incompatibility between both measurements which disfavours solar neutrino oscillations into sterile states.

In Fig. 2 we plot the allowed regions obtained by a χ^2 analysis of the total observed rates including also the theoretical uncertanties from the solar fluxes and interaction cross sections. In the table 2 we list the local minima in each of the allowed regions and the value of $\chi^2_{\rm min}$ in each local minimum. As seen in the figure for oscillations into active neutrinos we have four oscillation regimes which are compatible within errors with the experimental data: MSW small mixing angle (SMA), MSW large mixing angle (LMA), MSW low mass (LOW) and vacuum oscillations (VAC). Before including the SNO(CC) data, the best fit corresponded to the SMA solution, but after SNO the best fit corresponds to the LMA solution.

For oscillations into sterile neutrinos the parameter space is considerably smaller. Oscillations into sterile neutrinos would imply very similar neutrino fluxes in SNO and SK. Schematically, in presence of oscillations,

$$R_{\rm SNO/SK} \equiv \frac{\phi^{\rm SNO}_{^8B}}{\phi^{\rm SK}_{^8B}} = \frac{\langle P_{ee}\rangle_{\rm SNO}}{\langle P_{ee}\rangle_{\rm SK} + r\langle P_{ea}\rangle_{\rm SK}} \tag{1}$$

which has to be compared with the measured value $R_{\rm SNO/SK} = 0.754 \pm 0.069$. P_{ee} is the ν_e survival probability while P_{ea} is the oscillation probability into active neutrinos. By $\langle P\rangle_{\rm SNO\,(SK)}$ we denote the corresponding averaged oscillation probability at SNO (SK) taking into account the detector response function which accounts for energy threshold and resolution and the neutrino interaction cross section. $r \equiv \sigma_\mu/\sigma_e \simeq 0.15$ is the ratio of the the $\nu_e - e$ and $\nu_\mu - e$ elastic scattering cross-sections. Naively, ignoring the difference between

Solution	$\Delta m^2[\text{eV}^2]$	$\tan^2\theta$	χ^2_{min} (3 d.o.f)	g.o.f.
LMA	1.9×10^{-5}	3.0×10^{-1}	2.62	45%
SMA	7.7×10^{-6}	1.7×10^{-3}	4.40	22%
VAC	7.7×10^{-11}	0.33(3.0)	5.83	12%
Sterile VAC	9.2×10^{-11}	0.32(3.1)	10.1	1.7%
LOW	1.0×10^{-7}	6.1×10^{-1}	11.0	1.2%
Sterile SMA	5.0×10^{-6}	8.9×10^{-4}	22.8	0.4%

Table 2: Best-fit parameters for the analysis of total rates only, corresponding to Fig. 2. The sterile SMA solution does not appear in Fig. 2 because χ^2_{min} is too large.

the detectors response functions we see that oscillations into sterile neutrinos predict that Eq. (1) would be equal to ~ 1 which is ruled out at $\sim 3\sigma$. Eq. (1) also shows us that the difference between the solar neutrino fluxes measured in SNO(CC) and SK(ES) must come from transitions into active neutrinos, which would contribute only to SK(ES) through NC interactions. In summary, the combination of SNO(CC) and SK(ES) results strongly disfavours the hypothesis of pure $\nu_e \to \nu_s$ oscillations.

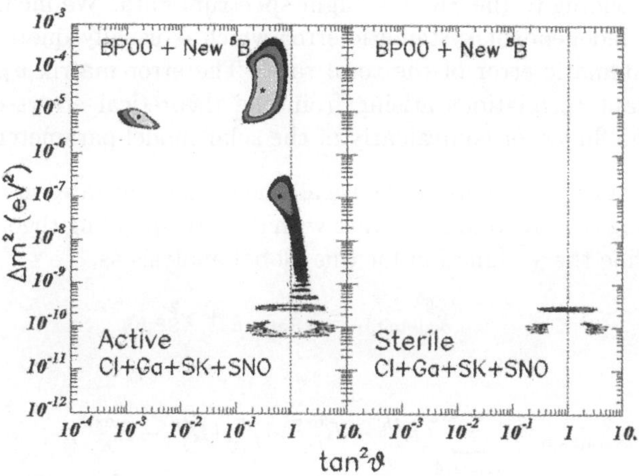

Figure 2: Allowed oscillation parameters (at 90, 95, 99 and 99.7% CL) from the analysis of the total event rates of the Chlorine, Gallium, SK and SNO CC experiments. The best fit point is marked with a star.

3 Global Analysis

The results shown in figures 3– 5 were derived combining the results from the total event rates and and the 2 × 19 bins of the (1258 day) SK [5] electron recoil energy spectrum measured separately during the day and night periods. In order to explore the robustness of the inferred allowed regions in neutrino parameter space, we obtain the permitted regions using calculations based upon three different plausible prescriptions for the statistical analysis that have been used in the published literature and which we label as follows.

(a) For our "standard" analysis, we adopt the prescription described in Refs. [7]. We do not include here the SK total rate, since to a large extent the total rate is represented by the flux in each of the spectral energy bins. We define the χ^2 function for the global analysis as:

$$\chi^2_{\rm global,a} = \sum_{i,j=1,41} (R_i^{th} - R_i^{\rm exp})\sigma_{G,ij}^{-2}(R_j^{th} - R_j^{\rm exp}), \tag{2}$$

where $\sigma^2_{G,ij} = \sigma^2_{R,ij} + \sigma^2_{SP,ij}$. Here $\sigma_{R,ij}$ is the corresponding 41×41 error matrix containing the theoretical as well as the experimental statistical and systematic uncorrelated errors for the 41 rates while $\sigma_{SP,ij}$ contains the assumed fully–correlated systematic errors for the 38×38 submatrix corresponding to the SK day–night spectrum data. We include here the energy independent systematic error which is usually quoted as part of the systematic error of the total rate. The error matrix $\sigma_{R,ij}$ includes important correlations arising from the theoretical errors of the solar neutrino fluxes, or equivalently of the solar model parameters.

(b) For the "all rates" analysis we use all the total event rates in the analysis, including the SK total event rate with its corresponding theoretical error. We define the χ^2 function for this global analysis as:

$$\chi^2_{\rm global,b} = \chi^2_{\rm Rates,b} + \chi^2_{\rm SP,b}, \tag{3}$$

where

$$\chi^2_{\rm Rates,b} = \sum_{i,j=1,4} (R_i^{th} - R_i^{\rm exp})\tilde{\sigma}_{R,ij}^{-2}(R_j^{th} - R_j^{\rm exp}), \tag{4}$$

$$\chi^2_{\rm SP,b} = \sum_{i,j=1,38} (\alpha_{\rm SP} R_i^{th} - R_i^{\rm exp})\tilde{\sigma}_{SP,ij}^{-2}(\alpha_{\rm SP} R_j^{th} - R_j^{\rm exp}). \tag{5}$$

In $\chi^2_{\rm Rates}$, we include the CC event rate measured at SNO [6], the Chlorine [3] and Gallium [4] event rates and the SK [5] total event rate as described in the previous section. The matrix $\tilde{\sigma}_{R,ij}$ contains the theoretical uncertainties, as well as the experimental statistical and systematic errors, for the total rates. In particular, $\tilde{\sigma}_{R,ij}$ includes the energy independent systematic error for the SK rate. On the other hand, the matrix $\tilde{\sigma}_{\mathrm{SP},ij}$ contains both the statistical as well as the systematic errors (both those that are correlated with energy and those that are uncorrelated) of the 38 spectral energy data; $\tilde{\sigma}_{\mathrm{SP},ij}$ does not include the theoretical error of the ^{8}B flux.

In order to avoid double counting, we allow a free normalization (for which no theoretical error is included) for the SK day and night electron recoil energy spectra. Following the usual procedure for this approach, we introduce an arbitrary normalization factor, α_{SP}, for the amplitude of the energy spectra. For each value of the oscillation parameters $\chi^2_{\mathrm{global},b}$ is minimized with respect to α_{SP}.

(c) The "free ^{8}B" analysis is identical to our standard analysis, described in (a) above, except that the ^{8}B neutrino flux is not constrained by solar model predictions. This freedom is implemented by multiplying the ^{8}B flux contribution to the R_i^{th} in eq. (2) by the normalization factor, f_{B}, and removing the theoretical ^{8}B flux errors from $\sigma_{R,ij}$. In this case, the resulting χ^2 function can also be written as

$$\chi^2_{\mathrm{global,c}} = \chi^2_{\mathrm{Rates,c}} + \chi^2_{\mathrm{SP,c}}, \tag{6}$$

where

$$\chi^2_{\mathrm{Rates,c}} = \sum_{i,j=1,3} (R_i^{th}(f_{\mathrm{B}}) - R_i^{\mathrm{exp}})\hat{\sigma}^{-2}_{R,ij}(R_j^{th}(f_{\mathrm{B}}) - R_j^{\mathrm{exp}}), \tag{7}$$

$$\chi^2_{\mathrm{SP,c}} = \sum_{i,j=1,38} (R_i^{th}(f_{\mathrm{B}}) - R_i^{\mathrm{exp}})\hat{\sigma}^{-2}_{SP,ij}(R_j^{th}(f_{\mathrm{B}}) - R_j^{\mathrm{exp}}). \tag{8}$$

In $\chi^2_{\mathrm{Rates,c}}$, we include only the CC event rates measured in the SNO, Chlorine, and Gallium experiments. For these three rates, the error matrix $\hat{\sigma}_{R,ij}$ only differs from $\tilde{\sigma}_{R,ij}$ due to the absence of the theoretical error for the ^{8}B neutrino flux. The error matrix $\hat{\sigma}_{SP,ij}$ only differs from $\tilde{\sigma}_{SP,ij}$ by the inclusion of the energy independent systematic error, which is usually quoted as part of the systematic error of the total rate.

For each value of the oscillation parameters, $\chi^2_{\mathrm{global,b}}$ is minimized with respect to f_{B}. Notice that the only difference in the "free ^{8}B" analysis when performed in terms of BP00 or in terms of BP00 + New ^{8}B is a shift in the corresponding best value of f_{B} for each value of the oscillation parameters $f_{\mathrm{B}} \to f_{\mathrm{B}}\left(\frac{5.05}{5.93}\right)$; the allowed regions are left unchanged.

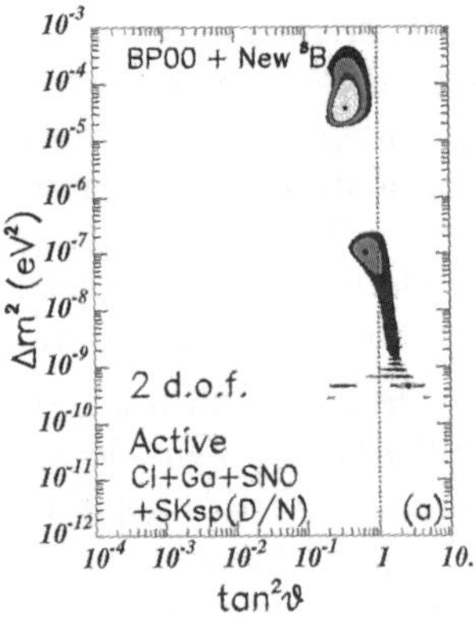

Solution	Δm^2	$\tan^2(\theta)$	$\chi^2_{\min}$
LMA	3.7×10^{-5}	3.7×10^{-1}	34.5
LOW	1.0×10^{-7}	6.7×10^{-1}	40.8
VAC	4.6×10^{-10}	2.5×10^{0}	42.3
Sterile VAC	4.7×10^{-10}	3.0×10^{0}	49.1
SMA	5.2×10^{-6}	1.8×10^{-3}	49.9
Sterile SMA	4.6×10^{-6}	3.4×10^{-4}	52.3

Figure 3: Allowed regions and best-fit parameters for the analysis of global analysis (a).

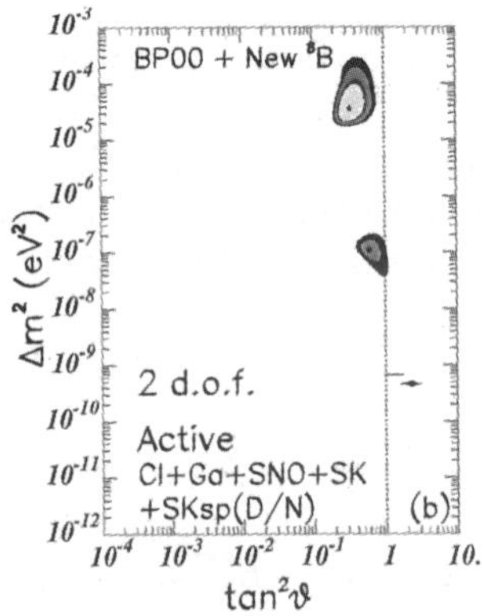

Solution	Δm^2	$\tan^2(\theta)$	$\chi^2_{\min}$
LMA	3.3×10^{-5}	3.5×10^{-1}	34.6
LOW	1.1×10^{-7}	6.1×10^{-1}	42.0
VAC	4.6×10^{-10}	2.3×10^{0}	45.5
SMA	5.2×10^{-6}	2.0×10^{-3}	51.2
Sterile VAC	4.6×10^{-10}	2.7×10^{0}	53.8
Sterile SMA	4.6×10^{-6}	3.9×10^{-4}	59.3

Figure 4: Allowed regions and best-fit parameters for the analysis of global analysis (b).

In constructing these regions we assumed that only active neutrinos exist. We derive therefore the allowed regions in χ^2 using only two free parameters: Δm^2 and $\tan^2\theta$. As discussed above, oscillations into pure sterile neutrinos are highly disfavored in the global analysis and are not included in our standard demarcation of the allowed regions.

The allowed regions obtained and the location of the best fit points , although different, do not depend dramatically on the adopted strategy. For any strategy the LMA solution is favored, but the LOW and VAC solutions are also allowed at a C.L. corresponding to 3σ. Other solutions such as oscillations into sterile neutrinos or the SMA solution for active neutrinos are disfavored at 3σ if the standard analysis strategy is adopted, but SMA is barely allowed at 3σ if the ^{8}B neutrino flux is unconstrained by the solar model predictions and uncertainties.

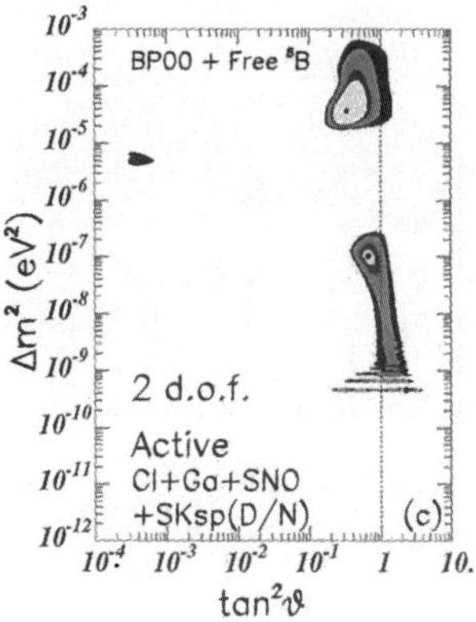

Solution	Δm^2	$\tan^2(\theta)$	f_B	$\chi^2_{\min}$
LMA	3.6×10^{-5}	3.3×10^{-1}	1.00	34.6
LOW	1.0×10^{-7}	6.7×10^{-1}	0.73	38.7
VAC	4.6×10^{-10}	2.3×10^{0}	0.59	39.5
SMA	4.9×10^{-6}	4.2×10^{-4}	0.44	45.3
Sterile SMA	4.6×10^{-6}	3.5×10^{-4}	0.42	48.5
Sterile VAC	4.6×10^{-10}	2.3×10^{0}	0.61	51.0

Figure 5: Allowed regions and best-fit parameters for the analysis of global analysis (c).

The allowed regions are reduced in size for strategy (a) and strategy (b) as compared to the regions derived using the 1998 standard value of $S_{17}(0)$. The most restricted allowed regions are obtained for strategy (b). As discussed in Ref.[7], the smallness of the allowed regions for this strategy results from the absence in the analysis of the ^{8}B theoretical error for the SK spectrum normalization. The largest allowed regions correspond to strategy (c), in which the ^{8}B neutrino flux is treated as a free parameter.

4 Predictions for SNO and Borexino

Next we present the predictions for SNO and Borexino of the currently allowed neutrino oscillation solutions. Some predictions for the SNO experiment are shown in Fig 6. All three analysis strategies yield essentially the same 3σ range for the neutral current to charged current double ratio predicted by the favored LMA, LOW, and VAC solutions, namely, $1.4 < [NC]/[CC] < 5.3$ (For analysis strategy (c),the upper limit extends to 6.2.) If the ^{8}B is allowed to vary freely, our analysis strategy (c), then the SMA solution for active neutrinos is allowed at the 3σ C.L. and for this solution $1.15 < [NC]/[CC] < 1.31$. Figure 6 shows that all of the values of [NC]/[CC] predicted by the allowed neutrino oscillation solutions are separated from the no oscillation value of $[NC]/[CC] = 1.0$ by more than the expected experimental uncertainty For the average difference in the CC day-night effect, $A_{\rm N-D}$(SNO CC), all three analysis strategies also yield essentially the same results. The predicted range of the day-night difference in charged current rates is $8.3^{+5.0}_{-5.6}(1\sigma)\%$ $[8.3^{+13.1}_{-8.6}(3\sigma)\%]$ and is predicted to be strongly correlated with the day-night effect for neutrino-electron scattering $A_{\rm N-D}$(SNO ES) which is bounded by the 3σ allowed range for SK, i. e.,

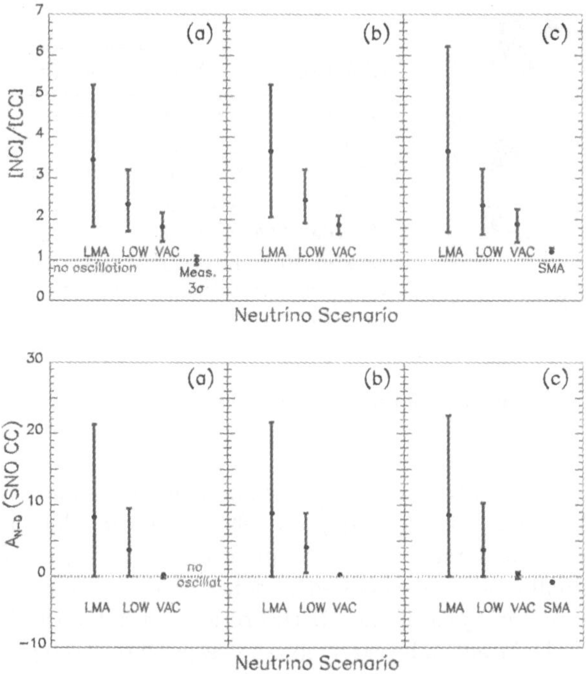

Figure 6: The double ratio [NC]/[CC] rate and the percentage difference between the night and the day rates at SNO.The figure shows the predictions obtained for recoil electrons with kinetic energies in the range 6.75 MeV $< T_e$. The tiny 3σ estimated measuring error in the left hand panel assumes the systematic uncertainties projected and 3000 CC events and 1200 NC events.

$A_{\mathrm{N-D}}$(SNO ES.) lies between 0% and 11%. Finally the current set of allowed oscillation solutions do not predict energy spectrum distortions that are large enough to be detected by SNO at a high level of significance.

Our predictions for the Borexino experiment are shown in Fig. 7 which shows that the currently LMA, LOW, and VAC oscillation solutions predict similar values for the neutrino-electron scattering rate for ^{7}Be neutrinos, $[^7\mathrm{Be}] = 0.655 \pm 0.035, 1\sigma (0.66 \pm 0.13, 3\sigma)$. The SMA solution, which is allowed at 3σ only in the "free ^{8}B analysis strategy predicts a much smaller value, $[^7\mathrm{Be}] < 0.34$. Figure 7 also shows that the LOW solution is the only currently allowed (at 3σ) neutrino oscillation solution that predicts a significant day-night variability in BOREXINO. If a difference between day and night rates is

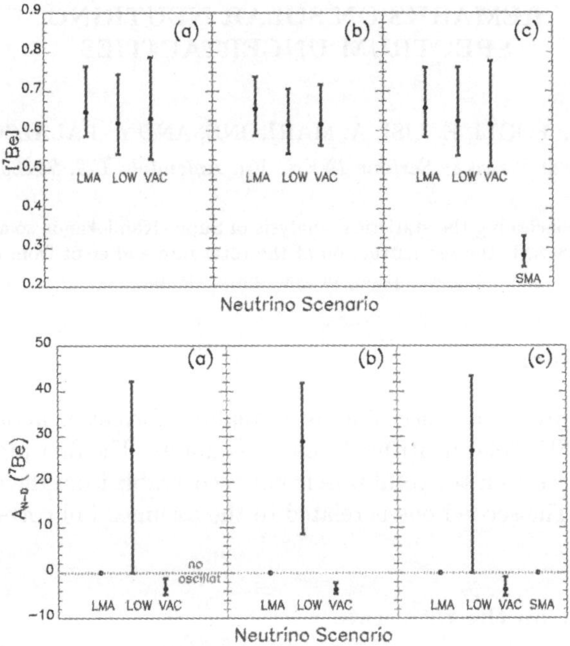

Figure 7: The ^{7}**Be** neutrino event rate and the percentage difference between the night and the day rates at Borexino. The figure shows the predictions obtained for recoil electrons with kinetic energies in the range 0.25 MeV $< T_e <$ 0.8 MeV.

observed, this will very strong evidence in favour of the LOW solution.

References

1. J.N. Bahcall, M.H. Pinsonneault, S. Basu, Astrophys. J. **555** (2001) 990.
2. A.R. Junghans et al., Phys. Rev. Lett. **88** (2002) 041101.
3. B.T. Cleveland et al., Astrophys. J. **496** (1998) 505.
4. W. Hampel et al. (GALLEX), Phys. Lett. **B 447** (1999) 127; M. Altmann et al. (GNO), Phys. Lett. **B 490**(2000) 16; V. Gavrin (SAGE), Nucl. Phys. Proc. Supp. **B 91**(2001),36.
5. S. Fukuda et al. Phys. Rev. Lett. **86** (2001) 5651.
6. Q.R. Ahmad et al., Phys. Rev. Lett. **87** (2001) 071301.
7. J.N. Bahcall, M.C. Gonzalez-Garcia, C. Peña-Garay, JHEP 08(2001)014; hep-ph/0111150.

REMARKS ON SOLAR NEUTRINO SPECTRUM UNCERTAINTIES

G.L. FOGLI, E. LISI, A. MARRONE*, AND A. PALAZZO

Dipartimento di Fisica & Sezione INFN, Via Amendola 173, 70126 Bari, Italy

Two issues concerning the statistical analysis of Super-Kamiokande solar neutrino data are discussed: the reconstruction of the total rate and error from the energy spectrum bins, and the correlations of systematic errors.

1 Introduction

There are two problems concerning the available information about the Super-Kamiokande (SK) solar neutrino data uncertainties. The first problem is that the published error on the total rate is not recoverable from the errors on the binned rates. The second one is related to the assumed full correlation of the spectral errors.

2 The error on the total rate

The relevant information about SK data and uncertainties is summarized in Tab. 1, which, for each energy bin, reports our estimate (with *hep* flux equal to 0) of the unoscillated event rate R_i (event/bin/day), the published observed rate f_i [in units of the standard solar model (SSM) expectations], the statistical error σ_i, and the systematic correlated and uncorrelated errors, $s^i_\pm$ and $u^i_\pm$. Table 2 shows how to recover various integral quantities from the binned distributions. Note, in particular, that uncorrelated errors are added quadratically, while (supposedly) fully correlated errors sum up linearly. Our calculated ratio of observed to expected rate of events is

$$\text{our estim.}: \quad \frac{\text{SK}}{\text{SSM}} = 0.455 \pm 0.008(stat)^{+0.006}_{-0.005}(sys)\,, \tag{1}$$

to be compared with the official published ratio

$$\text{published}: \quad \frac{\text{SK}}{\text{SSM}} = 0.459 \pm 0.005(stat)^{+0.016}_{-0.014}(sys)\,. \tag{2}$$

From Eqs. (1) and (2), a significant difference between the two statistical errors emerges, as well as a huge difference between the systematic errors.[a]

---

*SPEAKER

[a] In [1], an energy-independent and fully correlated error of +2.9% (−2.6%) is quoted.

Table 1. Binned rates and errors.

				SK published data					
bin	E_{min}	E_{max}	R_i	f_i	σ_i	s^i_+	s^i_-	u^i_+	u^i_-
	(MeV)		(event/bin/day)	(SK/SSM)		(%)			
1	5.0	5.5	10.094	0.445	0.046	0.2	0.2	3.9	3.1
2	5.5	6.0	9.177	0.446	0.025	0.2	0.2	1.7	1.6
3	6.0	6.5	8.250	0.444	0.019	0.3	0.3	1.3	1.4
4	6.5	7.0	7.328	0.446	0.015	0.5	0.6	1.4	1.4
5	7.0	7.5	6.427	0.472	0.016	0.8	0.8	1.3	1.4
6	7.5	8.0	5.559	0.492	0.016	1.0	1.1	1.3	1.4
7	8.0	8.5	4.739	0.474	0.017	1.4	1.3	1.3	1.4
8	8.5	9.0	3.977	0.447	0.017	1.7	1.7	1.3	1.4
9	9.0	9.5	3.282	0.459	0.018	2.1	2.0	1.3	1.4
10	9.5	10.0	2.659	0.464	0.020	2.5	2.3	1.3	1.4
11	10.0	10.5	2.112	0.451	0.021	3.0	2.7	1.3	1.4
12	10.5	11.0	1.643	0.415	0.022	3.4	3.2	1.3	1.4
13	11.0	11.5	1.249	0.464	0.026	3.9	3.6	1.3	1.4
14	11.5	12.0	0.927	0.431	0.029	4.5	4.2	1.3	1.4
15	12.0	12.5	0.669	0.430	0.033	5.1	4.8	1.3	1.4
16	12.5	13.0	0.470	0.491	0.042	5.8	5.4	1.3	1.4
16	13.0	13.5	0.320	0.439	0.048	6.5	6.2	1.3	1.4
18	13.5	14.0	0.211	0.615	0.067	7.4	7.0	1.3	1.4
19	14.0	20.0	0.326	0.503	0.050	10.7	9.5	1.3	1.4

Table 2. Total rates and errors.

	definition (event/bin/day)	numerical value
Total unoscillated rate	$R_0 = \sum_i R_i$	69.4
Total observed rate	$R = \sum_i R_i f_i$	31.6
Total stat. error of R	$\sigma = \sqrt{\sum_i (R_i \sigma_i)^2}$	0.59
Total uncorrelated ($\pm$) error of R_0	$u_\pm = \sqrt{\sum_i (R_i u^i_\pm/100)^2}$	0.47(0.41)
Total correlated ($\pm$) error of R_0	$s_\pm = \sum_i R_i s^i_\pm/100$	0.82 (0.79)
Total systematic ($\pm$) error of R_0	$\sigma_\pm = \sqrt{s^2_\pm + u^2_\pm}$	0.85(0.81)
SK/SSM ratio	$F = R/R_0$	0.455
Statistical error of F	$\sigma_{stat} = \sigma/R_0$	0.008
Systematic ($\pm$) error of F	$\sigma^\pm_{sys} = \sigma_\pm R/R_0$	0.006(0.005)

Table 3. Separate components of the correlated systematic errors.

bin	s_B (%)	s_E (%)	s_R (%)	s_{tot}^{our} (%)	s_{tot}^{SK} (%)
1	0.40	-0.07	-0.20	0.45	0.20
2	0.48	0.07	-0.20	0.52	0.20
3	0.57	0.22	-0.20	0.64	0.30
4	0.67	0.39	-0.19	0.80	0.55
5	0.78	0.58	-0.18	0.99	0.80
6	0.91	0.79	-0.15	1.21	1.05
7	1.05	1.03	-0.11	1.47	1.35
8	1.20	1.30	-0.05	1.77	1.70
9	1.36	1.60	0.03	2.10	2.05
10	1.55	1.93	0.13	2.48	2.40
11	1.75	2.30	0.27	2.90	2.85
12	1.97	2.71	0.44	3.38	3.30
13	2.21	3.16	0.67	3.92	3.75
14	2.47	3.66	0.95	4.52	4.35
15	2.74	4.21	1.29	5.19	4.95
16	3.04	4.80	1.70	5.94	5.60
17	3.35	5.45	2.20	6.77	6.35
18	3.69	6.14	2.80	7.69	7.20
19	4.46	7.95	4.67	10.25	10.10

Therefore, it is not possible to reconstruct the total statistical and systematic rate errors from the published binned rates and errors. To do this, further detailed information about error definition is needed from SK.

3 The correlation of systematic errors

The correlated systematic errors in each bin $s_{\pm}^i$ have been assumed to be nearly fully correlated by the SK collaboration.[2] To study such correlations, it is necessary to break down the total error in their independent components.[b] We separately analyze the three main sources of error:

1) The shape error s_B^i on the ^{8}B spectrum (as calculated in [3]);

2) The energy scale error S_E^i (simulated by shifting electron energies by 0.64%);

3) The resolution error S_R^i (simulated by changing the resolution width by 2%).

[b]Separate components of the correlated systematic error, as estimated by SK, are given in [1].

Table 3 shows our estimates of s^i_B,s^i_E and s^i_R, expressed as % of the rates R_i in the corresponding bins. The last two columns report our total systematic correlated error $(s^{our}_{tot})^i = \sqrt{{s^i_B}^2 + {s^i_E}^2 + {s^i_R}^2}$, and the symmetrized SK error, $(s^{SK}_{tot})^i = (s^i_+ + s^i_-)/2$. There is a reasonable agreement between the SK calculations and ours. However, once the correlation ρ_{ij} between $(s^{our}_{tot})^i$ and $(s^{our}_{tot})^j$ is calculated,

$$\rho_{ij} = \frac{s^i_B s^j_B + s^i_E s^j_E + s^i_R s^j_R}{(s^{our}_{tot})^i (s^{our}_{tot})^j}, \tag{3}$$

we found that $\rho_{ij} < 1$ significantly for $i \neq j$, in disagreement with the SK assumption $\rho_{ij} \simeq 1$. Only recently there has been an improvement in the way SK error correlations are treated.[1] This kind of information, we think, should be more deeply discussed in technical publications.

4 Conclusions

Known information about the SK solar ν data, errors and correlations, suggests that a detailed description of the SK uncertainties is highly desirable.

References

1. M. Smy, these Proceedings.
2. SK Collaboration, *Phys. Rev. Lett.* **86**, 5656 (2001).
3. J. Bahcall *et al.*, *Phys. Rev.* **C** 54, 411 (1996).

Table 4. Systematic error correlations (our estimate). The (symmetric) correlation matrix is shown only for odd energy bins, due to space limitations.

1	3	5	7	9	11	13	15	17	19	bin
1.00	0.87	0.69	0.56	0.45	0.37	0.30	0.24	0.17	0.07	1
	1.00	0.96	0.89	0.83	0.77	0.72	0.66	0.61	0.51	3
		1.00	0.98	0.96	0.92	0.89	0.85	0.80	0.72	5
			1.00	0.99	0.98	0.95	0.93	0.89	0.82	7
				1.00	0.99	0.98	0.96	0.94	0.88	9
					1.00	0.99	0.99	0.97	0.92	11
						1.00	1.00	0.99	0.95	13
							1.00	1.00	0.97	15
								1.00	1.00	17
									1.00	19

THE STATUS OF RESONANT SPIN FLAVOR PRECESSION

C. S. LIM

Dept, of Physics, Kobe University, Rokkodai-cho 1-1, Nada, Kobe 657-8501, Japan
E-mail: lim@phys.sci.kobe-u.ac.jp

The present status of Resonant Spin Flavor Precession (RSFP) scenario, originally proposed as a solution to the solar neutrino problem, is discussed. It is argued that the scenario (with Majorana-type transition magnetic moment) is capable of explaining all event rates of existing solar neutrino experiments, including the recently reported data from SNO. In addition, it is pointed out that, in the realistic three generation scheme, to accont for the atmospheric neutrino oscillation is also possible. A simple formula, which enables us to reduce the problem in the three generation scheme to that in two generation scheme, is given.

1 Introduction

1.1 Present Status of the Solar Neutrino Problem

It is the recent great achievement that the combined result of super-Kamiokande (SK) [1] and SNO charged current experiment (SNO-CC) [2] has clearly shown the presence of the (active) ν_μ and/or ν_τ components in the solar neutrino flux, and therefore some physics beyond the standard model, irrespectively of the solar model. In the most popular scenario of neutrino oscillation, i.e. MSW scenario [3], the allowed parameter region to account for all existing data has been rahter restricted now to, so called LMA (large mixing angle) solution.

One comment here is that in the LMA solution, the genuine feature or resonant matter oscillation, i.e. enough depletion of ν_e flux even for small mixing angle, which was the original motivation to advocate the MSW, doesn't seem to play a central role. In fact, the survival probability of ν_e in vacuum oscillatrion $S = 1 - \frac{1}{2}\sin^2 2\theta$ and that in the matter oscillation $S = \sin^2\theta$ just coincide for "maximal mixing" $\theta = \frac{\pi}{4}$.

1.2 Particle Physics Scenarios to Solve the Solar Neutrino Problem

Having such progress in experiment what will be theoretical particle physics scenario to solve the solar neutrino problem ? The MSW is the most promissing scenario. It just needs non-vanishing neutrino masses and the resultant flavor mixing. The small non-vanishing neutrinos masses are theoretically well-motivated, as suggested by the seesaw mechanism [4], though the large (or even maximal as indicated by the atmospheric neutrino oscillation) mixing is a great theoretical challenge.

There, however, still remain a few possibilities of "exotic" scenario, which is theoretically less motivated but on the other hand needs a dramatic modifications of the standard model. In this talk I would like to report on the present status of one of such exotic scenarios, i.e. Resonant Spin Flavor Precession (RSFP) scenario [5].

2 RSFP Scenario

The story started when R.Davis and collaborators claimed that there is an indication of the time variation of the solar neutrino flux, anti-correlated with sun-spot number (with a period of 11 years). As the sun-spot number is a measure of magnetic activity, it is natural to suspect that neutrino may have magnetic interaction. From such reasoning, Okun-Voloshin-Vysotsky (OVV) proposed a scenario of spin precession [6]. They assumed that the ν_e has an anormalous magnetic moment μ. Then the emitted state ν_{eL} experiences a spin-precession due to the magnetic field inside the sun, perpendicular to the direction of neutrino velocity, into a state ν_{eR}. The state ν_{eR}, being a "sterile state", escapes the detection and the spin-precession may account for the deficit (and the time variation) of solar neutrinos. The transition probability is given as $P(\nu_{eL} \to \nu_{eR}) = \sin^2(\mu B t)$ (t : the time of travel, B : the strength of the magnetic field). To get rough idea, for $B \sim 1(kG)$ the necessary magnetic moment is $\mu \simeq 3 \cdot 10^{-10}$ $(e/2m_e)$, which is much larger than what we get by incorporating ν_{eR} to the standard model, $\mu \simeq 3 \cdot 10^{-19}(\frac{m_\nu}{1eV})(\frac{e}{2m_e})$, and needs a real New Physics.

Unfortunately, this scenario has a few problems to be overcome in explaining the solar neutrino data. First, once the matter effect inside the sun is switched on, as only ν_{eL}, not ν_{eR}, gets the matter effect, it tends to prevent the spin-precession:

$$P(\nu_{eL} \to \nu_{eR}) = \frac{(2\mu B)^2}{a_\nu^2 + (2\mu B)^2} \cdot \sin^2 \left(\frac{\sqrt{a_\nu^2 + (2\mu B)^2}}{2} t\right), \qquad (1)$$

where a_ν (assumed to be constant, for brevity) is the matter effect and when it is larger than μB it clearly prevents the precession. Secondly, the survival probability is independent of the neutrino energy E_ν, and therefore does not fit to the existing data, having sensitivities to different E_ν, and showing defferent solar neutrino deficit rates. Finally, it also should be pointed out that the spin-precession is a sort of neutrino oscillation into a sterile state ν_{eR}, and is now disfavored by the combined data of SK and SNO-CC, as stated above.

Such problems can be solved once the spin-precession is generalized to RSFP. Spin flavor precession is the precession where not only spin but also flavor precess, e.g. $\nu_{eL} \rightarrow \nu_{\mu R}$, $\nu_{\tau R}$, for Dirac neutrinos. This time the matter effect, combined with the mass difference, causes the resonant neutrino oscillation. Just as in MSW case, the survival probability of ν_{eL} has E_ν dependence. Again, however, this type of RSFP is disfavored as $\nu_{\mu R}$, $\nu_{\tau R}$ are sterile states.

Just as in the Majorana-type mass term, the chiral partner can be anti-particle of ν_L, and "Majorana-type" spin precession (with $|\Delta L| = 2$), such as

$$\nu_{eL} \rightarrow \overline{\nu}_{\mu R},\ \overline{\nu}_{\tau R} \quad (\overline{\nu}_{\mu R} \equiv (\nu_{\mu L})^c, etc.), \tag{2}$$

is possible [5]. Now the converted final states are active states and are in complete consistence with the combined result of SK and SNO-CC. The 2×2 Hamiltonian to govern the time evolution, relevant for e.g. $\nu_{eL} \rightarrow \overline{\nu}_{\mu R}$ reads as

$$H = \begin{pmatrix} a_{\nu_e} & \mu B \\ \mu B & (\Delta m^2_{21}/2E_\nu) - a_{\nu_\mu} \end{pmatrix}, \tag{3}$$

where the net matter effect, $a_{\nu_e} + a_{\nu_\mu} = \sqrt{2}G_F(N_e + N_n)$ (N_e, N_n : the number densities of electron and neutron), is similar to that in MSW in view of the fact $N_n \sim (1/6)N_e$ inside the sun. The resonance, or level crossing, is realized at the point satisfying $(\Delta m^2_{21}/2E_\nu) = a_{\nu_e} + a_{\nu_\mu}$. There is a feature, not shared by the MSW scenario, i.e. the oscillation length at the resonance point, $1/(\mu B)$, is E_ν-independent. This becomes important when we discuss the characteristic energy spectrum of the survival probability in RSFP.

It is interesting to note that a Majorana neutrino cannot have a magnetic moment, while a Dirac neutrino may have: $\overline{N}\sigma_{\mu\nu} N = 0$, for $N^c = N$. This means that for Majorana neutrinos only "transition magnetic moment" $\mu_{\alpha\beta}$,

$$\mu_{\alpha\beta}\ \overline{N_\alpha}\ \sigma_{\mu\nu}\ N_\beta \quad (\alpha \neq \beta), \tag{4}$$

which connects different flavors, is allowed. Namely, for Majorana particle spin flavor precession is inevitable [7]. As has been already pointed out, to get enough magnetic moments for neutrinos, while keeping their masses small enough, is a theoretically very non-trivial task. The essential reason is that to get a magnetic moment we need a chirality-flip and the magnetic moment is easily proportional to the neutrino mass, which should be kept very small. Though there have been considerable efforts to construct models to solve this problem, it seems that no natural and satisfactory model has been put forward so far. So if the RSFP is established, it will clearly require a dramatic modification of the standard model.

It seems to be important that the RSFP scenario provides a unique clue to the nature of neutrinos, i.e. either Dirac or Majorana. Let us recall that in the scenario invoking to the ordinary neutrino oscillation due to flavor mixing, such as MSW, it is basically not possible to distinguish these two. This is simply because what we are observing is neutrino oscillations without chirality flip, $\nu_L \to \nu_L$, since an oscillation with chirality-flip is suppressed in its probability by a factor $(m_\nu/E_\nu)^2$ and is practically not observable. Thus, for Dirac neutrinos what we are seeing is a process with even number of mass term insertion such as $\nu_L \to \nu_R \to \nu_L$, while for Majorana case what we are seeing is $\nu_L \to (\nu_L)^c \to \nu_L$. Therefore it is not possible to say whether the intermediate state is ν_R (a sterile state) or $(\nu_L)^c$ (an active state). In the RSFP process, however, the situation completely changes; the magnetic field B does cause a chirality flip (spin-precession) without the chiral suppression factor. Thus if RSFP is due to Dirac-type magnetic moment, the oscillation is of the type of active $\to$ sterile, $\nu_L \to \nu_R$, which is now experimentally disfavored. On the other hand, if RSFP is due to Majorana-type (transition) magnetic moment, the oscillation is of the type of active $\to$ active, $\nu_L \to (\nu_L)^c$, and is completely consistent with the recent SK and SNO data, as we will see in some detail below. So if we ever observe anti-neutrinos from the sun it will be a good indication of the Majorana-type RSFP.

It will be worth noticing that the RSFP scenario provides an "ideal energy spectrum" of ν_e survival probability $P(\nu_e \to \nu_e)$. This spectrum seems to play an important role in the fit of this scenario to the data, though in the detailed fitting procedure other informations, such as the magnetic field profile inside the sun, become also relevant. In the MSW scenario (for relatively small mixing angles), it is well-known that the survival probability as the function of E_ν once goes down from 1 (smaller energies) to almost 0 (intermediate energies) and then recovers 1 (higer energies). On the other hand, in RSFP scenario with appropriate strength of magnetic field B, though it also once goes down from 1 to almost 0, it does not recover 1, but approaches approximately to 1/2 for higher energies [8]. Such difference stems from the difference in the adiabaticity conditions in these two cases:

$$\text{MSW}: \frac{\tan 2\theta}{\left(\frac{d\ \log N_e}{dx}\right)} \gg \frac{E_\nu}{\Delta m_{21}^2\ \sin 2\theta}, \tag{5}$$

$$\text{RSFP}: \frac{\left(\frac{\mu B}{\Delta m_{21}^2/E_\nu}\right)}{\left(\frac{d\ \log N_e}{dx}\right)} \gg \frac{1}{\mu B}, \tag{6}$$

with μ being the relevant transition magnetic moment. Namely for MSW case, the time-evolution of neutrino states is not adiabatic for higher energies, as we see in the above relation, and the enough transition $\nu_e \to \nu_\mu$, for instance in 2 generation scheme, is not allowed even if there is a level crossing. This is why the survival probability recovers 1. For the RSFP scenario, the situation is just opposite; the above equation indicates that for higher energies the evolution is adiabatic. Thus enough transition is expected. In fact when E_ν is large, the factor $\Delta m_{21}^2/2E_\nu$ and therefore the mater effects a_{ν_e}, a_{ν_μ} at the resonance point becomes negligible. Thus the Hamiltonian is well approximated by

$$H \simeq \begin{pmatrix} 0 & \mu B \\ \mu B & 0 \end{pmatrix}. \tag{7}$$

Namely the situation just reduces to the case of OVV, where there is no mass-squared difference and matter effect is not considered. Thus the time averaged probability just gives 1/2: $\overline{P(\nu_e \to \nu_e)} \simeq 1 - \overline{\sin^2(\mu B t)} = 1/2$.

3 Comparison with MSW Scenario

3.1 Post-SNO-CC Data

Having the new data of SNO-CC, it will be meaningful to compare the fitness of the RSFP scenario to all existing data with that of the most popular MSW scenario. I first would like to point out that the fact the solar neutrino flux observed at SK is larger than that of SNO-CC can be easily understood also in the Majorana-type RSFP scenario, since the converted final state, anti-neutrinos, are active states just as in MSW case and do contribute to the event rate at SK. For instance for $E_\nu = 10(MeV)$, the cross sections for the active states are $\sigma(\nu_\mu e \to \nu_\mu e) = 1.6 \times 10^{-44}(cm^2)$, $\sigma(\overline{\nu}_\mu e \to \overline{\nu}_\mu e) = 1.3 \times 10^{-44}(cm^2)$, which are comparable.

As was mentioned above, the RSFP has "ideal energy spectrum" of the survival probability. By ideal I mean that it roughly fit to the really observed survival rates of solar neutrinos reported by the 3 types of experiments. Namely Ga, Cl and νe experiments, which are sensitive to the solar neutrinos with lower, intermediate and higher energies, suggest distinct survival probabilities for different energies; high and almost 0 survival probabilities for pp and Be neutrinos, while the survival probability of the B neutrinos with higer energies being close to 1/2, roughly speaking, but not 1. Thus it is expected to be relatively easy for the RSFP scenario to fit to the existing data. In fact the post-SNO-CC analysis, including all data, done in the recent paper [9] tells us that the RSFP scenario has better "goodness of fit" of 22% than that of LMA-MSW solution of 14%. It has also been shown that the predicted recoil energy spectrum at SK detector is just flat, around 0.45, in complete agreement with the SK data [9].

3.2 Implications of RSFP to SNO-NC and Planned Low Energy Solar Neutrino Experiments

If the deficit of solar neutrino event rates are attributed to the neutrino oscillation into the active states ν_μ, ν_τ as in MSW, the flux observed by the SNO neutral current (NC) experiment will be the same as what we expect when there is no oscillation, and will be a good confirmation of the scenario. It is interesting to note that the Majorana-type RSFP scenario gives more or less similar prediction, since the converted anti-neutrino states, $\overline{\nu}_\mu$, $\overline{\nu}_\tau$, being active states, do contribute to the NC process, just as in MSW. It has been explicitly confirmed by the detailed analysis in [10]; i.e. the RSFP has been shown to give a similar prediction with that of LMA-MSW.

Now let us move to the discussion on the planned experiments aimed to detect the low energy pp and Be neutrinos [11]. This type of experiment will be quite useful clue to discriminate LMA-MSW and RSFP. The reason is rather simple. As I have pointed out at the introduction, in the LMA solution the mechanism of resonant matter oscillation does not play a central role, and it rather mimics the case of vacuum oscillation. Thus the survival probability has mild energy dependence and the predicted event rates for pp and Be neutrinos are roughly the same, i.e. $R(pp) \sim R(Be)$ (R: the ratio of the expected number of events with oscillation to the one without oscillation). As for the RSFP the situation is completely different. As we have already seen, RSFP shares the same energy spectrum of survival probability with that of SMA(small mixing angle) MSW solution, as far as the energies up to the intermediate range are concerned (though the SMA-MSW is disfavored anyway, as it fails to predict almost constant enegy spectrum for higer energies, which B neutrino is sensitive to). Namely, the probability shows rapid decrease from 1 to almost 0 when energy changes from lower to intermediate range. Thus we expect that the predicted $R(Be)$ is considerably lower than $R(pp)$, in clear contrast to the case of LMA solution. The detailed analysis of these ratios for lower energy solar neutrinos hs been given by H.Nunokawa [12]. We may say that such small $R(Be)$ is genuine nature of resonant matter oscillation. Thus if this small event rate of Be neutrino is ever confirmed, it may be a good indication of the RSFP.

4 Realistic RSFP Scenario to Accommodate the Atmospheric Neutrino Oscillation

4.1 To Accommodate the Atmospheric Neutrino Oscillation into RSFP

Although the RSFP scenario was originally proposed as an explanation of solar neutrino deficits, for the scenario to become realistic it should also be able to account for another puzzle, i.e. the deficit of the atmospheric (ν_μ) neutrino. Obviously the magnetic field of the earth is too weak for the spin-flavour precession to be relevant for the atmospheric neutrino deficit. Thus a natural guess is that the deficit of the atmospheric neutrino is attributed to the ordinary neutrino oscillation due to the flavor mixing. Thus we need a "hybrid scenario" with both of (transition) magnetic moment and flavor mixing angles [13]. Generalizing further to the 3 generation scheme, the Hamiltonian for the time-evolution is now 6 × 6 matrix in the basis of

$$\left(\nu_{eL}\ \nu_{\mu L}\ \nu_{\tau L}\ \overline{\nu}_{eR}\ \overline{\nu}_{\mu R}\ \overline{\nu}_{\tau R}\right)^t . \tag{8}$$

described by 9 parameters, $\mu_{e\mu}$, $\mu_{e\tau}$, $\mu_{\mu\tau}$, Δm^2_{21}, Δm^2_{31}, θ_1, θ_2, θ_3, δ (in literature, θ_1, θ_2, θ_3 are often written as θ_{12}, θ_{23}, θ_{13}).

4.2 A Reduction Formula

The Hamiltonian in 3 generation scheme is complicated having 9 parameters. Even if an experimental datum such as survival probability is given, it will lead to a 8-dimensional hypersurface in the parameter space, which is practically inpossible to analyze. Furthermore, though the experimental data are usually shown in the plane of parameter space $(\Delta m^2, \theta)$ assumimg simplified 2 generation scheme, it will be quite hard to say, for instance, which angle among $\theta_{1,2,3}$ the θ actually corresponds to.

However, it is encouraging to know that in the case of MSW scenario, it turns out that an (approximate) formula, which connects the survival probability for solar neutrino in 3 generation scheme S to the one in (effective) 2 generation scheme S_{eff}, thus realizing the reduction of 3 generation → 2 generation [14]:

$$S = \cos^4\theta_3 \cdot S_{eff} + \sin^4\theta_3, \tag{9}$$

where S_{eff} is the survival probability in an (effective) 2 generation scheme handled by the parameters Δm^2_{21} and θ_1, while the matter effect should be understood as re-scaled by the presence of θ_3 as $\cos^2\theta_3 \cdot a_\nu$. (This formula appeared only in a coference proceeding [14], and the detailed derivation was not given there. A paper to clarify this, including other related issues, will show up soon [15]). The important thing is such reduction is valid for arbitrary time-dependent matter effect and for arbitrary mixing angles, as far as a hierarchy in the mass-squared

$$\Delta m^2_{21},\ \ 2E_\nu a_\nu(0)\ \ll\ \Delta m^2_{31}, \tag{10}$$

is met, where $a_\nu(0)$ should be understood as the maximal value of the matter effect. It not only reduces the relevant parameters and make the analysis of the data very transparent, but also enables us to identify which angle is really handling the survival probability. Let us note that the hierarchy between two mass-squared difference, whic was assumed when this formula was first proposed by the analogy to the mass hierarchy in the quark sector, is a fact now; solar and atmospheric neutrino oscillations are sensitive to two distinct mass-squared differences, which should be identified with Δm^2_{21} and Δm^2_{31}, respectively. In view of the fact that θ_3 is rather small as suggested by the CHOOZ experiment, we know that the survival probabilty is well described by that in the effective 2 generation scheme, $S \simeq S_{eff}(\Delta m^2_{21},\ \theta_1,\ a_\nu)$.

I would like to poit out that a similar reduction formula is possible also for the survival probability of solar neutrino in the RSFP scenario, extended to the 3 generation scheme. Here I just give the result. For the details of the derivation see [16]. The reduction formula itself is exatly the same as Eq.(9). The difference is that now the effective survival probability S_{eff} should be calculated by use of the following 4×4 matrix with just 3 parameters $\hat{\mu}$, Δm^2_{21}, θ_1, and with the initial condition of state $(1,0,0,0)^t$,

$$H_{eff} = \begin{pmatrix} M+a & B\cdot\mu^\dagger \\ B\cdot\mu & M-a \end{pmatrix}, \tag{11}$$

where the 2×2 matrices are given as

$$M = \begin{pmatrix} \frac{\Delta m^2_{21}}{2E_\nu}\sin^2\theta_1 & \frac{\Delta m^2_{21}}{4E_\nu}\sin 2\theta_1 \\ \frac{\Delta m^2_{21}}{4E_\nu}\sin 2\theta_1 & \frac{\Delta m^2_{21}}{2E_\nu}\cos^2\theta_1 \end{pmatrix}, \tag{12}$$

$$a = \begin{pmatrix} \cos^2\theta_1 a_{\nu_e} + \sin^2\theta_1 a_{\nu_\mu} & 0 \\ 0 & a_{\nu\mu} \end{pmatrix},\quad \mu = \begin{pmatrix} 0 & \tilde{\mu}_{e\mu} \\ -\tilde{\mu}_{e\mu} & 0 \end{pmatrix}, \tag{13}$$

where $\tilde{\mu}_{e\mu} = \cos\theta_2\cos\theta_3\cdot\mu_{e\mu} + \sin\theta_2\cos\theta_3\cdot\mu_{\tau e} + \sin\theta_3\, e^{i\delta}\cdot\mu_{\mu\tau}$.

Again, the small θ_3 means that the survival probability is well described by S_{eff}, though the transition magnetic moment gets rescaling of $\mu_{e\mu}\to\tilde{\mu}_{e\mu}$, as we have seen above. The analysis, therefore, done by Miranda et al. for the case of 2 generation hybrid scenario [9] is just applicable., once their μ, Δm^2, θ is understood as our $\tilde{\mu}_{e\mu}$, Δm^2_{21}, θ_1. Their result tells us that the pure RSFP scenario can be extended to hybrid scenario with the mixing angle up to $\tan^2\theta_1 = 10^{-2}$.

In this scenario, the atmospheric neutrino deficit is attributed to the ordinary oscillation due to flavor mixing. If we neglect the matter effect of the earth, for brevity, the relevant transition probability is given by

$$P(\nu_\mu \to \nu_\tau)_{atm} = \sin^2 2\theta_2 \cos^4\theta_3 \sin^2(\frac{\Delta m^2_{31}}{4E}t). \tag{14}$$

5 Summary

We summarize below the issues reported at this conference.
1. The Resonant Spin Flavor Precession (RSFP) scenario has its characteristic "ideal energy spectrum" of ν_e survival probability, which approaches to 1/2, instead of 1 in MSW scenario, for larger neutrino energy. The spectrum works, roughly speaking, so that the RSFP scenario gets a large "goodness of fit" at the post-SNO-CC global analysis, providing a perfect recoil energy spectrum to fit to the SK data.
3. As for the ratio of the numbers of event with and without oscillation, R(pp) and R(Be), for pp and Be neutrinos, to be seen at the planned low energy solar neutrino experiments, the RSFP scenario predicts small R(Be) as the genuine feature of the resonant oscillation, in clear contrast to the LMA-MSW scenario, where R(Be) and R(pp) are comparable and sizable. Thus, planned low energy solar neutrino experiments, KAMLAND, BOREXINO, LENS, MOON, CLEAN, HERON, XMASS [11], is expected to provide very useful clue to clearly distinguish LMA and RSFP scenarios.
4. So far there is no indication of time variation of solar neutrino flux with the period 11 years, which motivated the original spin precession scenarios. However, if the magnetic field deeper inside the sun (not convective zone) plays a role for the RSFP, it is argued that there is no reason to expect such time variation.
5. We have also tried to make the RSFP scenario more realistic, namely to aaccount for not only the solar neutrino deficit but also that of atmospheric neutrino, by generalizing the RSFP scenario to the realistic 3 generation scheme, where generation mixings are also taken into account. It has been shown that a formula, which enables us to reduce the problem of neutrino oscillation in 3 generation scheme to that in simple 2 generation scheme, exists and the analysis based on the 2 generation scheme becomes meaningful. Such analysis shows that there remains a enough space in the parameter space to account for the solar neutrino data, while explaining the atmospheric neutrino oscillation relying on the usual oscillation due to the flavor mixing.

6. If the RSFP scenario is ever confirmed, it will need "real" New Physics, i.e. dramatic modification of the standard model.

Acknowledgments

The author would like to thank the organizer of the workshop for the kind invitation. This work was supported by the Grant-in-Aid for Scientific Research of the Ministry of Education, Science and Culture, No.12047219, No.12640275.

References

1. Y. Koshio, These proceedings.
2. S. Oser, These proceedings.
3. S.P. Mikheyev and A.Yu. Smirnov, *Sov.J.Nucl.Phys.* **42**, 913(1985); L. Wolfenstein, *Phys. Rev.* D **17**, 2369 (1978).
4. T. Yanagida, Proc. of Workshop on Unified Theory and Baryon Number in the Universe, ed. by O. Sawada and A. Sugamoto, KEK, Japan, 1979; M. Gell-Mann, P. Ramond and R. Slansky, in *Supergravity*, ed. by P. van Nieuwenhuizen and D.Z. Freedman, North Holland Pub. Co., 1979.
5. C.S. Lim and M.J. Marciano, *Phys. Rev.* D **37**, 1368 (1988); E. Kh. Akhmedov, PLB **213**, 64 (1988).
6. L.B. Okun, M.B. Voloshin, and M.I. Vysotsky, *Sov.J.Nucl.Phys.* **44**, 440(1986).
7. J. Schechter and J. Valle, *Phys. Rev.* D **24**, 1883 (1981).
8. C.S. Lim and H. Nunokawa, *Astropart. Phys.* **4**, 63(1995).
9. O.G. Miranda, C.Pena-Garay, T.I. Rashba, V.B. Semikoz and J.W.F. Valle, *Phys. Lett.* B **521**, 299 (2001).
10. O.G. Miranda, C.Pena-Garay, T.I. Rashba, V.B. Semikoz and J.W.F. Valle, *Nucl. Phys.* B **595**, 360 (2001).
11. M. Nakahata, These proceedings.
12. H. Nunokawa, in the proceedings of "International Workshop on Low Energy Solar Neutrinos", Tokyo, Japan, 4-5 Dec 2000, hep-ph/0105027(2001)
13. H. Minakata and H. Nunokawa, *Phys. Rev. Lett.* **63**, 121 (1989).
14. C.S. Lim, in the Proceedings of "BNL Neutrino Workshop", Upton, N.Y., Feb. 1987, BNL-39675, p111; A.Yu. Smirnov, in the Proceedings of Int. Symp. "Frontiers in Neutrino Astrophysics", Takayama, Japan, Oct 1992, p105; X. Shi and D. Schramm, *Phys. Lett.* B **283**, 305 (1992).
15. C.S. Lim, K. Ogure, H. Tsujimoto, a paper in preparation
16. A paper in preparation

STATUS OF THE KAMLAND EXPERIMENT

J. GOLDMAN (FOR THE KAMLAND COLLABORATION)
Research Center for Neutrino Science, Tohoku University
Sendai 980-8578 , Japan
goldman@awa.tohoku.ac.jp

We report here on the current status of KamLAND (Kamioka Liquid Scintillator Anti-Neutrino Detector), the first reactor based neutrino experiment with direct sensitivity to the Large Mixing Angle (LMA) region of neutrino-oscillations parameter space.

1 Introduction

The theory of neutrino oscillations has been proposed [1] as a possible explanation for the solar neutrino deficit measured by the Homestake chlorine experiment [2]. This anomaly was later confirmed by the SAGE [3], GALLEX [4], and Kamiokande [5] experiments and, more recently, by Super Kamiokande [6] and SNO [7]. These latter experiments limited the neutrino oscillation parameters to be within the LMA or LOW (Low Mass) region, with LMA being the more highly favored [8]. KamLAND is in an excellent position to investigate this region due to a combination of factors including a high $\overline{\nu}_e$ flux from Japanese commercial power reactors, 1 kiloton target mass, and a ratio, $L/E \sim (10^6\ eV^{-2})$, appropriate to LMA studies. If the oscillation parameters, $\sin^2 2\theta$ and Δm^2 lie within this region, KamLAND will be able to measure them (see Fig. 3). KamLAND will also make a measurement of the ν_e flux from ^{7}Be decay in the sun.

2 Experimental Layout

2.1 Detector

The KamLAND experiment is located in the Kamioka mining company's Mozumi mine at 2700 mwe under Mount Ikenoyama in Gifu prefecture, Japan. The detector (Fig. 1) contains an active volume 13m in diameter enclosed by a nylon/EVOH spherical balloon containing 1 kiloton of liquid scintillator (80% mineral oil and 20% pseudocumene with 1.5 g/l PPO dissolved in the mixture). The balloon is supported by a kevlar net attached at the chimney and base of the detector. Surrounding the balloon is an outer buffer of pure mineral oil followed by an acrylic spheroid functioning as a radon barrier.

Outside of the spheroid are 1325 Hamamatsu #R7250 Photomultiplier Tubes (PMT) and 554 Hamamatsu #R3602 PMTs which are attached to a stainless steel spherical shell. The total photocathode coverage afforded by these is 32%. External to the shell is a water cerenkov outer detector. It contains 225 #R3602 PMTs and acts as a cosmic ray muon veto.

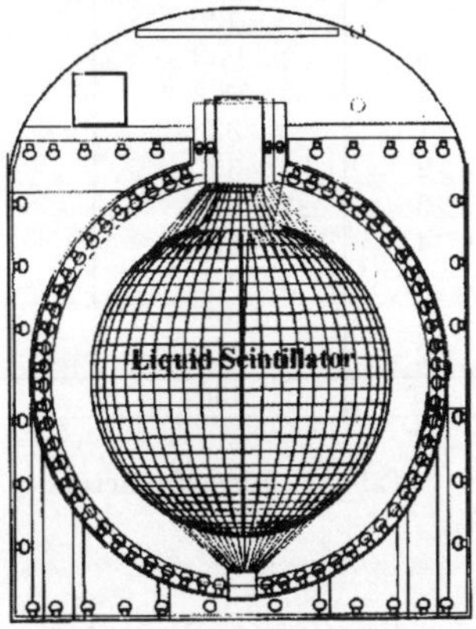

Figure 1. Diagram of the KamLAND detector

2.2 Reactor Neutrinos

KamLAND is well placed in that 86% of the incident reactor electron anti-neutrinos come from 175±35 km away resulting in an ideal baseline for LMA measurements. The flux incident on the detector is roughly $10^5\ \overline{\nu}_e/\mathrm{cm}^2/\mathrm{s}$ or 1075 $\overline{\nu}_e$ interactions/year throughout the detector volume (with no oscillations) under the assumptions of maximum power and that only the 16 nearest reactors contribute. A list of reactors and related information can be found in Table 1.

3 Physics Program

3.1 Reactor Program

KamLAND is sensitive to anti-neutrinos via the inverse beta decay reaction: $\overline{\nu}_e + p \rightarrow n + e^+$. The prompt positron is followed roughly 200 microseconds later by a neutron capture and the release of a 2.2 MeV photon. The primary backgrounds to the $\overline{\nu}_e$ signal are neutrons from muon spallation in the rock

Reactor Site	Distance (km)	# of reactors	Therm. Power (max) (GW)	Max. Flux (10^5 $\overline{\nu}_e$ /cm^2/s)	Max. Evt rate evts/kt-yr
Kashiwazaki	160	7	24.6	4.25	348
Ohi	180	4	13.7	1.90	154
Takayama	191	4	10.2	1.24	102
Hamaoka	214	4	10.6	1.03	84
Tsuruga	139	2	4.5	1.03	84
Shiga	81	1	1.6	1.08	89
Mihama	145	3	4.9	1.03	84
Fukushima-1	344	6	14.2	0.53	44
Fukushima-2	344	4	13.2	0.59	40
Tokai-II	295	1	3.3	0.17	14
Shimane	414	2	3.8	0.10	8
Ikata	561	3	6.0	0.08	7
Genkai	755	4	6.7	0.05	4
Onagawa	430	2	4.1	0.10	8
Tomari	784	2	3.3	0.02	2
Sendai	824	2	5.3	0.03	3
Total		51	130	13.1	1075

Table 1. List of reactors

surrounding the detector and natural radioactivity present in the scintillator. Particle identification (PID), timing, and spatial requirements are necessary to remove the neutron background as well as random coincidences due to ambient radioactivity. PID is particularly important for KamLAND as different particles of equal energies emit different amounts of light, due to quenching effects in the scintillator. KamLAND's Front End Electronics captures waveforms from each PMT thus allowing for single photoelectron counting and enhancing the experiment's ability to achieve the PID, spatial, and timing requirements. In addition to the use of this sophisticated electronics, low background levels of scintillator radioactivity (U$< 10^{-14}$ g/g, Th$< 10^{-14}$ g/g, ^{40}K$< 10^{-12}$ g/g, Rn< 0.5mBq/m^3) have been achieved. All scintillator and mineral oil were extensively purified during detector filling over the course of five months (May-September 2001). Figure 2 shows the positron energy spectrum compared with background from U, Th, and ^{40}K decay chains for several levels of radioactive element concentration. The current levels of radioactive contaminants in both scintillator and oil are lower than the threshold required for the reactor portion of the experiment. Note that, since the energy from the positron annihilation gamma rays are recorded in addition to the positron kinetic energy, the minimum visible energy of the inverse beta decay reaction is 1.022 MeV, which is above much of the background. The upper plot in Figure 3 shows the three year (theoretical) Δm^2 sensitivity region (LMA) visible to KamLAND ($1\times10^{-5} < \Delta m^2 < 2\times10^{-4}$ eV^2). The bands shown are allowed at 90%, 95%, 99%, and 3σ CL respectively. Limitations on the energy resolution make a precise measurement of the mixing angle parameters unfeasible for $\Delta m^2 > 2\times10^{-4}$ eV^2. The lower plot includes current solar data and is therefore slightly asymmetrical in $\tan^2\theta$ [9].

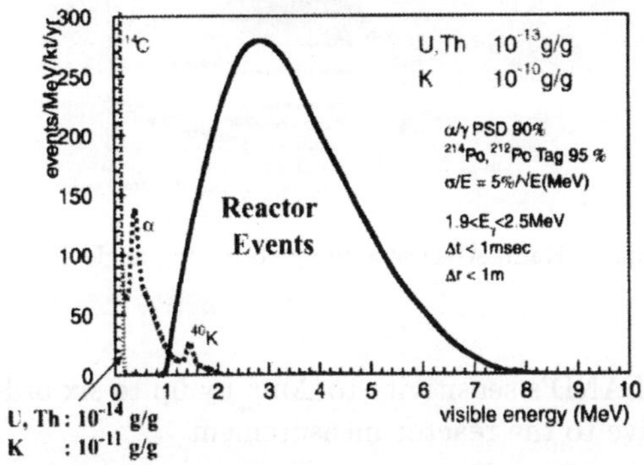

Figure 2. Positron energy spectrum from inverse β decay

3.2 Solar Program

The KamLAND solar program will require greater scintillator purity (U< 10^{-16} g/g, Th< 10^{-16} g/g, ^{40}K< 10^{-14} g/g, Rn< 0.01μBq/m^3) over the reactor phase and an aggressive purification program is scheduled. KamLAND is sensitive to 862 keV monoenergetic ν_e's from electron capture on ^{7}Be ($^7\text{Be}+e^-\rightarrow^7\text{Li}+\nu_e$). The signal for ^{7}Be $\nu_e e^-$ scattering will be a "Compton-like" edge at approximately 665 keV and solar neutrino observation efforts will therefore focus on this energy region. Although background levels are expected to be larger than the signal for this analysis despite the high degree of scintillator purity, the background will remain constant throughout the year, whereas the signal will change due to the eccentricity of the earth's orbit. The background may then be statistically subtracted away. This seasonal variation of the ^{7}Be flux is related to the SMA (Small Mixing Angle) and VAC (Vacuum Oscillations) solutions and, using subtraction, KamLAND will be able to see clear evidence for them, should they be correct, after three years of running. Additionally, if KamLAND can distinguish the day-night flux variation, it will be able to probe the LOW solution as well. The much longer baseline afforded by solar neutrinos in conjunction with the MSW [10] effect

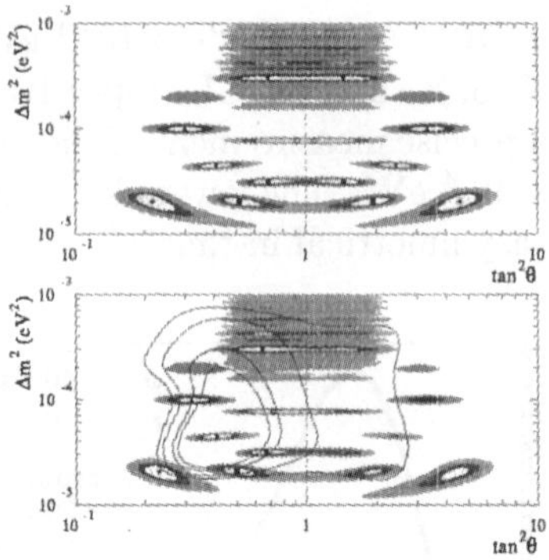

Figure 3. KamLAND sensitivity to mixing angle and Δm^2

will extend KamLAND's sensitivity to Δm^2 by up to six orders of magnitude downwards relative to the reactor measurement.

3.3 Current Status

As of January 22, 2002, the KamLAND experiment has begun normal 24 hour per day, 7 days per week data-taking. Most front-end electronics cards (currently 86%) have been installed and are fully operational. In addition, calibration efforts have begun with ^{65}Zn (1.116 Mev) and ^{60}Co (2.506 MeV) gamma emitters and pulsed LED and laser light sources. Figure 4 shows two preliminary muon candidates observed early in data taking. Outer (cerenkov) detector information was not available for these events but the left-hand figure contains what is likely an edge skimming muon (upper right-hand side of detector projection) while the right-hand figure is likely to be a through-going muon (entering at the top of the projection and exiting through the bottom). With several groups working simultaneously on calibration, we hope to have a great deal of information very soon relating to detector and PMT response and linearity. With the recent integration of the outer cerenkov detector, KamLAND is poised to begin taking neutrino data.

Acknowledgments

The KamLAND collaboration gratefully acknowledges the assistance of the Kamioka Mining and Smelting Company and the Mitsui Shipbuilding Company during and subsequent to construction of the experiment. The Kam-

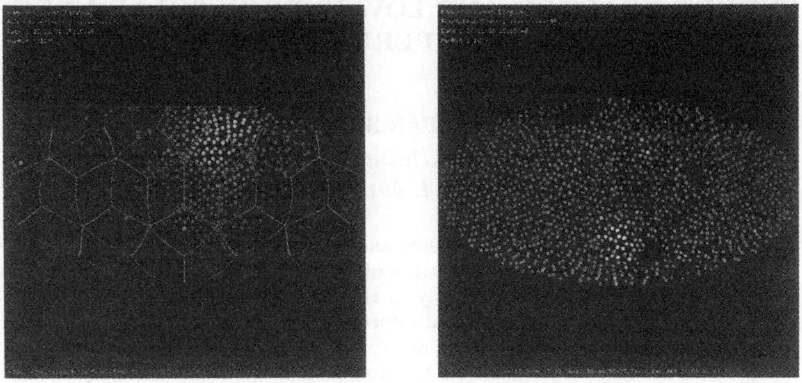

Figure 4. KamLAND muon candidate events

LAND experiment has been made possible by funding from the Japan Ministry of Education, Culture, Sports, Science, and Technology and the U.S. Department of Energy. The Japanese Society for the Promotion of Science has provided additional funding for research staff and other experimental expenses.

References

1. V.N. Gribov and B. M. Pontecorvo, *Phys. Lett.* B **28**, 493 (1969)
2. R. Davis, D. Harmer, and K. Hoffman, *Phys. Rev. Lett.* **20**, 1205 (1968)
3. SAGE Collaboration, Neutrino 2000, p.36, 2000
4. GALLEX/GNO Collaborations, Neutrino 2000 , p.44 (2000)
5. Kamiokande Collaboration, Y. Fukuda *et al.*, *Phys. Rev. Lett.* **77**, 1683 (1996)
6. Super-Kamiokande Collaboration, *Phys. Rev. Lett.* **86**, 5656-5660 (2001)
7. SNO Collaboration, Q.R. Ahmad *et al.*, nucl-ex/0106015
8. G.L. Fogli, E. Lisi, D. Montanino, and A. Palazzo, *Phys. Rev.* D **64**, 093007 (2001).
9. André de Gouvêa and C. Peña-Garay, *Phys. Rev.* D **64**, 113011 (2001)
10. See, for example: L. Wolfenstein, *Phys. Rev.* D **17**, 2369 (1978); S.P. Mikheyev and A. Yu. Smirnov, *Yad. Fiz.* **42**, 1441 (1985)

FUTURE REACTOR AND LOW ENERGY SOLAR NEUTRINO EXPERIMENTS

STEFAN SCHÖNERT
Max-Planck-Institut für Kernphysik,
Saupfercheckweg 1, 69117 Heidelberg, Germany

Sub-MeV solar neutrino experiments and long-baseline reactor oscillation experiments toe the cutting edge of neutrino research. The upcoming experiments KamLAND and BOREXINO, currently in their startup and final construction phase respectively, will provide essential information on neutrino properties as well as on solar physics. Future projects, at present under development, will measure the primary solar neutrino fluxes via electron scattering and neutrino capture in real time. High precision data for lepton mixing as well as for stellar evolution theory will become available in the future. This paper aims to give an overview of the upcoming experiments and of the projects under development.

1 Introduction

In the past, experiments with solar neutrinos and experiments with neutrinos from nuclear reactors probed different ranges of the parameter space for neutrino oscillations. Basically, the distances between source and detector in reactor neutrino experiments ($\lesssim 1$ km) were too short to probe the parameter space relevant to explain the solar neutrino deficit [1,2]. In the very near future, with the upcoming experiments KamLAND [3] and BOREXINO [4], it will be possible to study the parameter space of the large mixing angle (LMA) solution not only with solar, but as well with reactor neutrinos. Although BOREXINO is optimized for solar neutrino detection, it will be able to detect $\bar{\nu}_e$'s from far distant European nuclear reactors [5]. *Vice versa*, KamLAND is dedicated to reactor neutrino spectroscopy, however, if radio impurities are sufficiently low, it will also measure solar neutrinos.

Future projects, for example LENS[6] and XMASS[7], seek to measure the primary pp-, ^{7}Be- and CNO neutrino fluxes in real time via neutrino capture and neutrino-electron scattering respectively. Such measurements will provide important information on the flavor of solar neutrinos, on mixing parameters and on solar physics. If the LMA solution is realized in nature, Δm^2 will be determined with high accuracy using reactor neutrinos. However, the measurement of the mixing angle will be statistically limited. Precision measurements of the pp- and ^{7}Be fluxes by elastic scattering (ES) or neutrino capture (CC) could improve considerably the accuracy of the determination of the mixing angle[8]. If the "LOW", "(quasi-) VAC" or small mixing angle

(SMA) MSW solution is realized in nature, then only solar neutrino experiments can explore the parameter spaces. Full information of the solar neutrino spectrum, via both ES and neutrino capture will be essential in any of the mentioned scenarios.

2 Experiments and R&D projects

Table 1 lists sub-MeV solar as well as reactor neutrino experiments which soon become operational, as well as projects which are currently in an early stage of development.

Table 1. Listing of upcoming experiments and projects under development.

Project	Goals	Method	Technique (Target)
KamLAND [3]	1) $\bar{\nu}_e^{\text{reactor}}$	CC	LS (CH)
	2) ν^{sun} :^{7}Be, ^{8}B	ES	LS (CH)
BOREXINO [4]	1)ν^{sun}: ^{7}Be, pep, ^{8}B	ES	LS (CH)
	2) $\bar{\nu}_e^{\text{reactor}}$	CC	LS (CH)
LENS [6]	ν_e^{sun}: pp, ^{7}Be, CNO	CC	LS (CH+metal)
XMASS [7]	ν_x^{sun}: pp, ^{7}Be	ES	LS (Xe)
HERON [12]	ν_x^{sun}: pp	ES	LS (He)
CLEAN [11]	ν_x^{sun}: pp	ES	LS (He,Ne)
MOON [13]	ν_e^{sun}: pp, ^{7}Be	CC	hybrid/LS (CH+metal)
TPC [14,15]	ν_x^{sun}: pp, ^{7}Be	ES	TPC (He+CH)
GENIUS [16]	ν_x^{sun}: pp	ES	Germanium
LOW-C14 [17]	ν_x^{sun}: pp	ES	LS (CH)
Kr2Det[18]	$\bar{\nu}_e^{\text{reactor}}$: U_{e3}	CC	LS (CH)
TEXONO[19]	$\bar{\nu}_e^{\text{reactor}}$: μ_ν	ES	solid scint.

ES: Elastic scattering of neutrinos off electrons (CC+NC); CC: neutrino capture (charged current; CC); LS: liquid scintillator; CH: hydro-carbon.

2.1 ES and CC-detection of sub-MeV neutrinos

Two different types of interactions will be exploited for the detection of sub-MeV real-time measurements: $\nu_x e$–scattering (ES) which involves charged (CC) and neutral (NC) current interactions, and neutrino capture which only occurs via CC-interaction. The first is sensitive to all active flavors – albeit ν_μ, ν_τ interact with a probability of about 1/6 of ν_e – while the latter is sensitive purely to ν_e. Combining the results of an ES experiment with those of

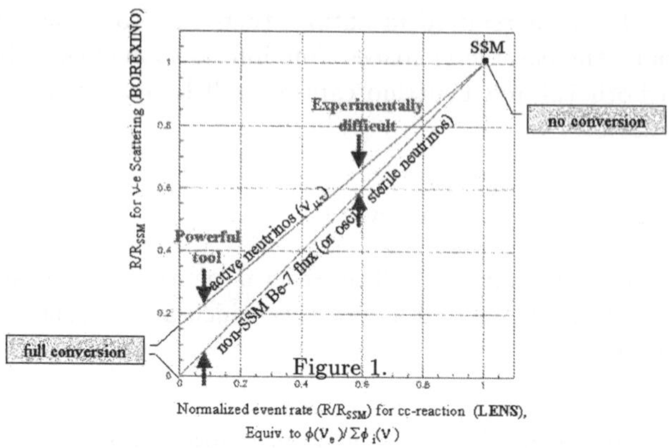

Figure 1.

a neutrino capture experiment provides a unique neutral current probe. Currently this provides the only possibility to study the contributions of ν_μ and ν_τ in the primary pp- and ^{7}Be neutrino fluxes. Electron scattering and neutrino capture experiments are therefore complementary, and should be realized in a coordinated strategy in order to maximize the potential physics output. For purposes of illustration only, Fig. 1 shows the expected relative interaction rates of ^{7}Be neutrinos in BOREXINO and LENS. It should be noted that modest suppression factors, as for example expected for the the LMA solution, challenge the accuracy reachable in the experiments. For instance, in the particular case indicated in Fig. 2 with "experimentally difficult", an error of $\lesssim 3\%$ is needed both for BOREXINO and LENS in order to obtain a signal for ν_μ, ν_τ appearance $\gtrsim 3\sigma$ CL.

2.2 BOREXINO and KamLAND

The BOREXINO (BX), displayed in Fig. 2, and the KamLAND (KL) detectors are similar in design. BX is optimized for solar neutrino detection, thus design priority has been given to high radiopurity, while KL is optimized for reactor neutrino detection where radiopurities are less stringent. Both detectors use liquid scintillators as ν-target and detection medium. Reactor-$\bar{\nu}_e$'s are detected via $\bar{\nu}_e p \rightarrow e^+ n$ where e^+ carries the energy information of the interacting neutrino. The $\bar{\nu}_e$-tag is given by a coincident e^+ – n signal which is correlated in space and separated in time with $\tau = 200\mu$sec. Solar neutrinos are measured via $\nu_x e^-$-scattering ($x = e, \mu, \tau$). The Compton-like recoil electron provides the energy information of the interaction. About 55 events/day/100 tons are expected in the "^{7}Be neutrino window" (i.e. recoil

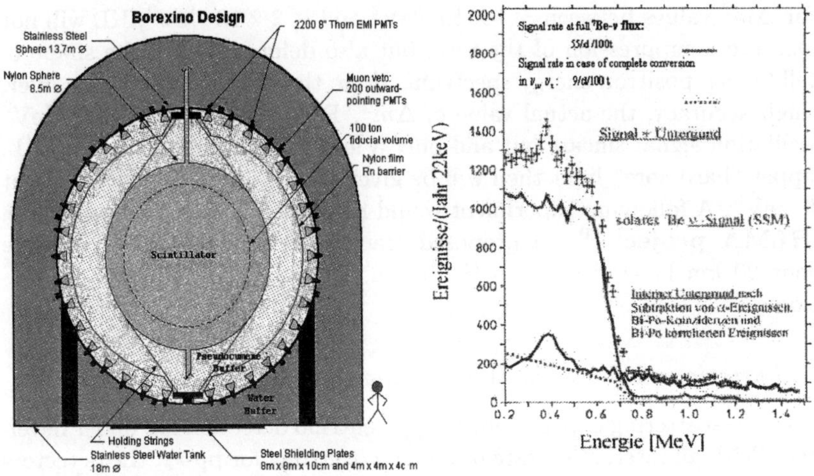

Figure 2. Schematic view of the BOREXINO detector at Gran Sasso, Italy, and expected electron recoil spectrum of ^{7}Be-ν's and background.

electrons with energies between 250-800 keV). Since $\nu_x e^-$–signals cannot be distinguished from radioactive background as β-decay events with the same energy, ultra-low backgrounds are required for neutrino detection. The signal and background spectra expected in BOREXINO are displayed in Fig. 2 assuming the uranium and thorium impurity levels at 10^{-16} g/g and potassium at 10^{-14} g/g. KL needs to meet similar radiopurity specifications to achieve solar-ν detection. However, variable time effects, such as day/night variations, can be observable even with a background larger than 10^{-16}g/g. KL data taking commenced the end of 2001 and the first results from reactor $\bar{\nu}_e$-measurements are expected in 2002. BX scintillator filling will be completed early 2003 and first solar neutrino results might be available during the same year.

The reactor signal in KL (and as well in BX) will be strongly suppressed in the case where LMA parameters drive solar neutrino oscillations. The oscillation length $L[\text{m}] = 1.24 \cdot E[\text{MeV}]/\Delta m^2[\text{eV}^2]$ would amount to 100 km for 3 MeV reactor–$\bar{\nu}_e$'s and $\Delta m^2 = 3.7 \times 10^{-5}$ eV^2. In case of no oscillations, KL will measure about 800 events per year in comparison to 30 events in BX. This obviously translates to a higher sensitivity with respect to $\sin^2 2\Theta$. The recent best fit value $\sin^2 2\Theta = 0.79$, however, should provide unambiguous signals in both experiments provided that background levels are as expected.

For Δm^2 values between $\sim 1 \times 10^{-5}\mathrm{eV}^2$ and $\sim 2 \times 10^{-4}\mathrm{eV}^2$, KL will not only observe a suppression of the rate, but also deformation of the shape of the well known positron energy spectrum. From the shape one can then infer, with high accuracy, the actual value of Δm^2. For values larger $2 \times 10^{-4}\mathrm{eV}^2$ the oscillation signal smears out and only a lower limit can be given by KL. The upper "hard core" limit then will be given by the CHOOZ experiment at $1 \times 10^{-3}\mathrm{eV}^2$. A follow-up experiment would be needed in this particular case. The **HLMA project** [20] - a proposed reactor experiment with a baseline of about 20 km located close to Heilbronn, Germany - would address this question.

2.3 ES experiments for sub-MeV ν detection

Various νe^-–scattering experiments for pp neutrino detection are under development. With an interaction rate of about 2/ton/day for pp-ν_e, the detectors need to have a target mass of about 10 tons to acquire high statistical accuracy. $^{14}\mathrm{C}/^{12}\mathrm{C}$ ratios at 10^{-18} in organic scintillators, as determined with the CTF of BX [21], prohibit the measurement of pp-neutrinos since the 156 keV β-decay endpoint significantly obscures the pp-energy range. If a low-^{14}C liquid scintillator with $^{14}\mathrm{C}/^{12}\mathrm{C} \lesssim 10^{-20}$ could be produced, as discussed in [17], a BX type detector could be realized for pp-ν detection. A promising approach to overcome the ^{14}C background is to avoid organic liquids and instead, to use liquified noble gas as a scintillator. Projects with helium, neon and xenon are under investigation. Argon and krypton are not suitable because of their long-lived radioactive isotopes. Tab. 2 summarizes features relevant for the design of pp-neutrino detection.

Table 2. Characteristics of liquid noble gases as ν-target and detector medium.

	He	Ne	Xe
Boiling Tmp.[K]:	4.2	27	165
Z:	2	10	54
Density [g/cm^3]:	0.125	1.20	3.06
Emission wl [nm]:	73	80	175
Photons/MeV:	22000	15000?	42000

High density liquid xenon, as proposed by the **XMASS** collaboration [7], provides efficient self-shielding. A geometry similar to that of BX, but smaller in size, could thus be realized. The scintillation photons can be detected with photomultiplier technology at liquid xenon temperatures. Backgrounds can

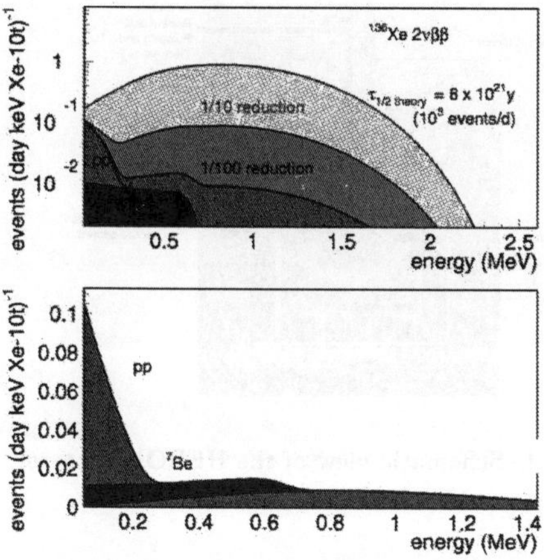

Figure 3. Neutrino recoil spectrum displayed together with the $2\nu - \beta\beta$–decay of ^{136}Xe. Isotope separation is required if $\tau_{1/2} \approx 8 \times 10^{21}$ years.

arise from ^{85}Kr β-decay and from $2\nu - \beta\beta$ decay of ^{136}Xe. The first isotope must be less than 4×10^{-15}g krypton/g. If the $\tau_{1/2}$ of ^{136}Xe is $< 8\times10^{23}$ years, then isotope separation is needed. The best limits have $\tau_{1/2} > 1.1\times10^{22}$ years [22]. Since theoretical calculations estimate $\tau_{1/2} \approx 8 \times 10^{21}$ years, the need for isotope separation is likely. Fig. 3 shows the spectrum of recoil electrons along with the $2\nu - \beta\beta$ decay of ^{136}Xe. A prototype detector containing 100 kg of liquid Xe is under construction. Photon detection will be achieved with newly developed low background photomultiplier tubes consisting of a steel housing and a quartz window. Milestones during this R&D phase will include the determination of the $2\nu - \beta\beta$ half-life and the optical properties of liquid xenon, such as the scattering length of scintillation light.

A similar concept is being pursued by the **CLEAN** project [11]. The target under discussion is neon (or helium) for the fiducial mass and liquid or solid neon as passive shielding buffer. Since neon scintillation occurs at 80 nm, a wavelength shifter is required in order to use photo multipliers for light detection.

About 20t of super-fluid helium as neutrino target is modelled in the **HERON** project [12]. Recoil electrons are detected via 1) prompt scintillation photons (80 nm) and 2) delayed phonon/rotons (1 meV; 10^8/MeV) on sapphire or silicon wafer calorimeters. About 2400 of these wafers are proposed to be located above the liquid phase. A schematic view of the detector is

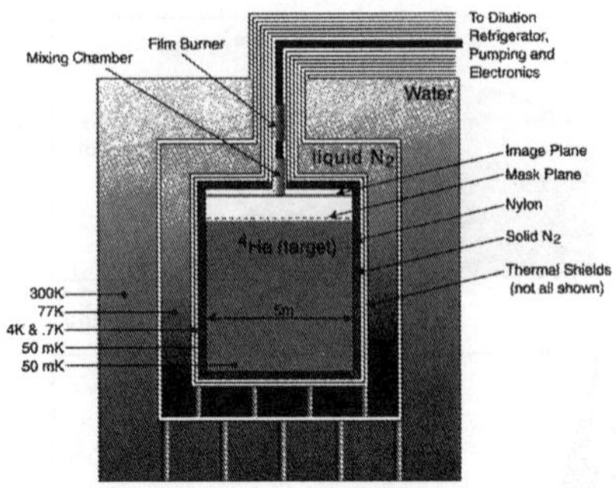

Figure 4. Schematic view of the HERON detector.

displayed in Fig. 4. The temperature rise after photon/phonon interactions is sensed by super-conducting transition edge thermometers or by magnetic micro-calorimeters. The goal is to obtain single (16 eV) photon sensitivity per wafer. Internal radioactive backgrounds are not expected to be a problem, since super-fluid helium is "self-cleaning" (i.e. gravity > kT or chemical binding energies) and impurities, if present, fall to the bottom. On the other hand, the low density of helium does not provide sufficient self shielding against external backgrounds. Therefore, high radiopurity shielding material, dewar and light sensors is imperative. Point-like events, such as νe^-–scattering, can possibly be distinguished from multiple scattering events typical for external γ–radiation by position reconstruction.

Full energy reconstruction of the neutrino energy in ES experiments requires the determination the angle that the recoil electron makes with respect to the incident neutrino direction. For this purpose **TPC's** for solar neutrino detection are being investigated by several groups. The pioneering research project HELLAZ was one such project. A follow up R&D project was initiated recently [14] that proposes substantial conceptual modifications: it is planned to use methane instead of isobutane to overcome the ^{14}C background, to decrease the operational pressure so as to increase the angular resolution, and to simplify the complexity of the electrodes in order to reduce the radioactive background. A further project under discussion is **Super-MUNU** [15]. The detector concept is based on the experience of the MUNU experiment that searched for a neutrino magnetic moment at the Bugey reactor. Four 50 m^3 modules, filled with CF_4 at 1-2 bars pressure are under discussion. As a prototype study, it has been proposed to operate the MUNU reactor detector in an underground laboratory in order to study radioactive background and detector performance (angular and energy resolution) at pp-ν energies.

2.4 Charged current experiments for sub-MeV ν_e real time detection

The electron flavor content of sub-MeV solar neutrinos will be probed via neutrino capture. Only a few promising candidate nuclei exist for real time detection. Either the transition must populate an isomeric excited state, or the final state needs to be unstable in order to provide a coincidence tag to discriminate against background events. Moreover, the energy threshold of the transition needs to be sufficiently low to have sensitivity to sub-MeV neutrinos. Tab. 3 lists nuclei which satisfy these condition.

Table 3. Candidate nuclei for real time solar neutrino detection via neutrino capture (CC).

	Abd.	$T_{1/2}$	Daught.	Q [keV]	E* [keV]	$T^*_{1/2}$
^{115}In	95.7%	$5 \cdot 10^{14}$a	^{115}Sn	128	116+498	3.26 μs
^{176}Yb	12.7%	stable	^{176}Lu	301	72	50 ns
^{160}Gd	21.9%	stable	^{160}Tb	244	75+64	6+60 ns
^{82}Se	9.4%	$1 \cdot 10^{20}$a	^{82}Br	173	29	10 ns
^{100}Mo	9.6%	stable	^{100}Tc	168	3209	15.8 s
^{71}Ga	39.9%	stable	^{71}Ge	404	175	79 ns
^{137}Ba	11.2%	stable	^{137}La	611	11	89 ns
^{123}Sb	42.7%	stable	^{123}Te	541	330+159	31 ns
^{159}Tb	100%	stable	^{159}Dy	543	177+121+56	9.3 ns

E* and $T^*_{1/2}$ refer to de-excitation energy and half life of the isomeric state populate in the transition. In the ^{100}Mo – ^{100}Tc transition, the ^{100}Tc ground state is β unstable with Q_β=3209 keV.

The **MOON** collaboration [13] favors the molybdenum isotope ^{100}Mo for solar neutrino detection. The threshold for neutrino capture is 168 keV and thus sensitive to pp-neutrinos. The GT strength to the 1^+ ground state in ^{100}Tc is measured to be $(g_A/g_V)^2 B(GT) = 0.52 \pm 0.02$ by EC and charge exchange reactions. An interaction rate of 1.1 and 3.3 events per day and 10 tons of ^{100}Mo is expected for pp- and ^{7}Be neutrinos. Fig. 5 displays the nuclear levels and transition involved in the neutrino capture process. The neutrino signature consists of a prompt electron and a delayed β-decay (τ = 15.8 s) with endpoint energy of 3.4 MeV. Two different design options for a detector based on 100 tons of natural molybdenum (9.6 tons of ^{100}Mo) exist: 1) a hybrid detector consisting of plastic scintillator bars interleaved by Mo-foils (0.05 g/cm^2) and read out with PMTs via wave-length shifting fibers, or 2)

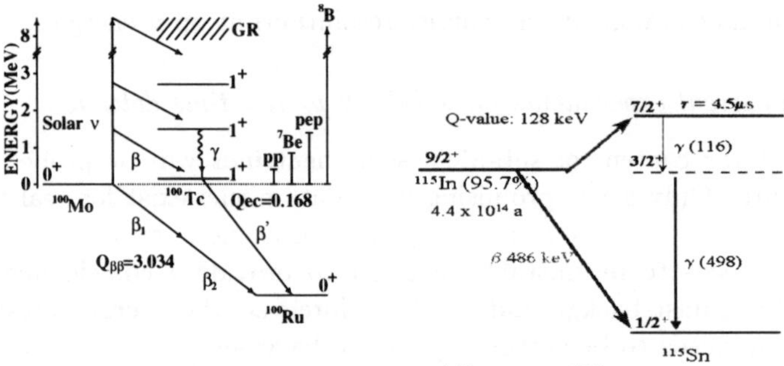

Figure 5. Level and transition scheme of ^{100}Mo and ^{115}In for solar neutrino capture.

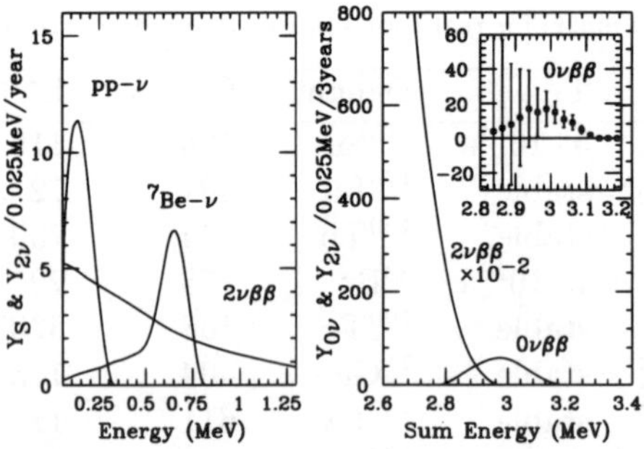

Figure 6. Energy spectrum for a possible detector with 3.3 tons of ^{100}Mo. The continuous line shows the expected background from $2 - \nu\beta\beta$ decay.

a similar detector geometry with liquid scintillators instead of plastic. Fig. 6 shows the expected spectrum together with the background of $2\nu - \beta\beta$ decay.

Various candidate nuclei listed in Tab. 3 are being investigated by the **LENS** collaboration [6]. These metals are being loaded up to 10% in weight in organic liquid scintillators. Ligands under study are carboxylic acids, phosphor organic compounds as well as beta-diketonates. Among the various nuclei, ^{115}In and ^{176}Yb are currently favored. Figs. 5 display the levels and transitions involved for neutrino capture in ^{115}In. ^{82}Se and ^{160}Gd are less promising because of challenges related to detection technique as well as to radioactive background.

The $B(GT)$ transition strengths of ^{176}Yb and ^{115}In have been measured via charge exchange reactions in order to estimate the interaction cross section.

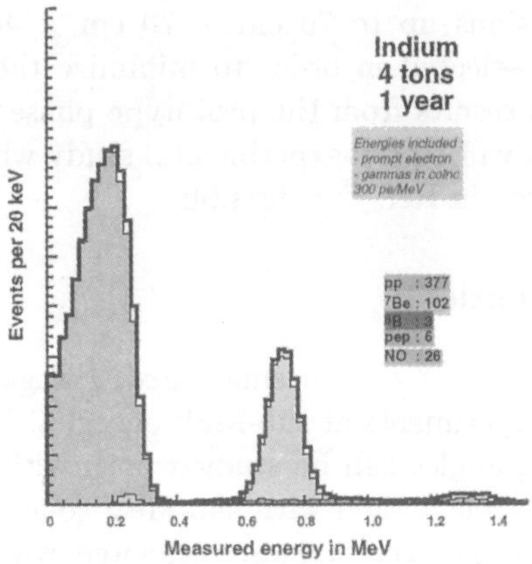

Figure 7. Estimated ν_e spectra of an In loaded liquid scintillator detector containing 4 tons of indium. The pp- and ^{7}Be rates are 377/year and 102/year.

For ^{176}Yb a $B(GT)$ value of 0.20 ± 0.04 (0.11 ± 0.02) was evaluated for the transition to the 194.5 keV (339 keV) level [23]. For ^{115}In the weak matrix element is $B(GT) = 0.17$ [24]. These values are sufficiently accurate to evaluate the target mass necessary for the LENS detector. However, to derive the neutrino fluxes, it is planned to use an artificial ^{51}Cr neutrino source of several MCi strength to calibrate the cross section at the few percent level. Fig. 7 display the expected spectra and interaction rates for 4 t of indium. The actual target mass for an indium detector will most likely be on the 10 t scale taking into account detector efficiencies and required statistical accuracy.

A high granularity is required in order to discriminate against background caused by random coincidences from ^{14}C, ^{115}In β-decay and other sources. This will be realized by a modular detector design. A basic module has a lengths of the order of the absorption length of the liquid scintillator which corresponds to a few meters. The modul area varies from 5×5 cm^2 to 20×20 cm^2 depending on metal loading, scintillator performance, choice of nucleus and neutrino source calibration optimization.

In order to study the detector performance as close as possible to the final detector geometry, the LENS Low-Background-Facility (LLBF) has been newly installed underground at the LNGS. A low-background passive shielding system with 80 tons of mass, located in a clean room, can house detector modules with dimensions up to 70 cm $\times$ 70 cm $\times$ 400 cm. All shielding materials have been selected in order to minimize the intrinsic radioactive contamination. First results from the prototype phase are expected in 2002. Of particular interest will be the experimental study whether pp-ν_e detection with ^{115}In, as proposed in Ref. [25] is feasible.

3 Summary and outlook

Solutions with $\Delta m^2 < 10^{-6}$ eV^2 *or* small mixing angles can be probed best with solar neutrino experiments at sub-MeV energies. Values of $\Delta m^2 > 10^{-6}$ eV^2 *and* large mixing angles can be studied both with long-baseline nuclear reactor neutrino experiments and with sub-MeV solar neutrinos. Large mixing, in particular at large Δm^2 values, is favored when combining all solar neutrino data. In particular, the LMA solution with $\Delta m^2 = 3.7 \times 10^{-5}$ eV^2 and $\tan^2 \Theta = 0.37$ [2] gives the best χ^2 values in global analysis. The exact value of the minimum, as well as the extension of the allowed parameter range, depend mainly on the treatment of the ^{8}B flux in the analysis. A large part of the 10^{-4} eV^2 range is allowed by the current data. Ultimately, large Δm^2 values are limited at 10^{-3} eV^2 by the CHOOZ reactor experiment [9].

Results of the upcoming experiments KamLAND and BOREXINO most likely will tell us which oscillation solution is realized in nature. The objective of the upcoming experiments will be to probe the mixing parameters with complementary techniques both with ES and CC reactions *with high precision.* For the LMA solution, reactor and solar neutrino experiments are complementary to determine mixing parameters: Δm^2 can be derived with high precision from reactor neutrino experiments while the mixing angle is measured best with solar neutrinos. A direct measurements of the primary solar neutrino fluxes will be essential to probe, with high accuracy, solar model predictions. Perhaps even new surprises await, if so, it would not be the first time in astroparticle physics – new experiments may find new phenomena.

References

1. T. Kirsten, Rev. Mod. Phys., 71(1999) 1213-1232.
2. recent golabal analysis can be found in: J.N. Bahcall, M.C. Gonzalez-Garcia, Carlos Pena-Garay, hep-ph/0111150v2; P.I. Krastev and A.Yu. Smirnov, hep-ph/0108177.
3. http://www.awa.tohoku.ac.jp/KamLAND/; Proposal for US Participation in KamLAND (1999) http://kamland.lbl.gov/ KamLAND.US.Proposal.pdf.

4. BOREXINO Collaboration, G. Alimonti et al., Astrop. Phys. 16, (2002) 205-234. Also at hep-ex/0012030.
5. S. Schönert, TAUP 1997, Nucl. Phys. B (Proc. Suppl.) 70 (1999) 195-198.
6. R.S. Raghavan, PRL Vol.78 No.19 (1997) 3618; LENS collaboration, Letter of Intent, Laboratori Nazionali del Gran Sasso (1999);
7. Y. Suzuki, LowNu2 workshop, Tokyo, 2000, http://www-sk.icrr.u-tokyo.ac.jp/neutlowe/, S.Moriyama, M.Yamashita, XENON01, 2001, http://www-sk.icrr.u-tokyo.ac.jp/xenon01/index.html
8. M. Nakahata, NOON2001, see this proceedings.
9. M. Apollonio et al., Phys. Lett. B 466 (1999) 415.
10. J.N. Bahcall, P.I. Krastev, A.Y. Smirnov, JHEP 0105 (2001) 015
11. D.N. McKinsey and J.M. Doyle, J. of Low Temp. Phys., 118, 153 (2000).
12. B. Lanou, LowNu2 workshop, Tokyo, 2000, http://www-sk.icrr.u-tokyo.ac.jp/neutlowe/, http://www.sns.ias.edu/ jnb/Meetings/ Lownu/lownuPres/pindex.html.
13. H. Ejiri et al., PRL 85 (2000) 2917-2910
14. G. Bonvicini et al., Contributed paper, Snowmass 2001, hep-ex/0109199, hep-ex/0109032.
15. Super-MUNU, http://isnwww.in2p3.fr/ munu/Supermunu/solar.html; C. Broggini et al., LowNu workshop, Sudburry (2000).
16. L. Baudis, H.V. Klapdor-Kleingrothaus, Proc. Beyond the Desert'99 (1999), astro-ph/0003435; H.V. Klapdor-Kleingrothaus, L. Baudis, G. Heusser, B. Majorovits, H. Paes, hep-ph/9910205.
17. E. Resconi, Dissertation Univ. di Genova (2000); S. Schönert and E. Resconi, to be published; R.S. Raghavan, unpublished.
18. Y. Kozlov, L. Mikaelyan and V. Sinev, NANP-2001, Dubna (2001), hep-ph/0109277
19. C.Y. Chang et al., Nucl. Phys. B (Procs. Suppl.) 66, 419 (1998), H.T-K. Wong and J. Li, NCTS workshop 2001, hep-ex/0201001.
20. S. Schönert, T. Lasserre, L. Oberauer, hep-ex/0203013.
21. G. Alimonti et al., BOREXINO collaboration, Phys. Lett. B 422 (1998) 349.
22. DAMA collaboration, TAUP01, this proceedings.
23. M. Fujiwara et al., PRL 85, 4442 (2000); M. Bhattacharya et al., PRL 85, 4446 (2000).
24. J. Rappaport et al. PRL 54, 2325 (1984).
25. R.S. Raghavan, hep-ex/0106054 (2001).

VALUE AND SENSITIVITY OF THE FUTURE LOW ENERGY SOLAR NEUTRINO EXPERIMENTS

M. NAKAHATA

Kamioka Observatory, Institute for Cosmic Ray research, University of Tokyo, Higashi-Mozumi, Kamioka-cho, Yoshiki-gun, Gifu, Japan, 506-1205
E-mail: nakahata@suketto.icrr.u-tokyo.ac.jp

The recent observations of ^{8}B solar neutrinos by Super-Kamiokande and SNO established the existence of solar neutrino oscillations. Combined with the ν_μ oscillation study by atmospheric neutrinos and accelerator neutrinos, the solar neutrino oscillations is quite valuable for constructing the neutrino mass matrix. The low energy solar neutrinos were measured only by a radiochemical method so far. Various low energy solar neutrino experiments which are able to perform real-time spectrum measurements have been proposed. Value and sensitivity of those experiments are reviewed.

1 Introduction

Solar neutrinos experiments so far have observed significantly smaller solar neutrino flux than the expectations from standard solar models (SSMs)[1]. The Super-Kamiokande(SK) has measured ^{8}B solar neutrinos by neutrino-electron(νe) scatterings and the observed flux was $(2.32\pm0.09)\times10^6/\text{cm}^2/\text{s}$.[2] The νe scattering is sensitive not only to electron neutrinos(ν_e) but also to muon neutrinos(ν_μ) and tau neutrinos(ν_τ). The observed flux of ν_e by SNO[3] using the $\nu_e + d \rightarrow e^- + p + p$ reaction was $(1.75\pm0.15)\times10^6/\text{cm}^2/\text{s}$ and it is significantly smaller than the flux measured by SK. The difference between SK and SNO results is naturally understood that it is due to the existence of ν_μ and ν_τ in solar neutrinos. Allowed region of neutrino oscillation parameters are discussed taking into account the fluxes measured by Homestake, GALLEX/GNO, SAGE, SNO, and energy spectrum and day/night variation by SK data.[4] The most favorite solution is in large mixing angle(LMA) region. But, if we relax the confidence level (C.L.) to 99.73% level (3σ), solutions in low Δm^2(LOW), small mixing angle(SMA) and just-so regions still exist. One of the major purpose of future solar neutrino experiments is to single out the solution and precisely determine the neutrino oscillation parameters. The low energy solar neutrino experiments are indispensable for this purpose as described below.

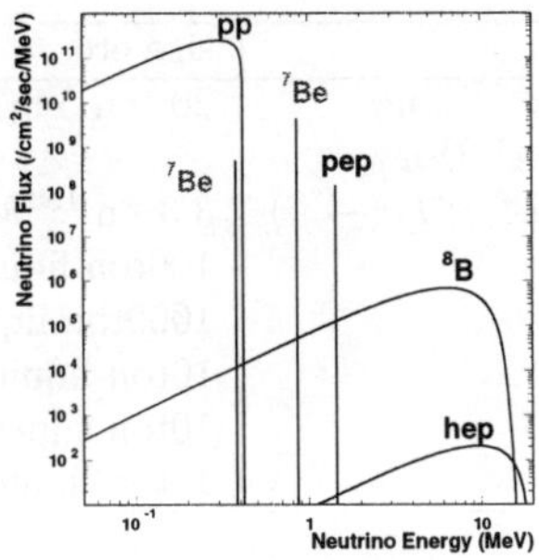

Figure 1: SSM spectrum.

2 Low energy solar neutrinos in SSM

It is well known that the energy source of the sun is a sequence of fusion reactions called pp-chain, which starts from $p+p \rightarrow d+e^{+}+\nu_{e}$ reaction. The energy spectrum of solar neutrinos calculated in SSM is shown in Fig.1. The most copious neutrinos are from $p+p \rightarrow d+e^{+}+\nu_{e}$ reaction (*pp* neutrinos). The second copious neutrinos are from $^{7}Be+e^{-} \rightarrow^{7} Li+\nu_{e}$ (^{7}Be neutrinos). The flux of those neutrinos are calculated in SSM[1] to be $5.95\times10^{10}(1.00 \pm 0.01)/\mathrm{cm}^2/\mathrm{s}$ for pp neutrinos and $4.77\times10^{9}(1.00 \pm 0.10)/\mathrm{cm}^2/\mathrm{s}$ for ^{7}Be neutrinos. The energy spectrum of *pp* neutrinos extends upto 0.42 MeV and ^{7}Be neutrinos are mono-energetic neutrinos with 0.861 MeV(90%) and 0.383MeV(10%). Those neutrinos are low energy components of the solar neutrinos. The error in pp neutrino calculation is only one percent because of the solar luminosity constraint. The quite small uncertainty in *pp* neutrino flux is a great advantage compared with other neutrinos, e.g. ~20% uncertainty in ^{8}B solar neutrinos, ~25% in atmospheric neutrinos, and ~3% in reactor neutrinos.

3 Low energy solar neutrino experiments

Table 1 shows proposed (or under construction) real time experiments aiming to detect *pp* and ^{7}Be solar neutrinos. They are classified into neutrino-electron scattering experiments (νe type) and charged current experiments (CC type). The CC type experiments can measure the energy spectrum of ν_{e} directly. The LENS project plan to use ~20 tons of Yb or In which are loaded in liquid scintillators. Delayed coincidence gamma rays are signature of the neuttrino signal. The MOON project use 3.3 tons of ^{100}Mo viewed by plastic scintillators. Fine segmentation is needed to tag neutrino signals by delayed beta decay signals.

Experiment	reaction	size etc.
LENS	$\nu_{\rm e}^{176}Yb \to e^{-176}Lu\gamma$	20 ton Yb(In)-loaded scintillator
	$\nu_{\rm e}^{115}In \to e^{-115}Sn\gamma$	
MOON	$\nu_{\rm e}^{100}Mo \to e^{-100}Tc(\to \beta)$	3.3ton ^{100}Mo foil + plastic scintillator
BOREXINO	$\nu e \to \nu e$	100ton liquid scintillator (^7Be only)
KamLAND	$\nu e \to \nu e$	1000ton liquid scintillator (^7Be only)
CLEAN	$\nu e \to \nu e$	10ton liquid Ne or He
HERON	$\nu e \to \nu e$	10ton super-fluid He
XMASS	$\nu e \to \nu e$	10ton liquid Xe

Table 1: Proposed (or under construction) low energy solar neutrino experiments.

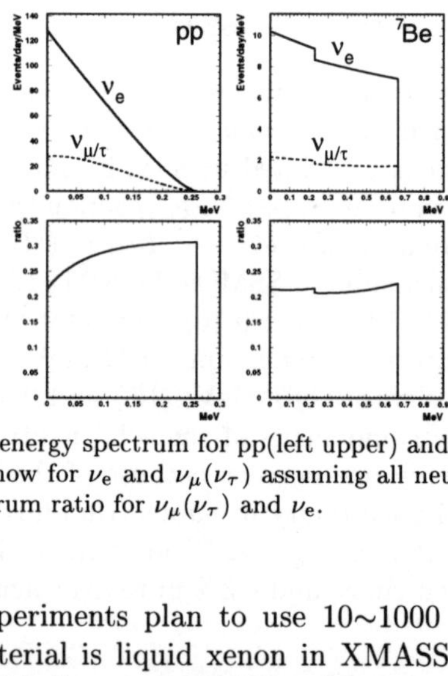

Figure 2: recoil electron energy spectrum for pp(left upper) and ^{7}Be(right upper) neutrinos. Solid and dotted lines show for $\nu_{\rm e}$ and $\nu_\mu(\nu_\tau)$ assuming all neutrinos are given types. The lower figures show spectrum ratio for $\nu_\mu(\nu_\tau)$ and $\nu_{\rm e}$.

The νe type experiments plan to use 10~1000 tons of detector material. The target material is liquid xenon in XMASS, liquid neon or helium in CLEAN, superfluid helium in HERON and liquid scintillator in KamLAND and BOREXINO. The recoil electron energy spectrum for pp and ^{7}Be neutrinos are shown in Fig.2. Although the spectrum is not the direct measurement of the neutrino energy spectrum, the event rate is quite high in the νe type experiments. The expected event rate is 10 events/day for pp neutrinos and 5 events/day for ^{7}Be neutrinos assuming SSM prediction. The cross section of $\nu_\mu(\nu_\tau) + e$ scattering is about 25 - 30% of $\nu_{\rm e} + e$ scattering for pp neutrinos and it is about 20% for ^{7}Be neutrinos. Therefore, combining the CC type experiments and νe type experiments, we can obtain the fluxes of $\nu_{\rm e}$ and $\nu_\mu + \nu_\tau$.

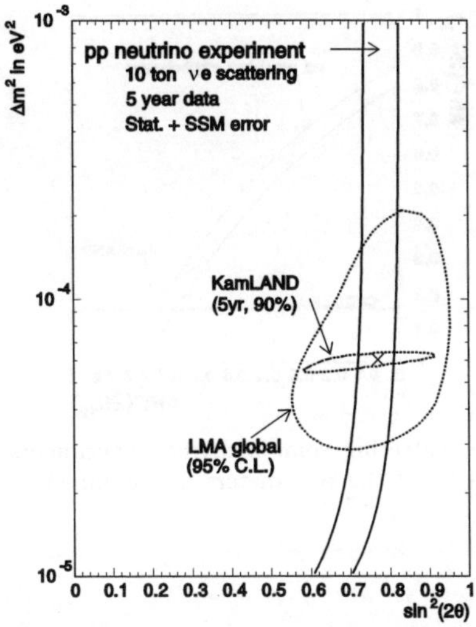

Figure 3: Sensitivity of low enery solar neutrino experiments and KamLAND reactor experiment for LMA solution. The dotted line shows 95% C.L. allowed region obtained by SK, SNO, Ga and Cl experiments. The solid band and dashed contour shows sensitivity of 10 ton νe type experiment and KamLAND reactor neutrinos for 5 year data taking.

4 Oscillation parameter determination

Because of the very precisely known flux of pp neutrinos, we can determine the neutrino oscillation parameters quite accurately. As described above, the solar neutrino experiments so far favors LMA solution. If it is the right solution, KamLAND is also able to observe neutrino oscillations by reactor $\bar{\nu}_e$ neutrinos. Figure 3 shows the sensitivity of solar neutrino experiments and KamLAND reactor neutrinos for 5 years' live time assuming the neutrino oscillation parameters are $\Delta m^2 = 6 \times 10^{-5}\mathrm{eV}^2$ and $\sin^2 2\theta = 0.77$. In this figure, 10 ton νe type solar neutrino experiment with 50keV threshold is assumed. Because of the large statistics of the νe scattering event rate even after neutrino oscillations(~12,000 events/5 years) and quite small error in predicted pp neutrino flux in SSM, precise determination of mixing angle is expected in the low energy solar neutrino experiments. For example, error of $\sin^2 2\theta$ is only 0.03 for $\sin^2 2\theta = 0.77$. Unfortunately, the low energy solar neutrino experiments are not sensitive to Δm^2 if the solution is in LMA region, because the neutrino energy is too low to satisfy matter resonance condition in the sun. On the other hand, KamLAND is quite sensitive to Δm^2, because the

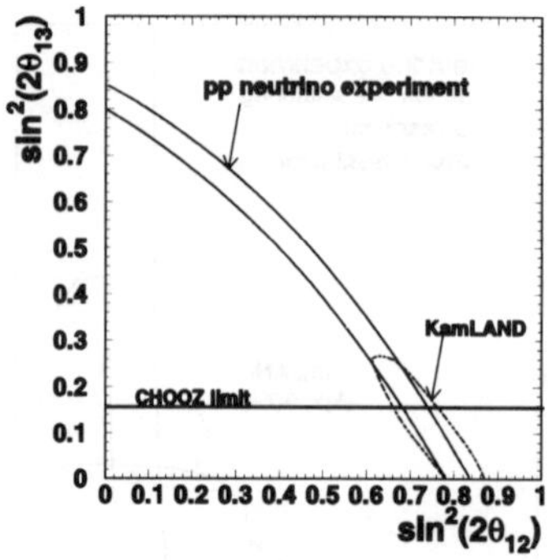

Figure 4: Sensitivity of νe scattering solar neutrino experiments and KamLAND reactor experiment. The real solution of the parameters are assumed to be $\sin^2 2\theta_{12} = 0.77$ and $\sin^2 2\theta_{13} = 0.078$ for $\Delta m^2 = 6 \times 10^{-5}\text{eV}^2$.

distance from the reactor to the detector is similar to the oscillation length in case of LMA solution. Thus, the combination of low energy solar neutrino experiments and KamLAND should be able to give a precise determination of oscillation parameters.

In reality, we need to consider three flavor neutrino oscillations. The following formula given by C.S.Lim et al.[5] is useful for discussing three flavor neutrinos in solar neutrinos and reactor neutrinos:

$$P^{(3)}(\nu_e \to \nu_e; A(x)) = (1 - |U_{e3}|^2)^2 P^{(2)}(\nu_e \to \nu_e; (1 - |U_{e3}|^2)A(x)) + |U_{e3}|^4,$$

where $P^{(2)}$ and $P^{(3)}$ are ν_e survival probabilities in case of two flavor and three flavor oscillations, respectively. U_{e3} is a component of the neutrino mass matrix and it is $\sin\theta_{13}e^{-i\delta}$ in standard parametrization. $A(x) \equiv \sqrt{2}G_F N_e(x)$ is the matter effect. Sensitivity of νe scattering solar neutrino experiments and KamLAND reactor experiment on $\sin^2 2\theta_{12}$ and $\sin^2 2\theta_{13}$ plane are shown in Fig.4 for an example of $\sin^2 2\theta_{12} = 0.77$, $\sin^2 2\theta_{13} = 0.078$ and $\Delta m^2 = 6 \times 10^{-5}\text{eV}^2$.

If the real solution of the oscillation parameters are in LOW region, low energy solar neutrinos will observe strong day/night asymmetry as shown in Fig.5. Hence, if the oscillation parameters are in LOW region, low energy solar neutrino experiments are able to determine the oscillation parameters precisely.

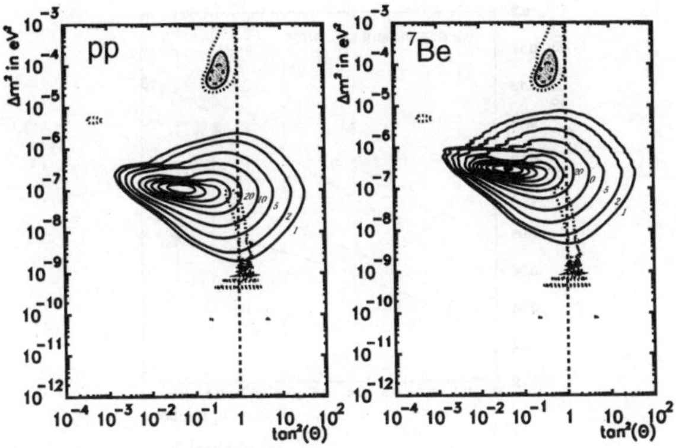

Figure 5: Iso-contour of day/night asymmetry for pp(left) and ^{7}Be(right) neutrinos, if the solution is in LOW region. The contours are for the asymmetry value, 2(night-day)/(night+day), of 1, 2, 5, 10, 20, 30, 40... from outside to inwards.

5 Unitarity test

The νe scattering type experiments observe not only ν_e but also ν_μ and ν_τ. On the other hand, CC type experiments detect only ν_e. Hence, combining those experimental data, we can get the total neutrino flux , $\phi(\nu_e+\nu_\mu+\nu_\tau)$. Because the total neutrino flux of pp neutrinos is precisely know in SSM to an accuracy of 1%, we can in principle discuss the consistency of the predicted neutrino flux and the experimentally obtained total flux down to this accuracy. This comparison gives a check of the unitarity of the neutrino mass matrix. If the predicted and observed total neutrino flux is different, it indicates the existence of new neutrinos which do not interact with matter (sterile neutrinos). Figure 6 shows the accuracy of total neutrino flux measurement as a function of fraction of ν_e arriving to the earth, assuming 10 ton νe scattering detector and 20 ton Yb CC type detector (173 pp events/year in SSM) both with 5 years' data. ν_e fraction is about 0.61 at the current best fit parameter ($\sin^2 2\theta = 0.77$). So, the accuracy of the total neutrino flux is about 7-10 % in future low energy experiments. If the CC type detector is 4 times larger (equivalent to 80 ton Yb detector), the accuracy will be 4-7 %.

6 Neutrino magnetic moment

If neutrinos have magnetic moment, the mechanism of the resonance spin flavor precession(RSFP) is able to explain the observed neutrino event rate and spectrum of SK, SNO, Ga and Cl experiments.[6,7] It is also suggested that the

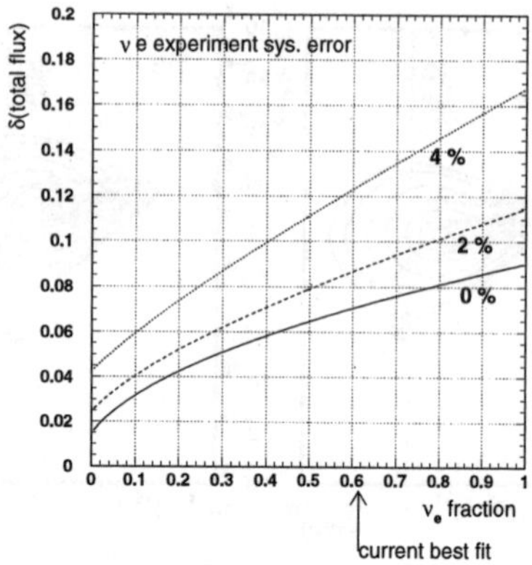

Figure 6: Accuracy of total flux measurement by 10 ton νe scattering and 20 ton Yb CC type experiment. The different lines show experimental errors in νe scattering experiment.

RSFP hypothesis give better fitting to the data compared with neutrino oscillation hypothesis.[7] The energy dependence of the neutrino survival probability in case of RSFP is quite different from that of neutrino oscillations. The survival probability is nearly 100% for pp neutrino energy region and ~20% for ^{7}Be neutrinos. The spectrum measurement of low energy solar neutrinos definitely answers to the RSFP hypothesis. The low energy νe scattering experiment is able to discuss neutrino magnetic moment using the shape of the energy spectrum. The existence of the electro-magnetic interaction term distorts the energy spectrum at low energy region, which enable us to discuss the magnetic moment down to $1 \times 10^{-11}\mu_B$.

7 Conclusion

The future low energy solar neutrino experiments are quite important in order to determine the neutrino oscillation parameters. If the real solution is in LMA region, the combination of low energy solar neutrino experiments and KamLAND give precise measurements of the oscillation parameters. If LOW is the solution, strong day/night asymmetry in pp and ^{7}Be neutrinos give precise determination of the oscillation parameters. The low energy solar neutrino measurements by νe scattering method and charged current method give precise measurement of the total neutrino flux. The accurately known flux of pp neutrinos enable us to discuss the unitarity of the neutrino mass matrix, which might be broken by an existence of sterile neutrinos. The low energy solar neutrinos are quite useful for the investigation of the neutrino magnetic moment.

8 Acknowledgement

The author would like to thank O.Yasuda for teaching me the formula of 3 and more flavour oscillations in solar neutrinos. The author also thank K.Inoue for the calculations of KamLAND sensitivities. The author also would like to thank J.Pulido and C.S.Lim for the discussion about RSFP. The calculation of neutrino oscillations and the estimate of the sensitivities were done by Y.Koshio.

References

1. J. N. Bahcall, M. H. Pinsonneault, S.Basu, *Astrophys. J.* **555**, 990(2001).
2. S. Fukuda et al., *Phys. Rev. Lett.***86**, 5651 (2001).
3. Q. R. Ahmad et al., *Phys. Rev. Lett.***87**, 71301 (2001).
4. Talk by M. Smy in this workshop.
5. C.-S. Lim, Proc. of the BNL Neutrino Workshop on Opportunities for Neutrino Physics at BNL, Upton, N.Y., February 5-7, 1987, ed. by M. J. Murtagh, p111; A. Yu. Smirnov, Proc. of the Int Symposium on Neutrino Astrophysics, Takayama/Kamioka 19 - 22 October 1992, ed. by Y. Suzuki and K. Nakamura, p.105.
6. C.-S. Lim, Proc. of the Int Symposium on Neutrino Astrophysics, Takayama/Kamioka 19 - 22 October 1992, ed. by Y. Suzuki and K. Nakamura, p.117.
7. J. Pulido and E. Kh. Akhmedov, *Astropart. Phys.* **13**, 227 (2000).

STATUS OF THE ATMOSPHERIC NEUTRINO STUDIES

M. D. MESSIER
High Energy Physics Laboratory
Harvard University
42 Oxford Street
Cambridge, MA 02138, USA
E-mail: messier@huhepl.harvard.edu

Studies of atmospheric neutrinos at a variety of detectors yield a consistent picture of neutrino oscillations, ie. that muon neutrinos oscillate to tau neutrinos with nearly maximal mixing and a mass splitting in the range of a few meV^2. In this article, emphasis will be given to the high-statistics Super–Kamiokande results, although results from the MACRO, and Soudan-II detectors will also be discussed.

1. Atmospheric Neutrinos

Atmospheric neutrinos are produced as the decay products of interactions of cosmic-ray primaries (p, He, etc.) with nuclei in the upper atmosphere. The primary decay chain is $\pi^+ \to \mu^+ + \nu_\mu$, $\mu^+ \to e^+\nu_e\bar{\nu}_\mu$. This simple decay chain yields a robust relationship between the number of muon neutrinos and electron neutrinos ($\nu_\mu/\nu_e \simeq 2$ at low energies) which is predicted with 5% uncertainty. In addition to the prediction of the ν_μ/ν_e ratio, a second robust prediction is that at high energies ($E_\nu > 1$ GeV) the neutrino fluxes are up-down symmetric based on simple geometric arguments. The observations of atmospheric neutrinos by several detectors have shown that neither of these simple, robust, expectations are true, providing conclusive evidence for neutrino oscillations. In particular, that muon neutrinos convert to tau neutrinos with a probability given by

$$P(\nu_\mu \to \nu_\tau) = 1 - \sin^2 2\theta \sin^2(1.27\Delta m^2 L/E), \tag{1}$$

where L is the distance from the point of neutrino production to the point of detection in km, E is the neutrino energy in GeV, θ is the angle between the neutrino mass and electroweak eigenstates, and Δm^2 is the difference in the squares of two mass eigenvalues measured in eV^2. The current experimental evidence reviewed in this article points to this mechanism of neutrino disappearance exclusive of all other mechanisms.

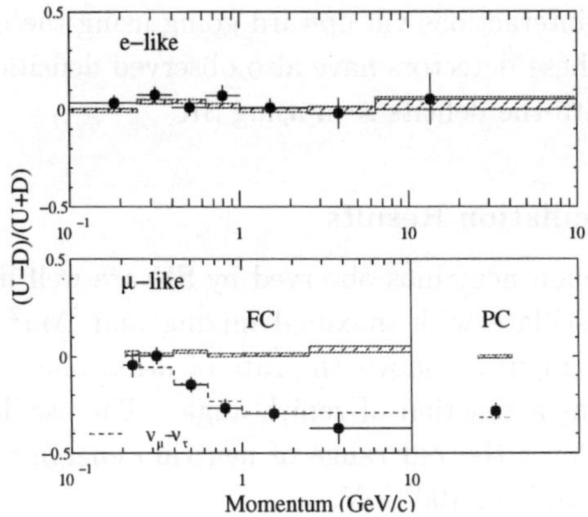

Figure 1. The zenith angle asymmetry of the atmospheric neutrino flux as measured by the Super–Kamiokande detector.

2. Experimental Detection of Atmospheric Neutrinos

Atmospheric neutrinos are detected by a variety of means. The largest sample has been accumulated using the Super–Kamiokande (SK) water Cherenkov detector[1,2,3]. This detector observes atmospheric neutrinos via charged-current (CC) and neutral-current (NC) interactions on its large (22.5 kiloton) water fiducial mass. These events are classified as fully-contained (FC) if all the event energy is deposited in the detector volume and partially-contained (PC) if particles deposit visible energy in the veto region which surrounds the detector inner volume. FC events are classified according to their multiplicity (single-ring and multi-ring), and neutrino flavor (e-like and μ-like) based on likelihood fits to the leading ring topology. SK has also isolated a sample of multi-ring events with enriched ($\simeq$35%) NC content. Partially-contained events are assumed to be 100% ν_μ CC events as the exiting particle is typically a muon. While typical neutrino energies of FC events is $\simeq$ 1 GeV, typical energies of PC events is higher, roughly 10-20 GeV.

SK also observes neutrino-induced upward-going muons which are produced via CC neutrino interactions on the rock surrounding the detector[4,5]. These muons either stop in the detector (upward-stopping) or enter and exit the detector (upward through-going). The difference in neutrino energies for the two samples is roughly 20 GeV vs 100 GeV.

Contained atmospheric neutrino interactions have also been observed by the Soudan-II iron-calorimeter[7,8,9]. The MACRO experiment, which was originally designed for magnetic monopole detection, also observed atmospheric neutrino interactions via upward-going using the detector tracking system[11,12,13]. These detectors have also observed deficits of muon neutrinos consistent with the deficits seen using SK.

3. $\nu_\mu \leftrightarrow \nu_\tau$ Oscillation Results

The deficits of muon neutrinos observed by SK are well fit if one assumes that neutrinos oscillate with maximal mixing and Δm^2 in the range of a few$\times 10^{-3}$eV^2. Figure 2 shows the rate of atmospheric neutrino events observed by SK as a function of zenith angle. The oscillation hypothesis fits the SK data over the full range of neutrino energies sampled by SK, from roughly 0.1 GeV to 100 GeV.

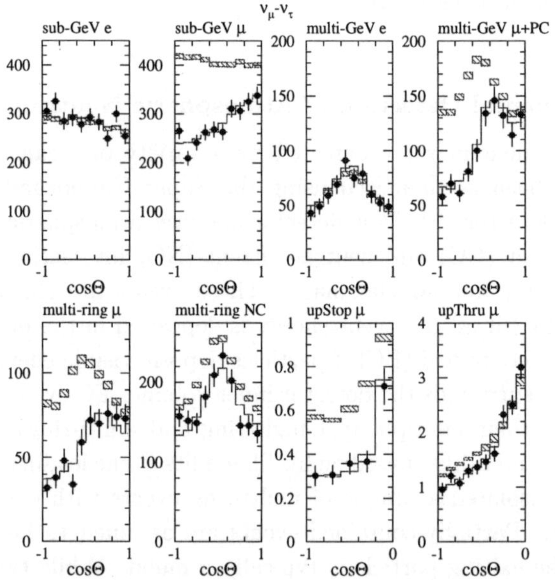

Figure 2. The zenith angle rates of the different atmospheric neutrino event samples measured by the Super–Kamiokande detector. The hatched boxes shows the expected rates in the absence of oscillations while the solid curve shows the expected results for best-fit neutrino oscillations of $\nu_\mu \leftrightarrow \nu_\tau$.

The MACRO upward-going muon data is shown in Figure 3. Although MACRO cannot measure the muon momentum directly, the muon momentum can be inferred by measuring the amount of multiple scattering using the relation $\theta_0 \simeq (13.6 \text{ MeV}/(\beta cp))z\sqrt{x/X_0}$[14]. Using this estimate, which is correlated with the neutrino energy, MACRO can divide the up-

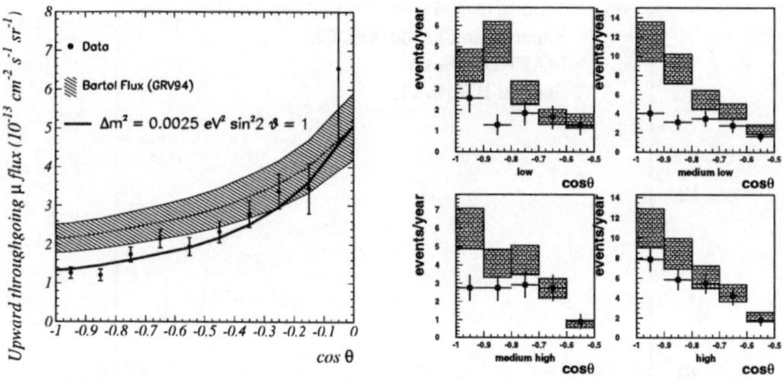

Figure 3. The rates of upward-going muons measured by the MACRO experiment

ward muon sample into bins of low, medium, and high average neutrino energy. The zenith-dependent and energy-dependent rates are consistent with neutrino oscillations with the parameters measured by SK.

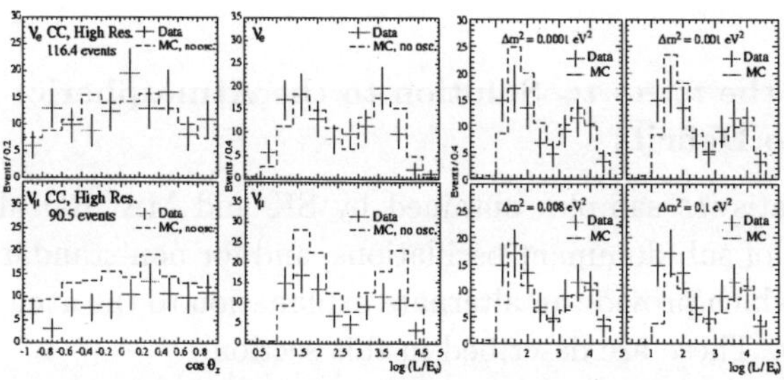

Figure 4. Soundan-II L/E fits to oscillations.

The rates of atmospheric neutrino observed using the Soudan-II are shown in Figure 4[10]. The figure shows the data plotted as a function of reconstructed using the "high-resolution" data sample – a sample of Q.E. events where the lepton and recoil proton are both measured giving roughly 20% resolution in L/E.

The oscillation parameters implied by the three experiments, SK, MACRO, and Soudan-II, are shown in Figure 5. All are consistent with the best limit which is obtained by SK, $\sin^2 2\theta > 0.90$ and $0.0015 < \Delta m^2 < 0.0038$ eV^2 at 90% confidence level.

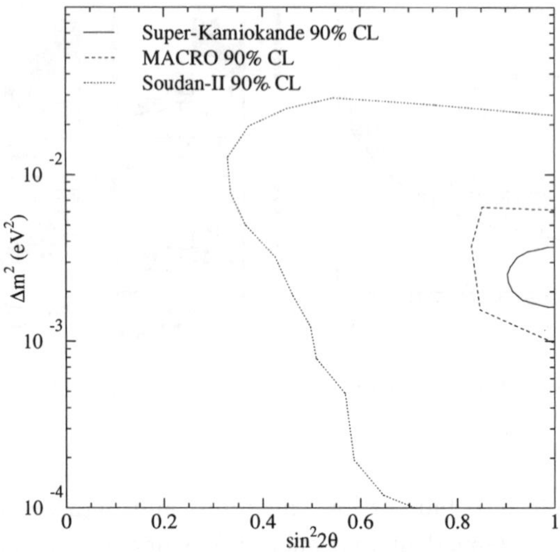

Figure 5. The allowed contours obtained from fits to the atmospheric neutrino deficits for the Super–Kamiokande, MACRO, and Soudan-II experiments.

4. Beyond the $\nu_\mu \leftrightarrow \nu_\tau$ Solution to the Atmospheric Neutrino Deficits

The high statistics samples obtained by SK and MACRO allow tests of the presence of sub-dominant oscillations, and for non-standard oscillation hypotheses which provide an alternate explanation to the atmospheric neutrino deficits. These are described in this section.

4.1. *Electron Neutrino Appearance*

If there is a significant coupling of muon neutrinos to electron neutrinos then one might expect to find resonant conversion of muon neutrinos to electron neutrinos as the neutrinos propagate through the Earth. For typical Earth densities, this would happen for multi-GeV neutrinos. SK has searched for evidence of electron neutrino appearance (measured using the energy and asymmetry distributions shown in Figure 6). No evidence is found limiting the parameter U_{e3}^2 to the values shown in Figure 7. The results are consistent with the more restrictive limits set by the CHOOZ experiment[15].

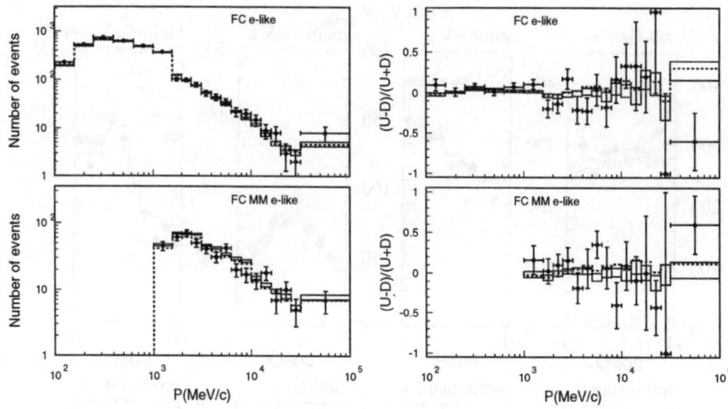

Figure 6. Left: The momentum distribution of electron neutrino events in Super–Kamiokande including multi-ring events (MM=multi-ring, multi-GeV). Right: The multi-GeV electron neutrino zenith angle asymmetry as measured by the Super–Kamiokande experiment. No evidence for a resonant conversion to electron neutrinos is found.

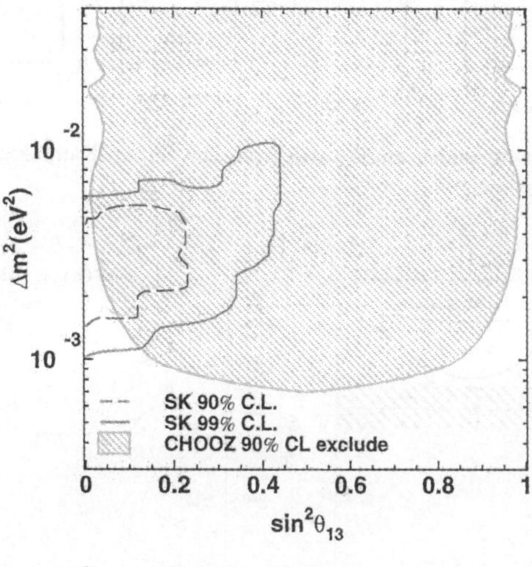

Figure 7. The limit on U_{e3}^2 obtained by the Super–Kamiokande experiment compared to the current limit from the CHOOZ experiment.

4.2. *Sterile Neutrinos*

4.2.1. *Limits on Sterile Neutrino Content*

Recently, there has been significant interest in the possibility that the atmospheric neutrino deficits could arise from oscillations into sterile neutrinos rather than tau neutrinos. There are three primary differences between the $\nu_\mu \leftrightarrow \nu_\tau$ oscillation hypothesis and the $\nu_\mu \leftrightarrow \nu_{sterile}$ hypothesis. Os-

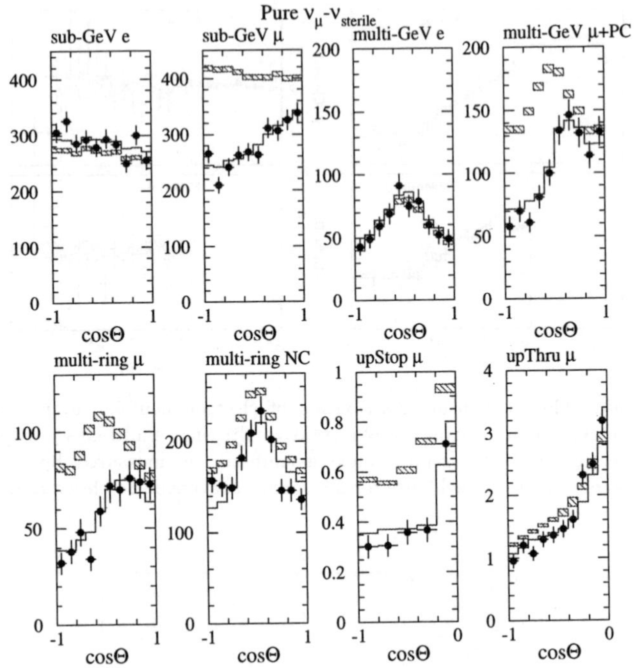

Figure 8. The best-fit zenith angle rates from the Super–Kamiokande experiment for $\nu_\mu \leftrightarrow \nu_{sterile}$ oscillations.

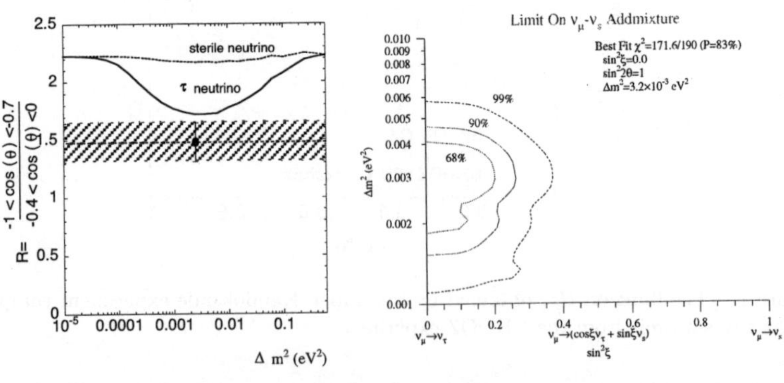

Figure 9. Left: The ratio of vertical to horizontal-going muons observed with the MACRO detector. The result is compared to expectations assuming $\nu_\mu \leftrightarrow \nu_\tau$ oscillations and $\nu_\mu \leftrightarrow \nu_{sterile}$ oscillations. Right: The limit on the sterile content of atmospheric neutrino oscillations using single and multi-ring FC events, PC events, and upward-going muons observed by the Super–Kamiokande detector.

cillations to sterile neutrinos should exhibit 1) a suppression of the NC interaction rate 2) a suppression of oscillations at high energies due to matter-effects, and 3) a suppression of the ν_τ interaction rates. The first of these two are discussed in this section, while the next section treats searches for ν_τ events.

The best fit assuming $\nu_\mu \leftrightarrow \nu_{sterile}$ oscillations have been compared to the SK atmospheric neutrino data [6]. As can be seen in Figure 8, the asymmetry introduced in the NC-enriched sample fits the data poorly. Also, the highest energy data (multi-GeV, and upward-going muons) is fit poorly as the matter suppression expected for $\nu_\mu \leftrightarrow \nu_{sterile}$ oscillations does not match the apparently fully-mixed rates implied by the data. The MACRO upward-going muon data (also shown in Figure 8) also disfavors the presence of matter-suppressed oscillations, and hence favors the $\nu_\mu \leftrightarrow \nu_\tau$ explanation over $\nu_\mu \leftrightarrow \nu_{sterile}$ [13].

Based on the poor quality of the fit to $\nu_\mu \leftrightarrow \nu_{sterile}$ SK can limit the amount of sterile content allowed in atmospheric neutrino oscillations, assuming that ν_μ mixes withe a state which is a superposition of ν_τ and $\nu_{sterile}$, given by $\cos\xi\nu_\tau + \sin\xi\nu_{sterile}$ [16]. SK obtains limit of 26% sterile content as shown in Figure 9.

4.2.2. *Search for ν_τ Appearance*

Assuming $\nu_\mu \leftrightarrow \nu_\tau$ oscillations, SK expects about $20\nu_\tau$ interactions to occur in its fiducial volume. These events are difficult to tag on an event-by-event analysis due large background of topologically similar deep-inelastic (DIS) neutrino interactions. Three studies have searched the SK data for the presence of ν_τ interactions. These searches combine several event variables using likelihood and neutral net techniques to identify events which are mostly likely to be ν_τ events. These analyses attempt to use that fact that ν_τ interactions are on average of higher multiplicity than DIS interactions and are expected to occur in a relatively narrow energy window. The analysis shown in the figure accepts ν_τ events with 51% efficiency. Figure 10 shows the comparison of the τ-like parameter for one analysis for upward and downward going data. The downward going data shows relatively good agreement, while the upward going data (where ν_τ events are expected) shows a slight deficit of events which are most likely to be ν_τ induced. Also shown in the figure is the zenith angle rates of the τ-like events. Based on a fit to the zenith angle rate of the τ-like events SK measures $92 \pm 35(stat.) \pm 18(\Delta m^2)^{+0}_{-15}(3-flavor)$ ν_τ interactions which is perhaps and upward fluctuation over the 74 events expected, and hints toward a non-zero

ν_τ interaction rate.

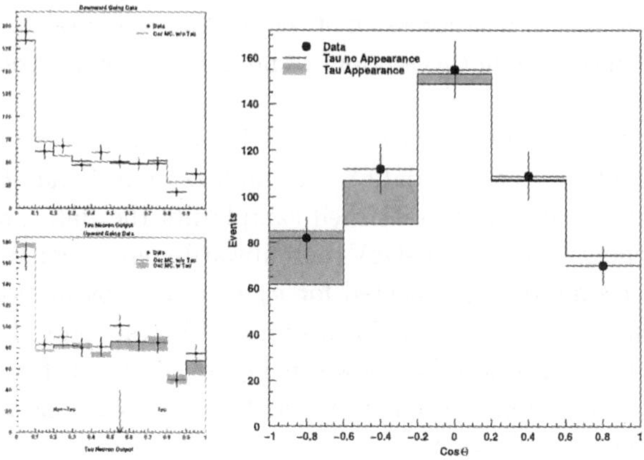

Figure 10. Left: Distribution of the τ-like parameter for downward-going (top) and upward-going (bottom) events. The upward-going data is shown without (lower curve) ν_τ interactions and with ν_τ interactions expected for the best-fit oscillation parameters. Right: The zenith angle rate for events with τ-like parameter larger than 0.55. Expected rates with and without ν_τ interactions are shown.

4.2.3. *Exotic Solutions*

Several "exotic" solutions to the atmospheric neutrino deficits observed by SK have been proposed in the literature, such neutrino decay [17], and violation of special relativity [18,19]. The high statistics obtained using SK as well as the large range of L/E probed by the detector has allowed SK to provide evidence for $\nu_\mu \leftrightarrow \nu_\tau$ oscillations over several other explanations by roughly 4 σ. Results are summarized in Table 1.

5. Conclusion

Observations of atmospheric neutrinos by the SK, MACRO, and Soudan-II detectors have provided a consistent picture that atmospheric muon neutrinos oscillation to tau neutrinos with oscillation parameters in the range of $\sin^2 2\theta > 0.9$, $0.0015 < \Delta m^2 < 0.0038$ eV^2 (SK, 90% C.L.). While the MACRO and Soundan-II experiments have completed their runs, the SK experiment expects to resume observations latter in 2002 following repairs.

References

1. Y. Fukuda *et al.*, *Phys. Lett. B* **433**, 9 (1998)

Table 1. Summary of tests using the Super–Kamiokande data of proposed solutions to the atmospheric neutrino muon neutrino deficits.

Mode	Best fit	χ^2	$P(\chi^2)$	$\Delta\chi^2$	σ
$\nu_\mu \leftrightarrow \nu_\tau$ $\sin^2 2\theta \sin^2(1.27\Delta m^2 L/E)$	$\sin^2 2\theta = 1.0$ $\Delta m^2 = 2.5 \times 10^{-3}$ eV^2	175.0	76%	0.0	0.0
$\nu_\mu \leftrightarrow \nu_e$ $\simeq \sin^2 2\theta \sin^2(1.27\Delta m^2 L/E)$	$\sin^2 2\theta = 0.97$ $\Delta m^2 = 5.1 \times 10^{-3}$ eV^2	261.7	0.04%	86.7	9.3
$\nu_\mu \leftrightarrow \nu_{sterile}$ $\simeq \sin^2 2\theta \sin^2(1.27\Delta m^2 L/E)$	$\sin^2 2\theta = 0.96$ $\Delta m^2 = 2.8 \times 10^{-3}$ eV^2	204.9	21.8%	29.9	5.5
$L \times E$ $\simeq \sin^2 2\theta \sin^2(\alpha L \times E)$	$\sin^2 2\theta = 0.90$ $\alpha = 5.4 \times 10^{-4}$ /GeV/km	261.2	0.05%	86.2	9.3
ν_μ Decay $\sin^4\theta + \cos^4\theta(1 - e^{-\alpha L/E})$	$\cos^2\theta = 0.48$ $\alpha = 3.9 \times 10^{-3}$ /GeV/km	269.5	0.01%	94.5	9.7
ν_μ Decay $(\sin^2\theta + \cos^2\theta e^{-\alpha L/E})^2$	$\cos^2\theta = 0.33$ $\alpha = 5.4 \times 10^{-4}$ /GeV/km	195.7	37%	20.7	4.5
No Oscillations	–	427.4	0%	252.4	15.9

2. Y. Fukuda *et al.*, *Phys. Lett. B* **436**, 33 (1998)
3. Y. Fukuda *et al.*, *Phys. Rev. Lett.* **81**, 1562 (1998)
4. Y. Fukuda *et al.*, *Phys. Rev. Lett.* **82**, 2644 (1999)
5. Y. Fukuda *et al.*, *Phys. Lett. B* **467**, 185 (1999)
6. S. Fukuda *et al.*, *Phys. Rev. Lett.* **85**, 3999 (2000)
7. W.W.M. Allison *et al.*, *Nucl. Instr. Meth.* **A 376**, 36 (1996); **A 381**, 385 (1996); W.P. Oliver *et al.*, *Nucl. Instr. Meth.* **A 275**, 371 (1989).
8. W.W.M. Allison *et al.*, *Phys. Lett.* **B 391**, 491 (1997); *Phys. Lett.* **B 449**, 137 (1999).
9. M. Sanchez, *Int. J. Mod. Phys. A* **16S1B**, 727 (2001).
10. W. A. Mann, in *Proc. of the 19th Intl. Symp. on Photon and Lepton Interactions at High Energy LP99* ed. J.A. Jaros and M.E. Peskin, Int. J. Mod. Phys. A **15S1**, 229 (2000)
11. M. Ambrosio *et al.*, Phys. Lett. B **478**, 5 (2000).
12. M. Spurio [MACRO Collaboration], Nucl. Phys. A **663**, 779 (2000).
13. M. Ambrosio *et al.*, *Phys. Lett. B* **517**, 59 (2001)
14. M. Sioli, arXiv:hep-ex/0201016.
15. M. Apollonio *et al.*, *Phys. Lett. B* **466**, 415 (1999)
16. G. L. Fogli, E. Lisi and A. Marrone, *Phys. Rev. D* **63**, 053008 (2001)
17. V. D. Barger, J. G. Learned, P. Lipari, M. Lusignoli, S. Pakvasa and T. J. Weiler, *Phys. Lett. B* **462**, 109 (1999)
18. S. L. Glashow, A. Halprin, P. I. Krastev, C. N. Leung and J. Pantaleone, *Phys. Rev. D* **56**, 2433 (1997)
19. S. R. Coleman and S. L. Glashow, *Phys. Lett. B* **405**, 249 (1997)

COSMIC RAY MEASUREMENTS FOR ATMOSPHERIC NEUTRINO WITH BESS-TEV

K. ABE

*for the BESS Collaboration**

Department of Physics, Graduate school of Science, The University of Tokyo, 7-3-1 Hongo, Bunkyo-ku, Tokyo 113-0033, Japan

To calculate the accurate atmospheric neutrinos spectra in a study of neutrino oscillation, it is essential to know the accurate primary cosmic ray fluxes up to about 1 TeV, as well as atmospheric muon, and secondary particle spectra at various altitudes which correlate directly to the hadronic interactions. To observe these fluxes, we upgraded the BESS detector to improve the rigidity resolution and muon/electron separation. Balloon experiment was performed to observe cosmic ray fluxes at Fort Sumner, New Mexico (cutoff rigidity is 4.2 GV), in September 2001. Because of gradually descendant of the balloon (pressure changed from 4.5 g/cm^2 to 33 g/cm^2 in 13.5 hours), it is difficult to calculate primary fluxes up to 1 TeV with only this flight data, but the date measured at various altitudes are very interesting for the hadronic interactions.

1 Introduction

For a more detailed study of neutrino oscillation phenomena observed in atmospheric neutrinos, it is essential to minimize the systematic errors in the predicted energy spectra of neutrinos. In order to improve the accuracy of the predictions, one has to know (i) the accurate primary cosmic ray fluxes, (ii) hadronic interactions, (iii) geomagnetic effect, and (iv) a structure of atmosphere. The precise measurements of the primary proton and helium energy spectra up to about 1 TeV are important to estimate atmospheric neutrino fluxes in a wide energy range between 0.1 and 100 GeV, which covers all the atmospheric neutrino events observed in Super-Kamiokande, i.e., "Sub-GeV events", "Multi-GeV events", "Stopping muons", and "Through-going muons". Because production and decay process of muons are accompanied by neutrino productions, the atmospheric muon flux is also crucial. The energy spectra of muons correlate directly to the hadronic interactions.

Although the energy spectrum up to around 1 TeV region has been measured by using calorimeters [1,2]. It is not trivial for calorimeters to determine the absolute energy of the incident particles, nor to distinguish going through

*BESS Collaboration formed with The University of Tokyo, High Energy Accelerator Research Laboratory (KEK), Kobe University, NASA Goddard Space Flight Center, University of Maryland, and The Institute of Space and Astronautical Science.

particles nor accidental tracks. On the other hand, magnetic spectrometer can determine the absolute magnetic-rigidity of incident particle only from its trajectory.

The BESS (Balloon-borne Experiment with a Superconducting Spectrometer) [3] has provided to observe proton and helium spectra up to 100 GeV and 50 GeV/nucleon with the overall uncertainties below ±5% and ±10%, respectively, from the flight data in 1998 [4]. Proton flux obtained from the BESS 98 and other experiments [5,6,7,1,2,8,9,10,11,12,13,14,15]. are shown in figure 1, BESS98 date is shown by closed circles.

The AMS has also provided proton spectra up to 200 GeV using spectrometer [15] which is shown in figure 1 by closed triangles . And agreement these two observations, BESS and AMS, are very well.

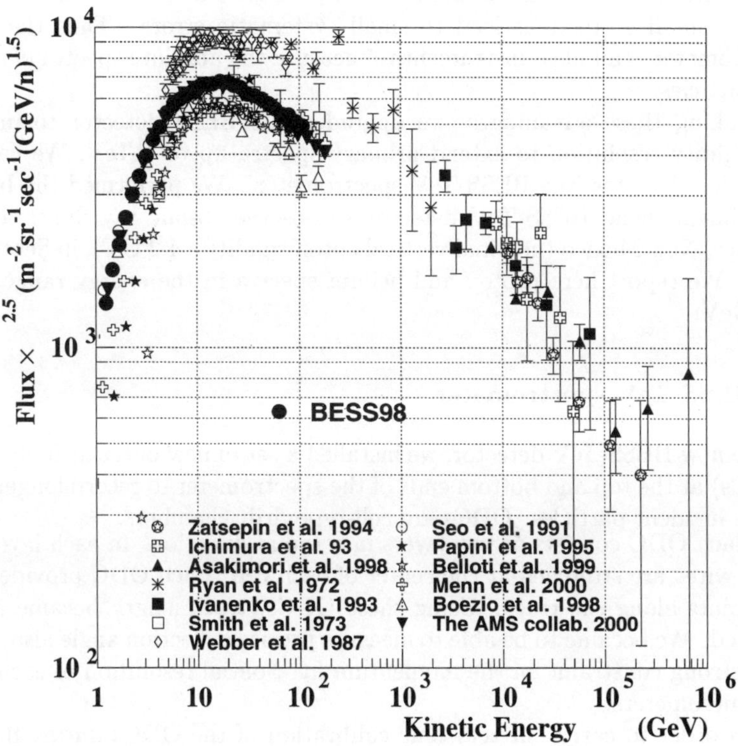

Figure 1. Proton fluxes measured by BESS in 1998, AMS in 1998 and other previous experiments

The BESS detector (the BESS 2000 detector is left one in figure 2.) is a high-resolution spectrometer with a large acceptance.

In the central region, a solenoidal superconducting coil provides a uniform magnetic field of 1 Tesla. The rigidity is determined using up to 28 hit points inside a JET-type drift chamber (JET) and two inner-drift-chambers (IDCs). A maximum detectable rigidity (MDR) of 200 GV was achieved. The outermost detector is a set of time-of-flight (TOF) hodoscopes. It provides the velocity ($\beta \equiv v/c$) and energy loss (dE/dx) measurements. The instrument also incorporates a threshold-type aerogel Cherenkov counter and lead (covered 1/5 of the whole acceptance) of 2 radiation length and acrylic plate.

Its simple cylindrical geometry makes an estimation of the geometrical acceptance reliable. Furthermore, the live time was measured exactly. An absolute flux is therefore calculated irrespective of normalization. The overall efficiencies in the off-line analysis was kept high and corrections were kept small. Small corrections lead to small systematic errors. Thus the BESS spectrometer is an ideal instrument to measure the absolute spectra of cosmic ray particles.

Taking these advantages, we upgraded the BESS detector to improve the rigidity resolution to extend primary spectra up to 1 TeV. We call this upgraded detector 'the BESS-TeV spectrometer'. We performed the balloon experiment using BESS-TeV detector to observe cosmic ray fluxes at Fort Sumner, New Mexico (nominal vertical cutoff rigidity : 4.2 GV), in September 2001. We report here proton and helium spectra in the energy range up to 500 GeV.

2 BESS-TeV spectrometer

In the new BESS-TeV detector, we installed a set of new outer drift chambers (ODCs) at the top and bottom ends of the spectrometer to record longer track of the incident particles. ODCs are cell-type drift chambers.

Each ODC consists of two layers in a radial direction. In each layer, two sense wires are stretched at the center of each cell. Each ODC provided four hit points along the track. Using these ODCs, track length became almost doubled. We became to be able to measure precise deflection angle also, which gave strong constraint on the momentum fit. Spatial resolution of each hit is 150 micrometers.

In order to carry out real-time calibration of the ODCs during flight, a scintillation fiber detector covers outer and inner surface of one cell of the ODC.

New JET and IDCs are also to be installed inside the super-conducting

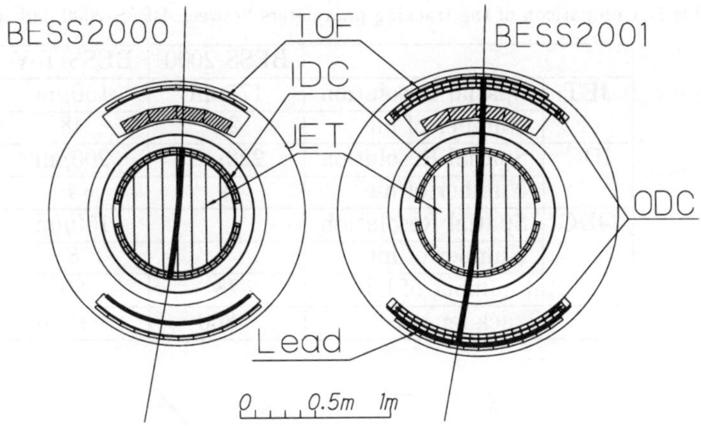

Figure 2. BESS 2000 and BESS-TeV2001.

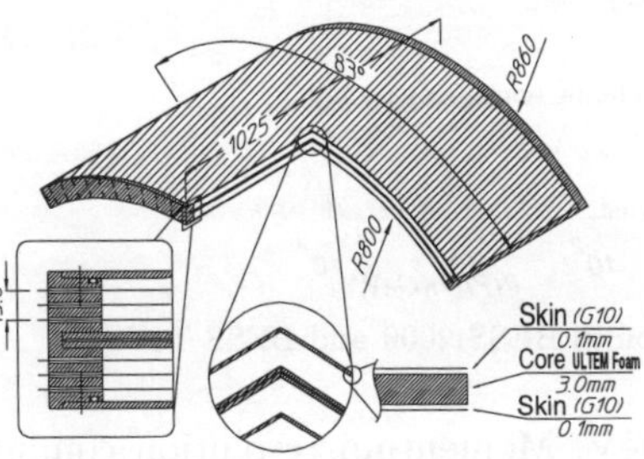

Figure 3. Schematic view of the outer drift chamber (ODC).

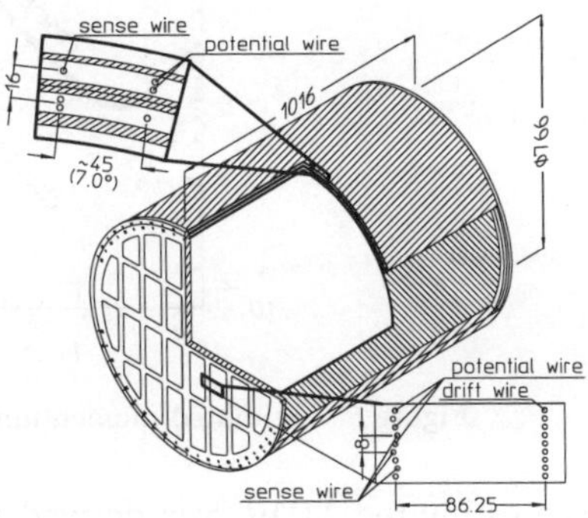

Figure 4. Schematic view of new JET and IDCs.

solenoid [a]. New JET/IDC have 20% better spatial resolution than that of old JET/IDC. And sampling points along the track becomes twice of old one. Table 1 shows the tracking parameters of the BESS-TeV detector in comparison with BESS-2000 detector.

Using the parameters in the table 1, a simple simulation was performed to estimate the expected performance of the BESS-TeV. As shown if figure 5 for the BESS 2000 configuration, the "estimated" MDR of 220 GeV was well consistent with the one from data. For the BESS-TeV configuration,

[a] New JET/IDC are to be installed in BESS-TeV 2002

Table 1. Comparison of the tracking parameters between BESS-2000 and BESS-TeV.

		BESS 2000	BESS-TeV
JET	Spatial Resolution	175μm	150μm
	Number of hit	24	48
IDC	Spatial Resolution	220μm	200μm
	Number of hit	4	4
ODC	Spatial Resolution		150μm
	Number of hit		8
Total number of hits		28	60
Track length		~800	~1600

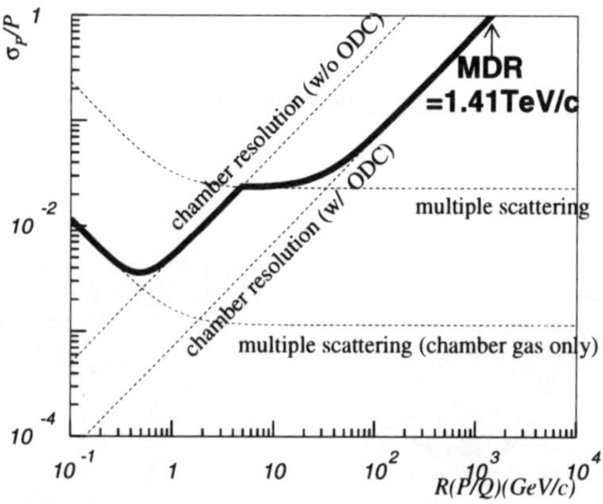

Figure 5. Estimated momentum resolution in BESS 2000 and BESS TeV.

the estimated MDR was derived as 1.4 TeV. Momentum resolution coming form the multiple scattering was estimated separately considering the material distribution of the magnet.

To perform electron/muon separation, a lead plate (covered 1/5 of the whole acceptance) of 2 radiation length was installed just above lower scintillator-hodoscopes. It functions as a electromagnetic shower counter.

3 BESS-TeV 2001 flight

3.1 Flight

The BESS-TeV 2001 campaign was performed at Fort Sumner, New Mexico (cutoff rigidity is 4.2 GV), in September 2001.

In the beginning of august, clues arrived at Fort Sumner and began to preparation for the flight. After one month, the preparation was finished and the detector became flight ready. The flight was carried out in the morning of September 24. The trajectory of the flight is shown in figure 6.
Figure 7 and 8 shows altitude and atmospheric pressure during the flight.

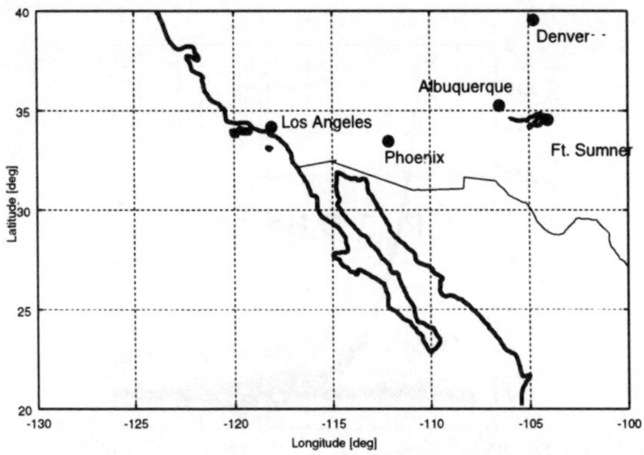

Figure 6. Flight trajectory for BESS-TeV 2001.

The balloon reached at the level altitude of 36 km (residual air of 5 g/cm^2), and floated for 2 hours. After that it gradually descended again down to 33 g/cm^2 in 7 hours. After about 16 hours from the launch, the ballooning was terminated in the midnight around Albuquerque, New Mexico. The BESS payload was safely recovered within one day.

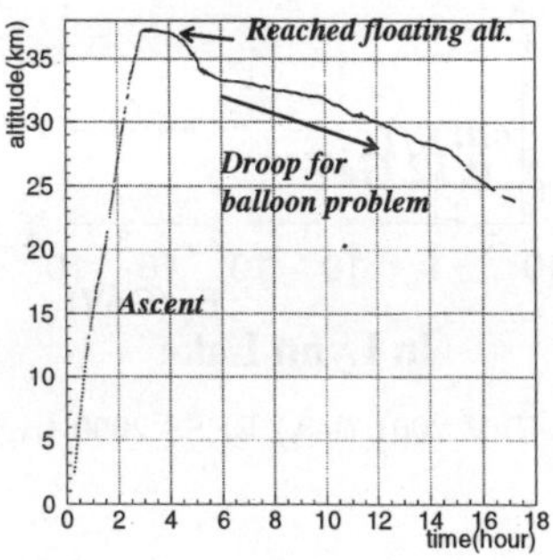

Figure 7. Altitude during the flight

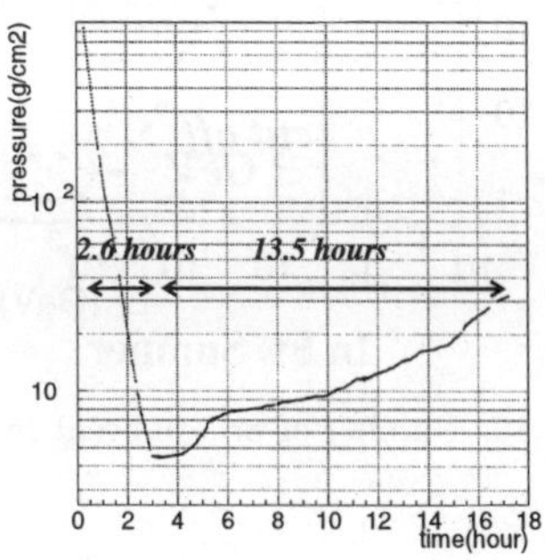

Figure 8. Pressure during the flight

3.2 Date

The date are now under analyzing, We pressed here very preliminary results. 2 hours date measured at around $5g/cm^2$ are plotted in figure 9. And we can easily identify each particles, in the figure.

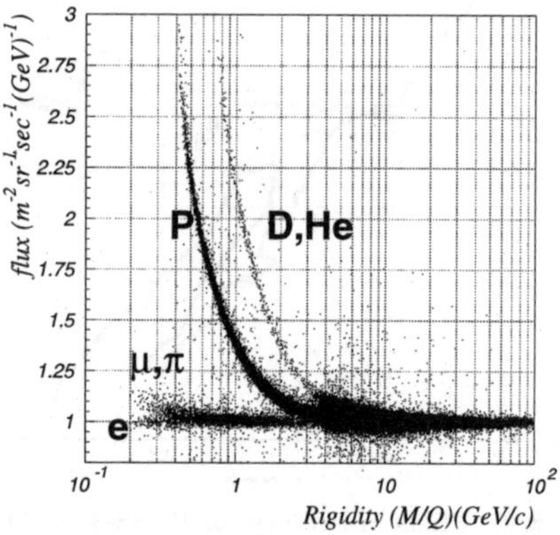

Figure 9. Identification plot

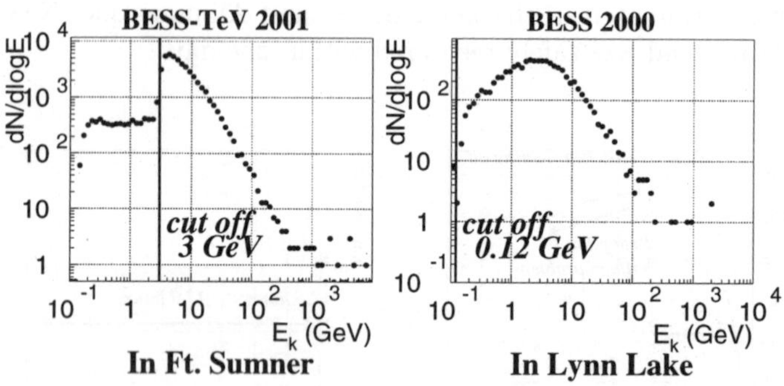

Figure 10. Number of proton event in BESS-TeV 2001 (left) BESS 2000 (right)

For primary cosmic ray flux, we can use only 2 hours data with residual atmospheric pressure around 5 g/cm^2. Because of large systematic errors from secondary fluxes, data with larger pressure is not suited for the calculation of primary fluxes. With only 2 hours data, it is difficult to calculate primary fluxes up to 1 TeV. We can calculate primary fluxes up to 20 GeV.

We can estimate the effect of magnetic cut off rigidity from this primary fluxes. Figure 10 shows a number of proton events measured in 2 hours date with residual atmospheric pressure around 5 g/cm^2, measured by BESS-TeV 2001 (left fig.) and BESS 2000 (right fig.). In each figure, we can see easily the effect of cut off, 3 GeV at Fort Sumner and 0.12 GeV, which is near the instrumental cut off, at Lynn Lake.

As for atmospheric muon and secondary particles, such as proton or helium fluxes, we took the very interesting data. During the balloon descended period, at various altitudes we took a large amount of muon data in comparison with the previous measurements. Pressure changed from 4.5 g/cm^2 to 33 g/cm^2 during the 13.5 hours measurement.

The counting rates of μ^- and μ^+ events at each altitudes are shown in figure 11. Energy ranges are from 0.5 Gev/c to 10 GeV/c for μ^-, and 0.5 GeV/c to 2 GeV/c for μ^+.
In the same way the counting rates of proton events at each altitudes are shown in figure 12. In this figure open square symbols shows the rates which energy are greater than cut off, that is these are primary rates, and the rates decrease as the pressure increases. While the rates which energy smaller than cut off (closed circles) increase as the pressure increases, because these are secondary.

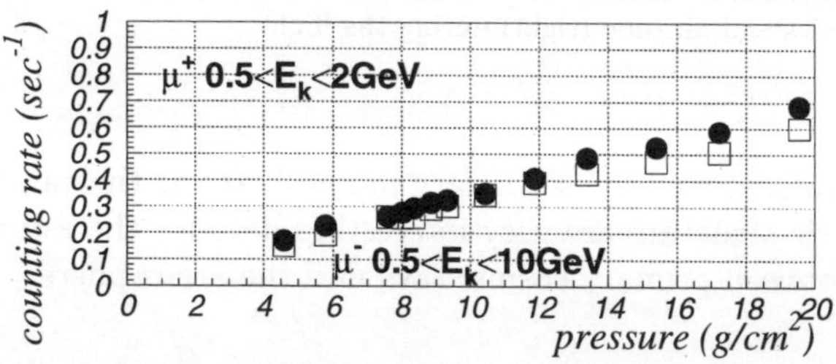

Figure 11. The counting rates of μ^- and μ^+ events observed during the flight.

Throughout the flight, we continuously monitored atmospheric pressure, temperature and altitude. And we know the accurate atmospheric structure in this flight time. Figure 13 shows pressure versus altitude and density of molecules versus altitude during the flight.

In short, the features of BESS-TeV 2001 are that, while it is difficult to calculate primary fluxes up to TeV region, the date primary fluxes up to 20

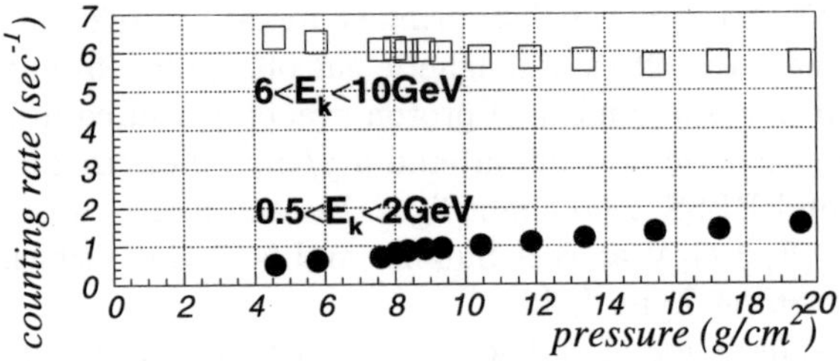

Figure 12. The counting rates of Proton events observed during the flight.

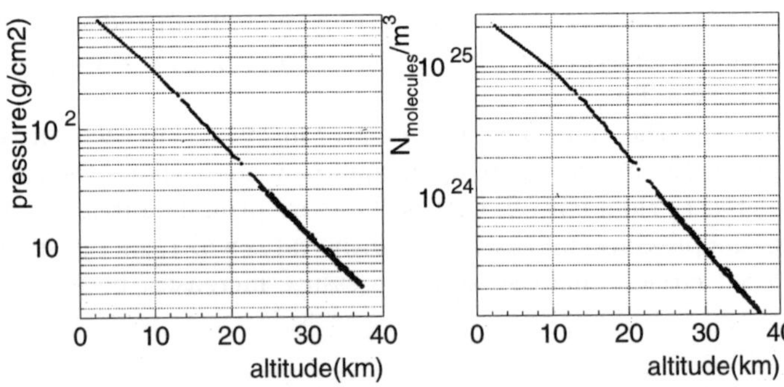

Figure 13. the correlation between pressure and altitude (left) and the correlation between density of molecules and altitude (right) during the flight

GeV, the geomagnetic effect, secondary muon fluxes at the various altitudes, and atmospheric structure are very interesting date for the estimation of the interactions between primary cosmic rays and the atmosphere.

4 Future prospect

Maintenance of each detectors have been finished, and integration for the BESS-TeV 2002 is started for the coming flight which is now scheduled in august 2002 at Lynn Lake, Canada (cut off rigidity 0.5 GV). 24 hours or more observation time is expected by a pre-turn-around flight to obtain enough statistics up to the maximum energy of 1 TeV.

References

1. M. J. Ryan et al., *Phys. Rev. Lett.* **28**, 1562 (1972).
2. I. P. Ivanenko et al., Proc. 23rd ICRC(Calgary) 2 (1993) 17.
3. Y. Ajima et al., *Nucl. Instrum. Methods* A **443**, 71 (2000).
4. T. Sanuki et al., *Astrophys. J.* **545**, 1135 (2000).
5. Zatsepin et al., *Yad. Fiz.* C **57**, 684 (1994).
6. M. Ichimura et al., *Phys. Rev.* D **48**, 1949 (1993).
7. K. Asakimori et al., *Astrophys. J.* **502**, 278 (1998).
8. L. H. Smith et al., *Astrophys. J.* **180**, 987 (1973).
9. W. R. Webber et al., Proc. 20th ICRC(Moscow) 1 (1987) 325.
10. E. S. Seo et al., *Astrophys. J.* **378**, 763 (1991).
11. P. Papini et al., Proc. 23rd ICRC(Calgary) 1 (1993) 579.
12. R. Bellotti et al., *Phys. Rev.* D **60**, 052002 (1999).
13. W. Menn et al., Proc. 25th ICRC(Durban) 3 (1997) 409
14. M. Boezio et al., *Astrophys. J.* **518**, 457 (1999).
15. The AMS collaboration, *Phys. Lett.* B **494**, 193 (2000).

THE PRIMARY PROTON SPECTRUM OF 0.4-30TEV DECONVOLVED FROM ATMOSPHERIC GAMMA-RAY SPECTRA AT BALLOON ALTITUDES

K.YOSHIDA

Faculty of Engineering, Kanagawa University,
3-27-1 Rokkakubashi Kanagawa-ku, Yokohama 221-8686, JAPAN
E-mail: yoshida@kit.ie.kanagawa-u.ac.jp

Y.KOMORI

Kanagawa Prefectural College,
1-5-1 Nakao Asahi-ku, Yokohama 241-0815, JAPAN
E-mail: yoshiko.komori@nifty.ne.jp

J.NISHIMURA

The Institute of Space and Astronautical Science,
3-1-1 Yoshinodai, Sagamihara 229-8510, JAPAN
E-mail: nisimura@icrr.u-tokyo.ac.jp

We have observed atmospheric gamma-rays for 30GeV to 10TeV for many years with the emulsion chamber at balloon altitudes. Since atmospheric gamma-rays observed at several g/cm^2 are produced almost by a single interaction of primary protons with atmospheric nuclei, we can deconvlve to the primary proton spectrum, referring to an appropriate hadronic interaction model. The proton spectrum estimated by our gamma-ray observations covers the energy range from 400GeV to 30TeV, filling a gap in the currently observed proton spectra.

1 Introduction

The Super-Kamiokande group has found evidence for oscillation of atmospheric neutrinos [1]. In the analysis of atmospheric neutrinos the expected neutrino fluxes obtained by Monte-Carlo calculations play a very important role, and their reliable calculations are necessary to determine the oscillation parameters in more detail. The uncertainty of the absolute flux calculation of atmospheric neutrinos comes mainly from the uncertainty in the primary cosmic-ray flux and the modeling of hadronic interactions.

Recently, the BESS and AMS groups have measured the precise primary proton spectra up to $\sim$ 100 GeV [2,3], giving consistent results with each other within an error of 5%. The helium component has been also measured by the both group with an error of about 10%. As for the hadronic interaction models, the BETS group has observed atmospheric gamma-rays in the energy

Table 1. List of balloon flights since 1977*.

Flight	Area (m^2)	Time (min)	Average Altitude(g/cm^2)	$S\Omega T$ (m^2 sr s)
1977	0.78	1760	4.5	6.55×10^4
1979	0.80	1680	4.9	1.41×10^5
1980	0.80	2029	7.8	9.52×10^4
1984	0.20	576	9.2	7.11×10^3
1985	0.40	940	9.4	1.32×10^4
1988	0.20	647	7.1	3.93×10^3
1996	0.20	2092	4.6	6.50×10^4
1998	0.20	1178	5.6	3.64×10^4

* See the reference [9] before 1977.

range from a few GeV to several 10 GeV at a mountain altitude of 2.77km and at balloon altitudes of $15-25$km [45]. They found that nuclear interaction codes with the Lund Fritiof V7.02 or dpmjet3 give fairly consistent results with the observed energy spectra and the altitude variation of the gamma-ray flux. Using the precisely observed primary cosmic-ray spectrum and the reliable nuclear interaction model, it is possible to calculate the better accurate neutrino flux around the $\sim$ 1GeV region.

Atmospheric neutrinos with higher energy of $\sim$ 100GeV, which are measured as upward through-going muons, are mainly produced from the interaction of $\sim$ 1TeV primary protons in the atmosphere. While the proton spectrum in the energy range from 10TeV to 1000TeV has been measured by several groups with reasonable agreement (e.g. JACEE Collaboration[6], RUNJOB Collaboration[7] and references therein), reliable measurements of the proton spectrum are poorly off in the energy range from 100GeV to 10TeV.

We have derived the atmospheric gamma-ray spectrum at an altitude of $4g/cm^2$, in which we included new gamma-ray data and improved the analysis method comparing with our previous paper[8]. Because of a scaling nature of the pion production cross section, we can deconvolve the proton spectrum from the gamma-ray spectrum in a semi-analityc way. In this deconvolution, we include the effect of the contributions of primary heliums and heavier nuclei, and also the minor contribution of gamma rays through η and K mesons.

In this paper, we present the primary proton spectrum of $0.4-30$TeV deconvolved from the atmospheric gamma-ray spectra observed at balloon altitudes of several g/cm^2, filling a gap of the observed spectra of primary cosmic rays in this energy region.

2 Atmospheric gamma-ray observations with ECC

We have observed primary electrons with the emulsion chamber on-board balloons from 1968 to 1998 [9] [10]. Simultaneously, we have also observed atmospheric gamma-rays at balloon altitudes of 4−9 g/cm^2 with a total observation time of 10.45 days. Table 1 summarizes the series of experiments since 1977.

The emulsion chamber consists of nuclear emulsion plates and lead plates which are stacked alternately. The typical area size and thickness is 40cm×50cm and $\sim$ 9 radiation length (r.l.). Total effective $S\Omega T$ is 6.45 m^2 sr day. Figure 1 shows typical configuration of the emulsion chamber in cross-sectional drawing from side view, developed for the primary electron observation. We can measure the electro-magnetic shower developed in lead plates using nuclear emulsion plates. From the detailed inspection of the starting point of the shower in the emulsions with a high position resolution

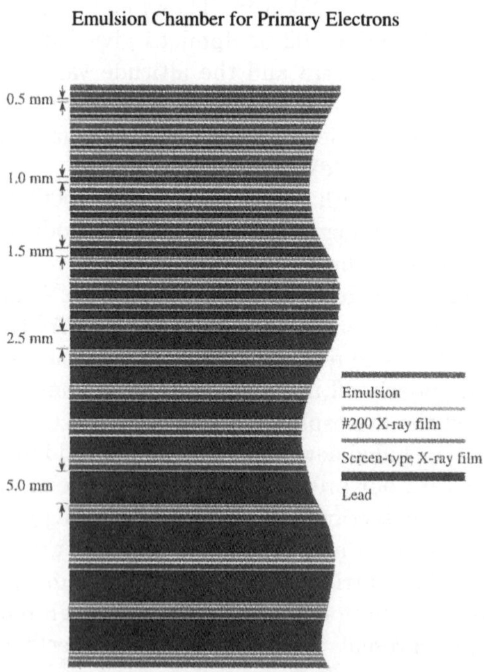

Figure 1. Typical configuration of the emulsion chamber in cross-sectional drawing from side view.

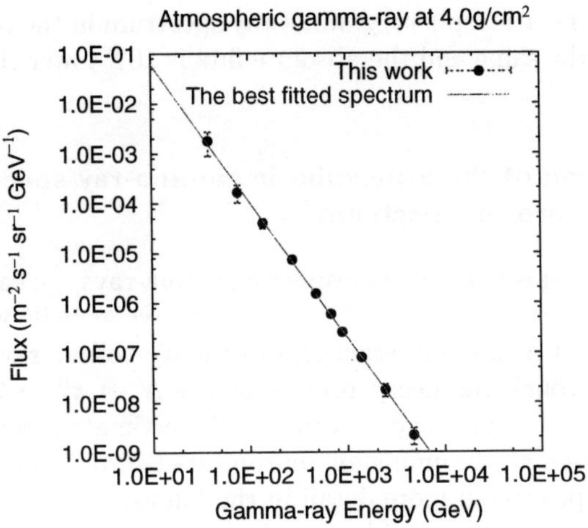

Figure 2. The atmospheric gamma-ray spectrum observed at an altitude of $4\mathrm{g/cm}^2$. The dash line shows the best fit power-law function with a spectral index of 2.74.

of 1μm, we can clearly identify such incoming particles as electrons, gamma-rays and other background particles which induce showers. In order to determine energies, we calculated the shower development for each chamber using Monte-Carlo simulations. We confirmed the simulation results comparing to the experimental results with FNAL electron beams.

In the derivation of the atmospheric gamma-ray spectrum, we took into accounts of the following corrections. As we selected gamma-ray events with a shower starting point within 3 r.l., we corrected the gamma-ray detection efficiency of $\sim$ 90%. The absolute flux of gamma-rays are affected by the statistical errors of the energy determination and enhanced in the case of the steep power-law spectrum. The enhancement factor due to the energy resolution of 12% at 100GeV is ~ 1.01. In this atmospheric gamma-ray flux, we subtracted the bremsstrahlung gamma-ray flux of $\sim 8 \times 10^{-6}$ (m^{-2} s^{-1} sr^{-1} GeV^{-1}) at 100GeV produced from primary electrons. The gamma-ray fluxes observed with each chamber are normalized to a $4\mathrm{g/cm}^2$ altitude.

The number of observed gamma-rays is 314 events in the energy range from 30GeV to 10TeV. Figure 2 shows the observed spectrum of atmospheric gamma-rays normalized at an altitude of $4\mathrm{g/cm}^2$, which is well represented by

$$F_\gamma(E) = (1.11 \pm 0.13) \times 10^{-4} (E/100\mathrm{GeV})^{-2.74 \pm 0.06} (\mathrm{m}^{-2}\mathrm{s}^{-1}\mathrm{sr}^{-1}\mathrm{GeV}^{-1}). \quad (1)$$

Comparing with the atmospheric gamma-ray spectrum in the reference [9], this spectral index is the same and the absolute flux is 10% lower than that in the reference [9].

3 Deconvolution of the atmospheric gamma-ray spectrum to the primary proton spectrum

From the observed spectrum of atmospheric gamma-rays, we can estimate the primary proton spectrum. Atmospheric gamma-rays at balloon altitudes are produced by almost a single interaction of primary cosmic rays in the atmosphere, mainly through the decay process of $p + N \to \pi^0 \to 2\gamma$. Therefore, the energy spectrum of primary protons can be estimated from atmospheric gamma-rays through π^0, assuming an appropriate hadronic interaction model. We describe this process in more detail in the following.

At first, primary protons incident to the top of atmosphere with a power-law spectrum of

$$F_p(E)dE_p = NE_p^{-\gamma}dE_p. \tag{2}$$

A mean free path length of a nuclear interaction and an absorption length of protons in the atmosphere are $\lambda = 90\text{g/cm}^2$ and $\Lambda = 110\text{g/cm}^2$, respectively. The flux of atmospheric gamma-rays observed at 4g/cm^2 is represented by

$$\Gamma(E_\gamma) = g(E_\gamma) \times \frac{1}{\lambda}\frac{\exp(-4/\Lambda) - \exp(-4\sigma/X_0)}{\sigma/X_0 - 1/\Lambda} = 0.0418g(E_\gamma), \tag{3}$$

considering the absorption of gamma-rays in the atmosphere (absorption coefficient of $\sigma = 0.7733$ and the radiation length of $X_0 = 36.7\text{g/cm}^2$), where $g(E_\gamma)$ is a production spectrum of gamma-rays generated in collisions between protons and atmospheric nuclei.

In the energy range over 100GeV, one can apply a scaling law for a production probability of π^0. A π^0 production rate in a single collision between protons and atmospheric nuclei can be represented by

$$f(E_\pi/E_p)d(E_\pi/E_p) \tag{4}$$

assuming a scaling law, where E_π is the energy of daughter π^0 and E_p is the energy of parent protons.

A production spectrum of π^0 generated in collisions between protons with a power-law spectrum and atmospheric nuclei is given by

$$\Pi(E_\pi)dE_\pi = \int_{E_\pi}^{\infty} dE_p E_p^{-\gamma} f(E_\pi/E_p)d(E_\pi/E_p).$$

Replacing as

$$\sigma_p = \int_0^1 x^{\gamma-1} f(x)dx$$

where $x = E_\pi / E_p$, $\Pi(E_\pi)dE_\pi$ is given by

$$\Pi(E_\pi)dE_\pi = \sigma_p E_\pi^{-\gamma} dE_\pi, \tag{5}$$

in which σ_p is a constant parameter independent of energies. We derive a numerical value of σ_p as 0.043 in the case of a spectral index $\gamma = 2.75$ by using a nuclear interaction code of Lund Fritiof V7.02. Here we include the contribution of η mesons to gamma-rays.

A production spectrum of gamma-rays from a single π^0 decay is given by

$$\frac{dn_\gamma}{dE_\gamma} = \frac{2}{E_\pi}. \tag{6}$$

By combining the equation (5) and (6), the energy spectrum of gamma-rays from a single collision between protons and atmospheric nuclei is given by

$$g(E) = \int_E^\infty \frac{dn_\gamma}{dE_\gamma} \Pi(E_\pi)dE_\pi = \frac{2\sigma_p}{\gamma} E^{-\gamma}. \tag{7}$$

From equation (2) and (7) the energy spectrum of gamma-rays from interactions of primary protons in the 4g/cm^2 atmosphere is represented by

$$\Gamma(E) = 0.0418 \times \frac{2\sigma_p}{\gamma} N E^{-\gamma}. \tag{8}$$

Using this formula, we can derive the primary proton spectrum from the observed atmospheric gamma-rays. Here we need to include the effect of the contribution due to the heavy primaries. We corrected the contribution of such heavier nuclei as He, C, N, and O, regarding the composed nuclei as independent particles. The flux of heavier nuclei with the same energy per a particle as protons is 5.5% for He and 0.8% for C,N,O of the flux of protons. Therefore the flux of heavier nuclei can be converted to the proton flux as follows:

$$4 \times F_{He} + 14 \times F_{CNO} = 0.22 \times F_p + 0.11 \times F_p = 0.33 \times F_p.$$

The proton flux derived in equation (9) is multiplied by $1/1.33 = 0.75$. We also corrected for the minor contribution of η and K mesons. The correction for η meson is included in the calculation of σ_p. The deconvolved primary proton spectrum is represented in the following,

$$F_p(E) = 23.9 \frac{\gamma}{2\sigma_p} \times 0.75 \Gamma(E). \tag{9}$$

Because of a scaling nature of nuclear interactions, the spectral index of protons in this spectrum (9) is the same as that of gamma-rays.

As shown from equation (9), one proton produces gamma-rays with the amount of

$$\frac{2\sigma_p}{\gamma} \times 10.75 = 0.042.$$

Hence, the energy of the corresponding parent protons is $(1/0.042)^{1/(\gamma-1)} =$ 6.1 times of that of the gamma-rays. Then we can approximately assume that atmospheric gamma-rays are produced on average from primary protons with the sixfold energy.

Figure 3 shows the proton spectrum deconvolved from the atmospheric gamma-ray spectrum at 4g/cm^2. The proton spectrum is well represented by

$$F_p(E) = (5.7 \pm 1.3) \times 10^{-2}(E/100\text{GeV})^{-2.74\pm0.06}(\text{m}^{-2}\text{s}^{-1}\text{sr}^{-1}\text{GeV}^{-1}) \quad (10)$$

in the energy range from 400 GeV to 30 TeV.

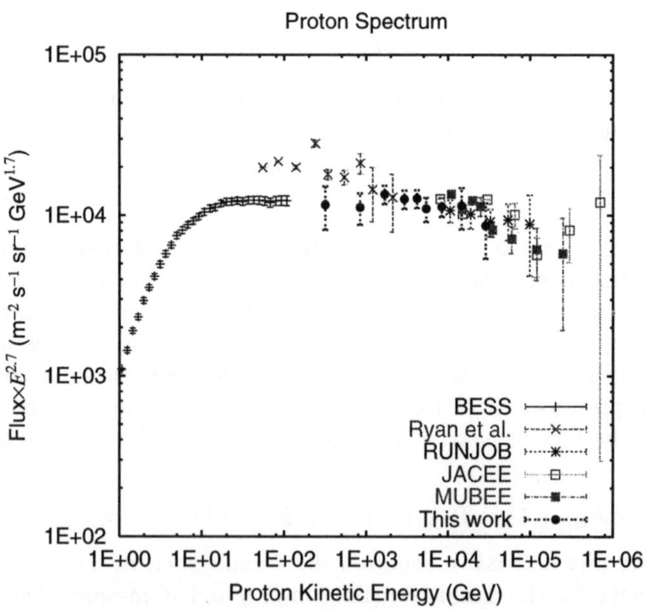

Figure 3. The proton spectrum deconvolved from our atmospheric gamma-ray spectrum, compared to the spectra with other groups.

4 Summary and Discussion

We estimated the primary proton spectrum from 400GeV to 30TeV, which is deconvolved from the atmospheric gamma-ray spectrum in a semi-analytic approach. In this deconvolution we include the effect of the contributions of primary heliums and heavier nuclei, and also include the contribution of gamma-rays through η and K mesons. As shown in figure 3, our estimated proton spectrum is almost consistent with the extension of the proton spectrum observed with the BESS and AMS instrument. While in the energy range from 100GeV to 10TeV there are no enough data in the currently observed proton spectra, our estimated proton spectrum fills this gap.

Although it is desirable to derive the proton spectrum with higher statistical precision, it is difficult to observe gamma-rays in a few hundred GeV energy region with the emulsion chamber because of the difficulty of the scanning. In order to overcome the difficulty of the microscope scanning to pick up low energy showers, we are trying to apply the automatic scanning method which has been successfully developed by Nagoya group to observe ν_τ with nuclear emulsions. An atmospheric gamma-ray observation in the energy range from 100GeV to 1TeV is also planned to be performed with PPB-BETS, which is launched in January of 2003 from the Showa station in the Antarctica.

References

1. Y. Fukuda *et al*, *Phys. Rev. Lett.* **81**, 1562 (1998).
2. T. Sanuki *et al*, *Astrophys. J.* **545**, 1135 (2000).
3. AMS collaboration, *Phys. Lett.* B **490**, 27 (2000).
4. S. Torii *et al*, *Proc. of the 2nd International Workshop on Neutrino Oscillations and their Origin*, 79 (2001).
5. K. Kasahara *et al*, *Proc. of 27th ICRC*, 966 (2001).
6. JACEE Collaboration, *Astrophys. J.* **502**, 278 (1998).
7. RUNJOB Collaboration, *Astrop. Phys.* **16**, 13 (2001).
8. T. Kobayashi, Y. Komori, K. Yoshida, and J. Nishimura, *Proc. of the 2nd International Workshop on Neutrino Oscillations and their Origin*, 87 (2001).
9. J. Nishimura *et al*, *Astrophys. J.* **238**, 394 (1980).
10. J. Nishimura *et al*, *Adv. Space Res.* **26**, No.11, 1823 (2001).
11. S. Torii *et al*, *Astrophys. J.* **559**, 973 (2001).

CALCULATIONS OF THE ATMOSPHERIC ν FLUXES

Paolo Lipari
Dipartimento di Fisica, Università di Roma I
P. A. Moro 2, 00185 Roma, Italy
E-mail: paolo.lipari@roma1.infn.it

Atmospheric neutrinos are produced by cosmic ray interactions in the Earth's atmosphere. The existence of discrepancies between the measurements and calculated predictions for these flxues can be interpreted as evidence for the existence of flavor oscillations. In this work we discuss systematic uncertainties in the calculation of the prediction, and argue that the evidence for new physics in neutrino propagation is very robust, and that the determination of the allowed intervals for the oscillation parameters (Δm^2 and $\sin^2 2\theta$) in the $\nu_\mu \leftrightarrow \nu_\tau$ hypothesis is free from significant biases.

1 Introduction

The comparison of the experimental measurements of the atmospheric ν fluxes performed by Super–Kamiokande (SK) [1] and other underground detectors [2] with predictions based on the standard model shows large discrepancies that have been interpreted as evidence for neutrino oscillations or some other form of "new physics". In this contribution we will discuss recent progress in the calculation of the atmospheric ν fluxes, give estimates about the size of the uncertainties in the prediction, and discuss possible consequences in the interpretation of the data.

The calculation of the atmospheric ν fluxes is based on three fundamental elements: (i) a description of the primary cosmic ray (c.r.) fluxes, (ii) a model for particle production in hadronic interactions, and (iii) a calculation scheme to obtain the neutrino fluxes that reach a certain detector.

2 The primary cosmic ray fluxes

The fluxes of primary particles reaching the Earth can be measured with detectors placed at high altitude on balloons or satellites. Several measurements of the proton flux are plotted as a function of energy after multiplication by the factor $E_{\rm kin}^{2.7}$ in fig.1. Our knowledge of the flux has improved dramatically in recent years. The most recent measurements of AMS [3] and BESS [4] have relatively small uncertainties and are in very good agreement with each other (to within 5%). The reasons behind the large discrepancies with earlier results, are not understood, but are likely to be the consequence of systematic errors. The data of Caprice [5] is approximately 20% lower in normalization. It appears

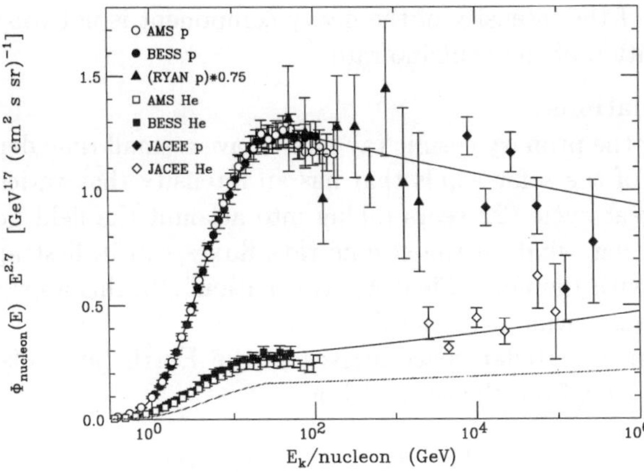

Figure 1: Measurements of the cosmic ray proton and Helium fluxes. including higher energy measurements.

reasonable to consider the proton flux as well determined up to a maximum energy of $E_0 \sim$ 100–200 GeV. The measurements of the p flux performed with magnetic spectrometer stop at a rigidity $p/Z \sim 100$ GV, due to the detector resolution and limited exposure. This is sufficient for the calculation of the rates of sub–GeV and multi–GeV events in a ν detector, when only a modest extrapolation is required. However for the calculation of up-going muon events, produced by higher energy neutrinos, one needs an extrapolation up to $E_0 \sim 10^5$ GeV, that results in a significant uncertainty of order 20% for the absolte normalization of the ν–induced muon rate/ Direct measurements up to an energy $E_p \lesssim 3 \times 10^5$ GeV, have been performed with emulsion chambers on balloon as for example by JACEE[6]. At still higher energies the results of EAS detectors provide a constraint, limited by lack of knowledge of the chemical composition.

The Helium component accounts for approximately $\sim$ 20 % of the atmospheric ν event rate. Both the AMS and BESS experiments have also measured the Helium flux. In this case the results show a 10% discrepancy when plotted as a function of rigidity. There is some indication that the Helium c.r. flux is a little flatter than the p one, a result of significant interest for cosmic ray physics. In the near future the knowledge of high energy c.r. will improve significantly. The BESS experiment will collect additional data with an improved detector, while the AMS detector will have have a very long exposure aboard the International Space Station.

Cosmic rays with $(Z \geq 3)$ account for $\sim$ 10% of the ν event rate, therefore uncertainties of the intensity of the heavy component is not too important in the determination of the neutrino rate.

Solar Modulations

At low energy the primary cosmic ray fluxes have a small time dependence due to the effects of the solar winds that has an intensity that varies periodically with a a 11 year cycle (22 years taking into account the field polarity). The effect of the solar wind on the cosmic rays fluxes can in first approximation be described with the Force Field Approximation". In this approximation the effects of the solar wind is modeled as a potential. A cosmic ray particle with energy $E_{\rm i.s.}$ in interstellar space arrives at the Earth (at a distance of one Astronomical Unit from the sun) with a

$$E_\oplus \simeq E_{(\rm i\ s)} - |Z|\ V_{\rm wind}(t) \tag{1}$$

It follows that the flux $\Phi_\oplus(E,t)$ in the vicinity of the Earth is related to the insterstellar flux $\Phi_{\rm i.s.}$ as:

$$\Phi_\oplus(E;\ t) \simeq \frac{p_\oplus^2}{p_{\rm i.s.}^2}\ \Phi_{\rm i.s.}[E + |Z|\, V_{\rm wind}(t)] \tag{2}$$

In recent years the BESS experiment has performed measurements of the cosmic rays fluxes in 4 balloon flights in 1997, 1998, 1999 and 2000, while the solar acivity was increasing from minimum to maximum. The results of these measurements are shown in fig. 2. In the figure the result is compared with fits based on the force–field approximation (equation (2). The value of the potential $V_{\rm wind}$ is found to be well correlated with the counting rates of neutron monitors, and therefore the flux at time t can be estimated from the available neutron monitor rates. A prediction of the neutrino fluxes at the SK site for different times is shown in fig. 3. Small (few percent) variation in the rate of sub–GeV events are expected).

Geomagnetic effects

The geomagnetic field introduces an anisotropy in the c.r. fluxes since it forbids low rigidity particles from reaching the vicinity of the Earth. The effect is stronger in the Earth's magnetic equatorial region, and for west–going particles and is reflected in the angular distribution of atmospheric neutrinos. A quantitative confirmation of the validity of the calculation of geomagnetic cutoffs has been obtained with the data of a ten days flight of the Anti Matter Spectrometer (AMS) aboard the Space Shuttle in June 1998. The AMS detector has measured the cosmic ray flux at an altitude of $\sim$ 400 Km over a large fraction of the Earth surface. The results for the proton flux[3] are shown in fig. 4. The

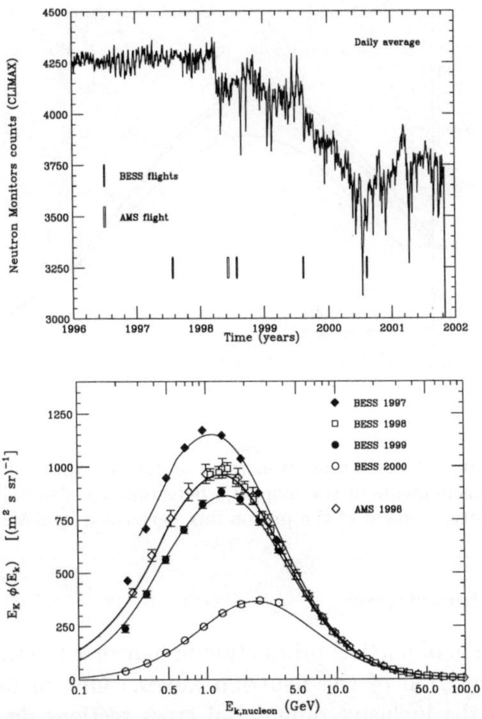

Figure 2: The top panel shows the neutron monitor rate for the CLIMAX detector. The time for different measurements of the magnetic detectors are also indicated. The bottom panel compares the measurement of the proton flux performed by AMS and BESS.

10 different sets of points correspond to the measurements of the vertical p flux taken when the detector was in different regions of magnetic latitude. Inspecting fig. 4 one can notice that: (i) the fluxes measured in different geographycal regions are equal (within errors) for kinetic energy $E_k \gtrsim 15$ GeV, (ii) a magnetic "cutoff" is present at a momentum that is largest when the detector is near the geomagnetic equator (the cutoff appears smooth because the data is integrated over a finite range of magnetic latitude and zenith angle), (iii) it is also evident the presence of non vanishing fluxes of sub–cutoff particles. The subcutoff particles are generated as secondary products of primary cosmic ray interactions in the atmosphere, (see for example [7,8] for more discussion). A conservative estimate of the rate of neutrino events due to subcutoff protons is of order 10^{-2} events/(kton year).

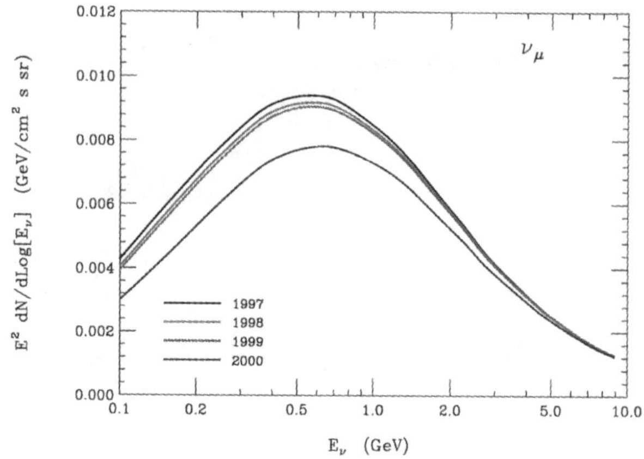

Figure 3: The top panel shows the neutron monitor rate for the CLIMAX detector. The time for different measurements of the magnetic detectors are also indicated. The bottom panel compares the measurement of the proton flux performed by AMS and BESS.

3 Hadronic Interactions

A precise description of particle production in hadronic interactions is essential for a precise calculation of the neutrino fluxes. One in fact needs a good representation of the inclusive differential cross sections $d\sigma_{p\to x}/dE_x(E_x, E_p)$ for a primary particle p to produce a secondary of type x in an interaction with an air nucleus. Some of the most important issues are: (i) The amount of energy transfered to a "leading nucleon" (that is of order $\sim$ 40–50%); (ii) how the energy is shared between pions of different charges: π^+, π^- and π° (neutral pions decay into photons and are a negligible source of neutrinos, while charged pions are the dominant ν source, with π^+ being the source of ν_e and π^- of $\overline{\nu}_e$); (ii) the shape of the inclusive differential spectra for π^+ and π^- production; (iv) the relative importance for the production of kaons ($K^\pm$, K_L) and pions. Note that K^+ can be produced in association with strange baryons (Λ K^+), (Σ K^+) and are significantly more abundant than K^-. Uncertainties in the hadronic interaction modeling is now probably the largest source of uncertainty for the prediction of the ν fluxes (see [19] for more discussion), however the dominant effect is about the absolute normalization.

4 Method of calculation: 3D versus 1D

The first calculations of the atmospheric ν fluxes have been performed in the so called "one–dimensional" approximation, where all secondary particles pro-

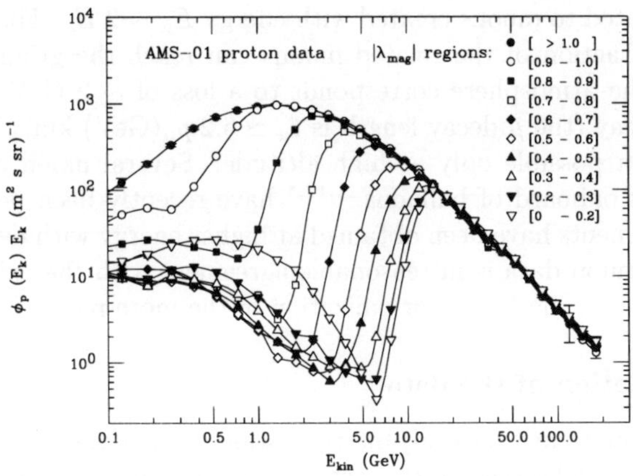

Figure 4: Cosmic ray proton flux measurements of the AMS detector[3] averaged for a detector position in ten regions of magnetic latitude. The error bars are shown only for one set of measures ($|\lambda_{\rm mag}| > 1$ radiant).

duced in a shower are considered as collinear with the primary particle. In this approximation one needs to study only the very small subset of primary particles, with trajectories that intersect the detector of interest, with an enormous reduction of the computational difficulty of the problem. The 1D approximation introduces a distortion of the angular distribution of low energy neutrinos. A 3D calculation, predicts an enhancement of the flux of horizontal ν's that can be understood as a simple geometrical effect[18,10]. This effect becomes negligible for large E_ν when the angle $\theta_{0\nu}$ becomes small. The 3D enhancement of the flux of low on the horizontal plane for low energy neutrinos, is difficult to observe directly because of the poor experimental angular resolution of the existing detectors. However since it implies a different pathlength distribution for the neutrinos, it can have observable consequences for the interpretation of the data, in particular for the determination of the Δm^2 allowed region. Preliminary analysis of the SK data suggest that the effect on the allowed region is only a modest shift.

5 The muon flux constraints

The fluxes of atmospheric muons are strictly related to the neutrino ones, because almost all ν's are produced either in association with, or in the decay of $\mu^\pm$. Precise measurements of the muons offer then a way to check directly the ν fluxes. Simple kinematical considerations show that neutrinos of energy E_ν

are closely related to muons created with energy $E_\mu \sim 2\,E_\nu$. For $E_\nu \lesssim 2$ GeV, only a small fraction of the related muons can reach the ground because of energy loss (the atmosphere corresponds to a loss of $\sim$ 2 GeV for a vertical muon) and decay (the μ decay length is $\ell_\mu \simeq 6.2\, p_\mu(\mathrm{GeV})$ km). The measurements are then possible only at high altitude. Several muon measurements with detectors onboard of balloons [11,12,4] have recently been performed, and other measurements have been obtained at higher energy with detectors at the ground. The muon data is in reasonable agreement with the existing calculation, and supports the lower normalization of the more recent calculations.

6 Interpretation of the data

The simplest (and most natural) interpretation of the data is in terms of two–flavor $\nu_\mu \leftrightarrow \nu_\tau$ oscillations. In this case the survival probability for ν_μ and $\overline{\nu}_\mu$ has the form:

$$P_{\nu_\mu \to \nu_\mu} = 1 - \sin^2 2\theta \; \sin^2 \left[\frac{\Delta m^2}{4} \frac{L}{E_\nu} \right], \tag{3}$$

is a function of the ratio L/E_ν and depends on the two parameters $\sin^2 2\theta$ and Δm^2. The expression (3) takes simple (and Δm^2 independent) forms when the ratio L/E_ν is much larger or much smaller than $|\Delta m^2|^{-1}$. For small L/E_ν the probability is approximately unity, while for large L/E_ν, averaging over the fast oscillations, one has: $\langle P_{\nu_\mu \to \nu_\mu} \rangle \sim 1 - \sin^2 2\theta/2$. The "critical value" of L/E_ν that separates the two behaviours of the probability corresponds to its first minimum and has the value:

$$\left(\frac{L}{E_\nu} \right)_{\mathrm{crit}} = \frac{2\pi}{|\Delta m^2|} \simeq 412 \left(\frac{3 \times 10^{-3}\ \mathrm{eV}^2}{|\Delta m^2|} \right) \frac{\mathrm{Km}}{\mathrm{GeV}} \tag{4}$$

The ν pathlength L is strongly correlated with the ν zenith angle θ_z (the correlation is not perfect because the ν creation point is not known), so to a reasonably good approximation one can consider the oscillation probability as a function of $\cos\theta_z$ and E_ν. For $E_\nu \sim 1$ GeV, that is for the sub–GeV and multi–GeV events, equation (4) corresponds to a zenith angle close to the horizontal, and therefore to a situation where down–going ν_μ are not influenced by oscillations $D \simeq D_0$, while up–going neutrinos are suppressed by a factor $U \simeq U_0\,(1 - \sin^2 2\theta/2)$. On the other hand, from spherical shape of the Earth and the assumption that osmic rays are isotropic (or equivalently that their flux is identical in different points of the Earth) one can deduce that the ν–fluxes are up–down symmetric:

$$\phi_{\nu_\alpha}(E_\nu, \theta) = \phi_{\nu_\alpha}(E_\nu, \pi - \theta) \tag{5}$$

and therefore $U_0 \simeq D_0$. The comparison of the up–going and down–going rate, allows to determine the mixing parameters after correcting for finite angular resolution effects. This assumption of c.r. isotropy is violated by geomagnetic effects that are non negligible for low rigidity particles. The validity of the treatment of geomagnetic effects has been tested reproducing the proton measurements of AMS in different regions of the Earth (see fig. 4) and studying the East–West effect for atmospheric neutrinos [9]. Note that for the Kamioka site geomagnetic effects predict an *excess* of upgoing neutrinos of order $\sim 2\%$ in the multi–GeV range of sign opposite to the large effect ($\sim -50\%$) observed and attributed to neutrino oscillations. Smaller effects on the no–oscillation up/down asymmetry are due to the presence of mountains above the detector site, and possible variations of the average atmosphere profile in different geographycal positions. Note that the angular resolution in reconstructing the neutrino direction is controlled by the physical μ–ν angle, whose distribution is determined by the ν interaction cross section. The average $\langle\theta_{\mu\nu}\rangle$ shrinks with energy and for multi–GeV and a small effect on the measured U/D ratio. We can conclude that the estimate of the mixing parameter $\sin^2 2\theta$ is solidly based and the size of the allowed interval is controlled by statistical errors.

To determine accurately the value of Δm^2 in the range indicated by the data with neutrinos below few GeV, one needs to study the angular distribution close to the horizontal plane. In this region small uncertainties in the neutrino direction are reflected in large uncertainties in the pathlength L so that the oscillation pattern become invisible, and one can only measure the oscillation probability averaged over a broad range of L. Increasing Δm^2 corresponds to a larger suppression, since neutrinos with shorter pathlength can oscillate. In order to extract the value of Δm^2 one needs a good prediction of the energy and angular distribution of the non–oscillated flux. In this case a very important constraint is obtained from the rate of e–like events, since the fluxes of $(\nu_\mu + \overline{\nu}_\mu)$ and $\nu_e + \overline{\nu}_e$ are closely related because they are produced by the chain decay of the same parent mesons. The dominant sources of uncertainties then become the ratio π^+/π^- and K/π. After the decay of the muon, π^+ is a source of ν_e while π^- produces $\overline{\nu}_e$ (both $\pi^\pm$ produce a pair $\nu_\mu + \overline{\nu}_\mu$), and since $\sigma_\nu > \sigma_{\overline{\nu}}$ a larger ratio π^+/π^- corresponds to a larger rate of e–like events (and a smaller $(\mu/e)_0$ ratio). The effect of uncertainties on the K/π ratio is reduced because in kaons generate higher energy ν_μ in the first generation decays like $K^+ \to \mu^+\nu_\mu$ (no kinematical suppression as in the decay $\pi^+ \to \mu^+\nu_\mu$ since $(m_\mu/m_K)^2$ is small), but also direct ν_e in decays such as $K^+ \to \pi^\circ e^+\nu_e$. The two effects have different signs on the $(\mu/e)_0$ ratio. The same reasons that are at the origin of the difficulties in the estimate of Δm^2 for an interpretation of the data, in terms of $\nu_\mu \leftrightarrow \nu_\tau$ oscillations, can explain the possibility to explain the data

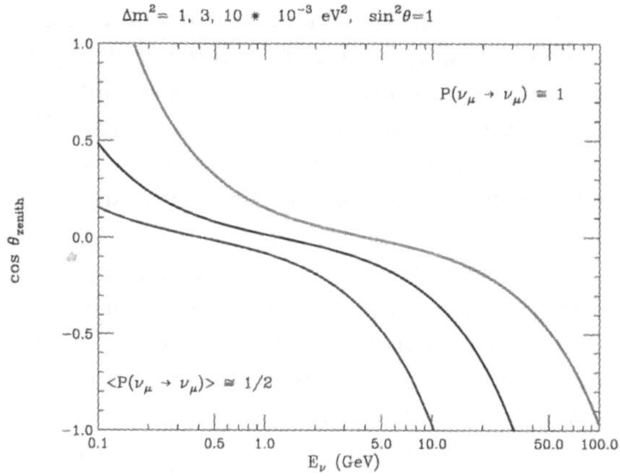

Figure 5: Curves in the plane $(E_\nu, \cos\theta_z)$ that correspond to the first (highest energy) minimum of the survival probability. The three curves correspond to $\Delta m^2 = 1$, 3 and 10 in units of 10^{-3} eV^2.

with a survival probability that decreases monotonically with increasing L/E_ν, instead of the oscillating form. Physical models that result in a probability $P_{\nu_\mu \to \nu_\mu}$ with this form have been constructed as a form of neutrino decay [13], or as decoherence [14]. An an illustration in fig. 5 we show the curves in the plane $(E_\nu, \cos\theta_z)$ that corrresponds to the first (lowest L/E_ν) minimum of the oscillation probability.

Several calculations of atmospheric neutrinos have been performed. The fluxes obtained by the Bartol group [15], Honda et al [16,17] and Fluka [18] have been compared with the experimental data of SK. The differences between the results of these calculations have been studied in detail and their origin is understood, and can be traced to: (a) differences in the choice of the input c.r. primary spectrum, (b) differences in the description of particle production in hadronic interactions, (c) inclusion or not of the approximation of "collinearity" of the secondary particles in the shower (or 3D versus 1D in the jargon of the field). This understanding implies the modeling of the showers (that is the description of energy losses, kinematics of particle decays, muon polarization effects and so on) is essentially equivalent in the three independent codes, giving confidence of their correctness. The effect in the determination of the allowed interval is found to be small.

7 Summary

In the last two years there has been some significant progress in the calculation of the atmospheric neutrino fluxes. New "second generation" calculations are under development and will soon become available. The main lessons of these critical investigations has been the confirmation that the evidence for new physics in the atmospheric ν data is robust. For the two–flavor oscillation scenario that is the simplest and probably most successful explanation of the data, no large theoretical bias in the parameter dermination has been found.

1. Super-Kamiokande coll. Phys.Rev.Lett. 81, 1562 (1998). M. Messier, these Proceedings.
2. Kamiokande Coll. Phys. Lett. B 205, 416 (1988); Phys. Lett. B 280, 146 (1992); Phys. Lett. B 335, 237 (1994). IMB Coll., Phys.Rev.Lett. 66, 2561 (1991); Phys.Rev. D 46, 3720 (1992). Soudan 2 Coll. Phys.Lett. B 449, 137 (1999). MACRO Coll. Phys.Lett. B 434, 451 (1998).
3. The AMS Coll., Phys.Lett. B 472, 215 (1999).
4. Bess Coll. Ap.J. 545, 1135 (2000). [astro-ph/0002481]
5. G. Barbiellini, *et al.*, Proc. 25th ICRC (Durban) vol. 3 (1997) 369.
6. JACEE Collaboration, Ap.J. 502, 278 (1998).
7. Paolo Lipari, Astrop.Phys. to appear (2001). [astro-ph/0101559]
8. P.Zuccon et al., astro-ph/0111111.
9. SK Collab. Phys.Rev.Lett. 82, 5194 (1999). [hep-ex/9901139].
10. Paolo Lipari, Astropart.Phys. 14,153 (2000), [hep-ph/0002282]; Astropart.Phys. 14, 171 (2000), [hep-ph/0003013].
11. M.P. De Pascale *et al.*, J.Geophys. Res. 98, 3501 (1993).
12. J. Kremer *et al.*, Phys. Rev. Lett. **83**, 4241 (1999).
13. V. Barger, J.G. Learned, P. Lipari, M. Lusignoli, S. Pakvasa, & T.J. Weiler, Phys.Lett. B 462, 109 (1999). [hep-ph/9907421]
14. E. Lisi, A. Marrone and D. Montanino, Phys.Rev.Lett. 85, 1166 (2000).
15. G. Barr, T.K. Gaisser and T. Stanev, Phys. Rev. D 39, 3532(1989); V. Agrawal et al, Phys.Rev. D 53, 1314 (1996). P. Lipari, T.K. Gaisser and T. Stanev, Phys.Rev. D 58, 073003 (1998).
16. M. Honda, T. Kajita, K. Kasahara & S. Midorikawa, Phys. Rev. D 52 (1995) 4985.
17. M. Honda, T. Kajita, K. Kasahara and S. Midorikawa, Phys. Rev. D **64**, 053011 (2001) [hep-ph/0103328].
18. G. Battistoni et al., Astrop. Phys. 12, 315 (2000), (hep-ph/9907408). See URL: www.mi.infn.it/~battist/neutrino.html
19. T. K. Gaisser and M. Honda, hep-ph/0203272.

OSCILLATION-ENHANCED SEARCH FOR NEW INTERACTION WITH NEUTRINOS

JOE SATO

Research Center for Higher Education, Kyushu University, Ropponmatsu, Chuo-ku, Fukuoka, 810-8560, Japan
E-mail: joe@rc.kyushu-u.ac.jp

We study how well we can observe the effect of new physics in neutrino oscillation experiments.

1 Introduction

Precision measurement on neutrino mass difference and mixing angles for lepton will be done in neutrino oscillation experiments of next generation (see, for example, [1]). According to those study we will know the mass square difference and the mixing angle relevant to the atomospheric neutrino anomaly with 3% and 1% accuracy, we can measure U_{e3} down to $O(10^{-2})$, and also we can observe the CP violation effect in near future.

Till today the main concern about the next generation experiments has been on these parameters, mixing angles and the mass square difference. Therefore there arise a question whether we can mesure only these parameters in those experiments. There are several suggestions that we may see the effect of flavor-violating new physics[2].

In this talk we consider the potential power of the next generation experiments to observe the effect of new interactions[3].

2 Measurement of Neutrino Oscillation

2.1 *Idea*

To see how the effect of new physics enter the oscillation phenomenon, first we remind ourselves what we really observe in oscillation experiments. Though we discuss only a neutrino factory to make the argument concrete, same discussion is followed in oscillation experiments with a conventional beam.

All we know in a neutrino factory is that the muons, say, with negative charge decay at an accumulate ring and wrong sign muons are observed in a detector located at a length L away just after the time L/c, where c is the light speed. This is depicted schematically in Fig.1.

Since we know that there is the weak interaction process, we interpret

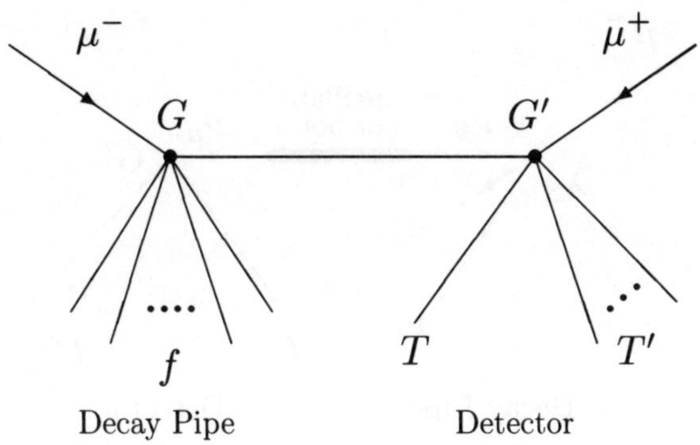

Figure 1. What we really see in a neutrino factory.

such a wrong sign event as the evidence of the neutrino oscillation, $\bar{\nu}_e \to \bar{\nu}_\mu$, which is graphically represented in Fig.2.

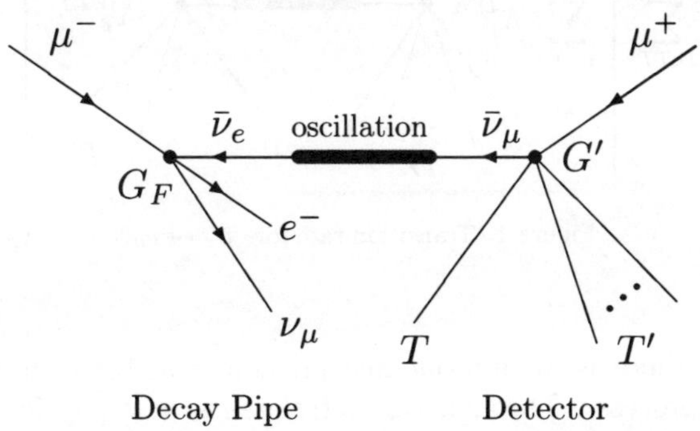

Figure 2. Standard interpretation of a wrong sign event.

Now if there is a flavor-changing exotic interaction, *e.g.*,

$$\epsilon(\bar{e}\gamma_\mu\mu)(\bar{\nu_\mu}\gamma^\mu\nu_\alpha), \quad \alpha \neq e, \tag{1}$$

then we will have the same signal of a wrong sign muon, whose diagram is shown in Fig.3, just like that caused by the weak interaction and the neutrino oscillation. We cannot distinguish these two kinds of contribution. The quantum mechanics tells us that in this case, to get a transition rate, we first sum up these amplitudes and then square the summation. In general, we have to calculate the graph fig.4 to get the transition rate.

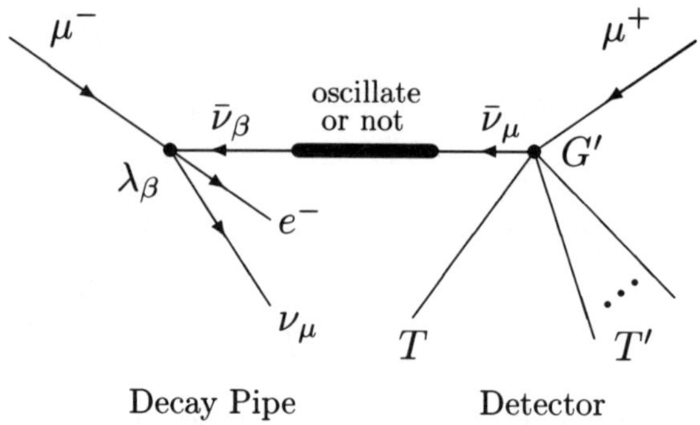

Figure 3. Diagram which gives same signal as that given by Fig.2.

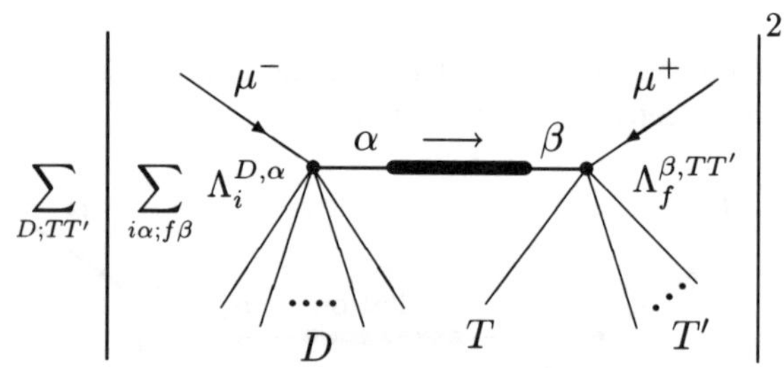

Figure 4. Transition rate for "$\bar{\nu}_e \rightarrow \bar{\nu}_\mu$".

Therefore there is an interference phenomenon between several amplitudes in this process. Through this interference new physics effect gives a contribution to the transition rate of order of not $O(\epsilon^2)$ but $O(\epsilon)$. We get an enhancement of the effect of new physics, that is, we can make oscillation-enhanced search for new interactions with neutrinos.

2.2 Parameterization

Here we consider how to parameterize new physics effects.

First we note that the amplitude for "neutrino oscillation" can be divided into three pieces: (1) Amplitude relevant with decay of a parent particle denoted as A^C_α, here C describes the type of interactions. For μ decay, as we will see in eq.(3) and (4), there are two types of interactions, $C = L, R$ while for π decay we do not need this label. α distinguishes the particle species which easily propagates in the matter and make an interaction at a detector. (2) Amplitude representing a transition of these propagating particles, which are usually neutrinos, from one species α to another/same β , denoted as $T_{\alpha\beta}$. (3) Amplitude responsible for producing a charged lepton l from a propagated particle β at a detector, stood for by $D^I_{\beta l}$. Here I denotes an interaction type. Using these notations we get the probability to observe a charged lepton $l^\pm$ at a detector as

$$P_{\mu^-\to l^+(l^-)} = \left| \sum_{\alpha\beta CI} A^C_\alpha T_{\alpha\beta} D^I_{\beta l^\pm} \right|^2 . \tag{2}$$

Therefore we can consider the effect of new physics separately for decay, propagation and detection processes.

First we consider the decay process of parent particles. For a neutrino factory, the exotic decays of muons which are $\mu^- \to e^-\nu_\alpha\bar{\nu}_e$ and $\mu^- \to e^-\nu_\mu\bar{\nu}_\beta$ can be amplified by the interference. This fact and Lorentz invariance allow only two kinds of new interactions in this process. For a wrong-sign mode, the allowed two interactions are the $(V-A)(V-A)$ type,

$$2\sqrt{2}\lambda_\alpha(\bar{\nu_\mu}\gamma^\rho P_L\mu)(\bar{e}\gamma_\rho P_L\nu_\alpha), \quad \alpha = \mu, \tau, \tag{3}$$

which has the same chiral property as the weak interaction but violates the flavor conservation, and the $(V-A)(V+A)$ type,

$$2\sqrt{2}\lambda'_\alpha(\bar{\nu_\mu}\gamma^\rho\nu_\alpha)(\bar{e}\gamma_\rho P_R\mu), \quad \alpha = e, \mu, \tau. \tag{4}$$

The latter has different chiral property from the former, so that it gives different energy dependence to the transition rate. However in the latter case there is a very strong chiral suppression for the interference and hence eventually there is no effect on the oscillation phenomena.

In the case of the $(V-A)(V-A)$ type exotic interaction, we can introduce the interference effect by treating the initial state of oscillating neutrino as the superposition of all flavor eigenstates. On the $\mu^- \to \mu^+$ process, we can take initial neutrino $\bar{\nu}$ as

$$\bar{\nu} = \bar{\nu}_e + \epsilon_\mu\bar{\nu}_\mu + \epsilon_\tau\bar{\nu}_\tau, \tag{5}$$

where $\epsilon_\alpha = \lambda_\alpha/G_F$. This simple treatment is allowed only for the $(V-A)(V-A)$ type interaction because of the same interaction form as the weak interaction except for difference of the coupling constant and the flavor of antineutrino. In this case we can generalize the initial neutrino for any flavor, using Y. Grossman's source state notation [4], as, [a]

$$\nu_\beta^s = U_{\beta\alpha}^s \nu_\alpha, \quad \alpha, \beta = e, \mu, \tau,$$

$$U^s \equiv \begin{pmatrix} 1 & \epsilon_{e\mu}^s & \epsilon_{e\tau}^s \\ \epsilon_{\mu e}^s & 1 & \epsilon_{\mu\tau}^s \\ \epsilon_{\tau e}^s & \epsilon_{\tau\mu}^s & 1 \end{pmatrix}. \tag{6}$$

We can include the total exotic effect into the oscillation probability as

$$P_{\nu_\alpha^s \to \nu_\beta} = \left| \langle \nu_\beta | e^{-iHL} U_{\alpha\gamma}^s | \nu_\gamma \rangle \right|^2. \tag{7}$$

This treatment is also valid for the effect on the ν_μ oscillation.

For π decay the situation is much simpler. In the presence of new physics there may be a flavor violating decay of π such as $\pi^- \to \mu^- \nu_\alpha (\alpha = e, \tau)$. This effect changes the initial ν state;

$$\nu_\mu \longrightarrow \nu_\mu^s = \epsilon_{\mu e}^s \nu_e + \nu_\mu + \epsilon_{\mu\tau}^s \nu_\tau. \tag{8}$$

In this case we do not have to worry about the type of new physics which gives a flavor changing π decay at a low energy scale. Due to kinematics, the energy and the helicity of the decaying particles, μ and ν are fixed.

Next we consider the propagation process. Exotic interactions also modify the Hamiltonian for neutrino propagation as [5],

$$H_{\beta\alpha} = \frac{1}{2E_\nu} \left\{ U_{\beta i} \begin{pmatrix} 0 & & \\ & \delta m_{21}^2 & \\ & & \delta m_{31}^2 \end{pmatrix} U_{i\alpha}^\dagger + \begin{pmatrix} \bar{a} + a_{ee} & a_{e\mu} & a_{e\tau} \\ a_{e\mu}{}^* & a_{\mu\mu} & a_{\mu\tau} \\ a_{e\tau}{}^* & a_{\mu\tau}{}^* & a_{\tau\tau} \end{pmatrix}_{\beta\alpha} \right\}, \tag{9}$$

where $\bar{a}$ is the ordinary matter effect given by $2\sqrt{2} G_F n_e E_\nu$, $a_{\alpha\beta}$ is the extra matter effect due to new physics interactions, that is defined by $a_{\alpha\beta} = 2\sqrt{2}\epsilon_{\alpha\beta}^m G_F n_e E_\nu$. Note that to consider the magnitude of the matter effect, the type of the interaction is irrelevant since in matter particles are at rest and hence the dependence on the chirality is averaged out.[6]

Finally we make a comment about new physics which affect a detection process. To consider this process we need the similar treatment to that at the decay process, that is, we have to separate contribution of new interactions following the difference of the chirality dependence. However to take into account new physics at a detector, the parton distribution and a knowledge about hadronization are necessary. Though we may wonder whether we can parameterize the effect of new physics at the detector g/G_T as ϵ^d like ϵ^s. It is expected that ϵ^d has a complicated energy dependence due to the parton distribution for example in a energy region of a neutrino factory. Consider

---

[a] U^s is not necessarily unitary.

the case that there is an elementary process from lepton flavor violating new physics including strange quark. To parameterize its effect we need both its magnitude and the distribution function of strange quark in nucleon which will show the dependence on the neutrino energy (more exactly, the transfered momentum from neutrino to strange quark). They are beyond our ability and hence we do not consider them further in this paper, though new physics which can affect the decay process have contribution to the detection process too.

3 Analysis

Let us discuss the feasibility to observe the signal induced by new physics. We will deal with a neutrino factory. For details see ref[3]. For the numerical calculation we use the parameters,

$$\sin\theta_{12} = \frac{1}{2}, \quad \sin\theta_{23} = \frac{1}{\sqrt{2}}, \quad \sin\theta_{13} = 0.1,$$
$$\delta m^2_{21} = 5\times 10^{-5}, \quad \delta m^2_{31} = 3\times 10^{-3}, \delta = \frac{\pi}{2}, \tag{10}$$

and take $|\epsilon| = 3\times 10^{-3}$, which is a reference value for the feasibility to observe the effect by using the method of the oscillation enhancement. Except for $\epsilon^{s,m}_{e\mu}$ and $\epsilon^{s}_{\mu e}$, the constraints of the processes of charged lepton have not forbidden this magnitude of ϵ's.

Here, we consider the case that there are only $(V-A)(V-A)$ type new interactions in the lepton sector. In this case the "oscillation probability" is given by eq.(7) in this situation.

The analytic expression of the probability for $\nu_e \to \nu_\mu$ given in the Appendix A of ref.[3]. It shows that the effect due to $\epsilon^m_{\mu\tau}$ and $\epsilon^{s,m}_{\alpha\alpha}$ are irrelevant since these terms are proportional to $\sin^2 2\theta_{13}\times\epsilon$ in the high energy region, so it is difficult to observe their effects. Therefore here as an example we consider the contribution of $\epsilon^{s,m}_{e\mu}$[b]. It can be represented analytically in the

---

[b]The flavor changing processes between muon and electron, *e.g.*, $\mu\to e\gamma$, $\mu\leftrightarrow e$ conversion, are strictly constrained from experiments, and these constraints are related to $\epsilon^{\ s}_{e\mu}$[7]. Therefore, the magnitude of $\epsilon^{\ s}_{e\mu}$ has very severe bound. However here we ignore these constraints.

high energy region as

$$\Delta P_{\nu_e^s\to\nu_\mu}\{\epsilon_{e\mu}\} = 2s_{23}s_{2\times 13} \times$$

$$\left[\left(s_\delta \mathrm{Re}[\epsilon^s_{e\mu}] - c_\delta \mathrm{Im}[\epsilon^s_{e\mu}]\right) \times \left\{1 - \frac{2}{3}\left(\frac{\bar{a}}{4E_\nu}L\right)^2 + \frac{2}{3}\left(2c_{2\times 13} - 3c^2_{23}c^2_{13}\right)\right.\right.$$

$$\left.\times\left(\frac{\bar{a}}{4E_\nu}L\right)\left(\frac{\delta m^2_{31}}{4E_\nu}L\right)\right\}\left(\frac{\delta m^2_{31}}{4E_\nu}L\right) \tag{11a}$$

$$-\left(c_\delta \mathrm{Re}[\epsilon^s_{e\mu}] + s_\delta \mathrm{Im}[\epsilon^s_{e\mu}]\right) \times \left[\left\{1 - \frac{1}{3}\left(\frac{\bar{a}}{4E_\nu}L\right)^2\right\}\left(\frac{\bar{a}}{4E_\nu}L\right)\right.$$

$$\left.-\left\{1 - 2s^2_{23}c^2_{13} - \left(1 - c^2_{13}\left(2 - \frac{4}{3}c^2_{23}\right)\right) \times \left(\frac{\bar{a}}{4E_\nu}L\right)^2\right\}\left(\frac{\delta m^2_{31}}{4E_\nu}L\right)\right]\left(\frac{\delta m^2_{31}}{4E_\nu}L\right) \tag{11b}$$

$$+\, 2c^2_{23}\left(s_\delta \mathrm{Re}[\epsilon^m_{e\mu}] + c_\delta \mathrm{Im}[\epsilon^m_{e\mu}]\right)\left(\frac{\bar{a}}{4E_\nu}L\right)\left(\frac{\delta m^2_{31}}{4E_\nu}L\right)^2 \tag{11c}$$

$$+\, 2\left(c_\delta \mathrm{Re}[\epsilon^m_{e\mu}] - s_\delta \mathrm{Im}[\epsilon^m_{e\mu}]\right) \times \left\{1 - \frac{1}{3}\left(\frac{\bar{a}}{4E_\nu}L\right)^2\right.$$

$$\left.\left.+\left(c^2_{23}s^2_{13} + \frac{2}{3}s^2_{23}c_{2\times 13}\right)\left(\frac{\bar{a}}{4E_\nu}L\right)\left(\frac{\delta m^2_{31}}{4E_\nu}L\right)\right\}\left(\frac{\bar{a}}{4E_\nu}L\right)\left(\frac{\delta m^2_{31}}{4E_\nu}L\right)\right]. \tag{11d}$$

where $s_{2\times ij} \equiv \sin 2\theta_{ij}$. Since we can suppose that E_ν is proportional to E_μ in a neutrino factory, E_μ and L dependence of the sensitivity to each term can be approximated as

$$\chi^2(11a) \propto \left\{1 - \frac{2}{3}\left(\frac{\bar{a}}{4E_\mu}L\right)^2\right\}^2 \times E^2_\mu/L,$$

$$\chi^2(11b, 11d) \propto \left\{1 - \frac{1}{3}\left(\frac{\bar{a}}{4E_\mu}L\right)^2\right\}^2 \times E^2_\mu,$$

$$\chi^2(11c) \propto E_\mu, \tag{12}$$

In Fig.5 we see that the sensitivity is indeed understood by the high energy behavior of the transition rate.

In the realisitic situation, we probably do not exactly know the value of the theoretical parameters. Once the uncertainties for those values are introduced, the sensitivities shown in Figs.5 may be spoiled completely. The $\epsilon_{e\mu}^{s,m}$ effect can be absorbed easily into the main (unperturbed) part of the oscillation by adjusting the theoretical parameters since the effects have the same energy dependence as the main part has.

Indeed, taking into account these uncertainties, some of the sensitivities to $\epsilon_{e\mu}^{s,m}$ are completely washed out. Therefore, we have to look for the terms whose energy dependence differ from that of the main oscillation term in the high energy region. Contribution for the transition probability labeled (11a), (11b) and (11d) depends on $1/E_\mu$. Consequently, the sensitivities to the terms must be robust against the uncertainties of the theoretical parameters. Since in the high energy region there is no $1/E_\mu$ dependence of the oscillation probability they can be distinguished from the main oscillation part by observing the energy dependence. The claim mentioned above are confirmed numerically by Fig.5 and 6. By comparison of these graphs, we can see that the sensitivities to observe the contribution of (11a), (11b) and (11d) do not suffer from the uncertainties. Incidentally, we note that though the uncertainties wreck the sensitivity to (11c) since it is proportional to $1/E_\mu^2$, the $\epsilon_{e\mu}^m$ second order term brings constant contribution for energy and this signal does not vanish.

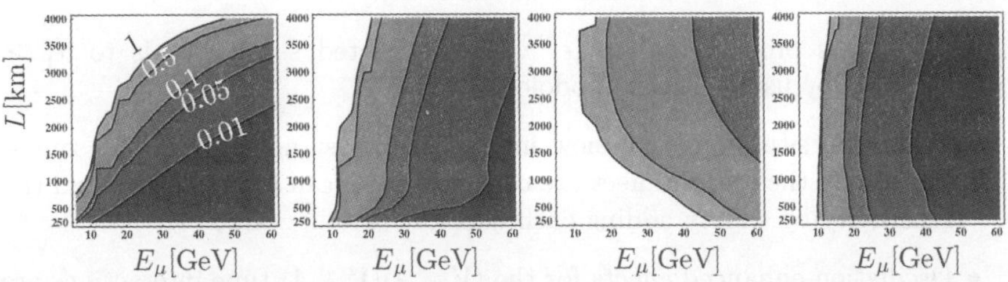

Figure 5. Contour plots of the required $N_\mu M_{\rm det}$ to observe the new physics effects concerning $\epsilon_{e\mu}^{s,m}$ at 90% C.L. in $\nu_e \to \nu_\mu$ channel when there is no uncertainty for theoretical parameters. From left to right: $(\epsilon_{e\mu}^s, \epsilon_{e\mu}^m) = (3.0\times10^{-3}, 0)$, $(3.0\times10^{-3}i, 0)$, $(0, 3.0\times10^{-3})$, $(0, 3.0\times10^{-3}i)$. Each plot corresponds to the sensitivities to eq.(11a)~eq.(11d) respectively.

4 Summary

In this talk we study how well we can reach the flavor changing interaction in neutrino oscillation experiments. In general, we can summarize it as follows.

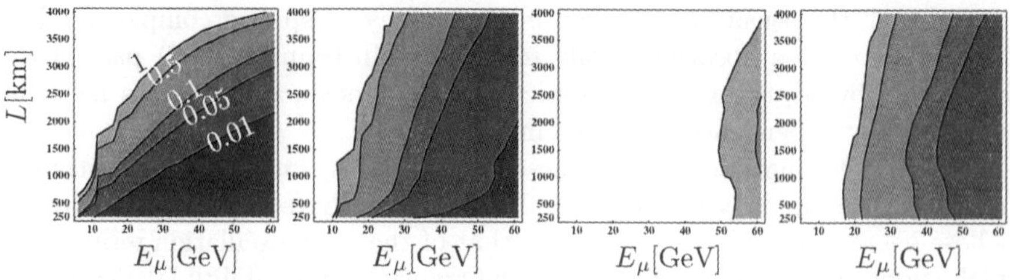

Figure 6. Same as Fig.5, but here each parameter has 10% uncertainty.

For a neutrino factory:

- In $\nu_\alpha \to \nu_\beta$, ($\alpha = e, \mu$, $\beta = e, \mu, \tau$) appearance channel, the observable effects of new $(V-A)(V-A)$ interactions come only from $\epsilon^{s,m}_{\alpha\beta}$. The others are too small or too vulnerable against the adjustment of the theoretical parameters. Note that δ and ϵ's phase are correlated. Namely the measured values are a certain combination of δ and ϵ.
- In $\nu_\mu \to \nu_\mu$ disappearance channel, we can measure $\epsilon^{s,m}_{\mu\tau}$ depending on their phase. In other words, the signal includes information of the phase. Furthermore, there is no correlation between δ and ϵ, so the measurement tells us directly the phase concerning the lepton-flavor violating process. In $\nu_e \to \nu_e$ disappearance channel, we can not get anything for new interactions in the oscillation enhanced way.
- The χ^2 is proportional to $|\epsilon|^2$. The expected sensitivity is to $|\epsilon| \geq \mathcal{O}(10^{-4})$ by using this methodology.
- When the situations that new interactions exist not only in the source but also in the matter effect are considered, we can easily understand the sensitivity by simply adding each effect.
- Oscillation-enhanced effects for the $(V-A)(V+A)$ type interactions are strongly suppressed by m_e/m_μ, so we can not get an advantage over a direct measurement.

For an upgraded conventional beam:

- We do not have to care the types of new interactions in the source. The analyses for the feasibility are similar to that of $(V-A)(V-A)$ type for a neutrino factory. In the assumed energy and baseline region, there is no sensitivity to the new effect in matter.

- The ϵ's for a conventional beam have different dependence from those for a neutrino factory on new interactions. Therefore, the comparison between two methods makes clear the species of new physics.

References

1. JHF-Kamioka project, Y. Itow *et al.*, hep-ex/0106019. Neutrino Factory and Muon Collider Collab., D. Ayres *et al.*, physics/9911009; C. Albright *et al.*, hep-ex/0008064;
2. M.C. Gonzalez-Garcia, Y. Grossman, A. Gusso, and Y. Nir, Phys. Rev. **D64** (2001) 096006 [hep-ph/0105159]; A. M. Gago, M. M. Guzzo, H. Nunokawa, W. J. C. Teves, and R. Zukanovich Funchal Phys. Rev. **D64** (2001) 073003 [hep-ph/0105196]; J. W. Valle, and P. Huber, hep-ph/0108193.
3. T. Ota, J. Sato, and N. Yamashita hep-ph/0112329.
4. Y. Grossman, Phys. Lett. **B359** (1995) 141 [hep-ph/9507344].
5. see A. M. Gago *et al.* in ref[2].
6. S. Bergmann, Y. Grossman, and E. Nardi, Phys. Rev. **D60** (1999) 093008 [hep-ph/9903517].
7. S. Bergmann, and Y. Grossman, Phys. Rev. **D59** (1999) 093005 [hep-ph/9809524];

Solar and Atmospheric Three- and Four-Neutrino Oscillations

M. C. Gonzalez-Garcia

Theory Division, CERN, CH-1211 Geneva 23, Switzerland,
Instituto de Física Corpuscular, Universitat de València - C.S.I.C.
Edificio Institutos de Paterna, Apt 22085, 46071 València, Spain,
and C.N. Yang Institute for Theoretical Physics
State University of New York at Stony Brook Stony Brook,NY 11794-3840, USA
email:concepcion.gonzalez-garcia@cern.ch

1 Introduction

There are three pieces of evidence for neutrino masses and mixing: solar neutrinos, atmospheric neutrinos and the LSND results[1] which can be individually interpreted in the framework of two-neutrino oscillations. The results of these partial analysis are summarized in Fig. 1 where we show the ranges of masses and mixing implied by these signals at 90 and 99% CL for 2 d.o.f., as well as relevant constraints from negative searches in laboratory experiments. The three pieces of evidence correspond to three values of mass-squared differences of different orders of magnitude. Consequently, there is no consistent explanation to all three signals based on oscillations among the three known neutrinos. As a consequence, if CPT is conserved, one must either assume that one of the experimental results is not due to oscillations (for instance LSND) and live with the three standard neutrinos, or enlarge the particle content of the standard model by introducing a fourth sterile neutrino.

In this talk we present the update of our *combined* analysis of the neutrino oscillation solutions of both the solar and the atmospheric neutrino problem, in the framework of three [2] and four-neutrino mixing [3]. We include in our analysis the most recent solar neutrino rates of Homestake , SAGE, GALLEX and GNO as well as the 1258-day Super-Kamiokande data sample, including the recoil electron energy spectrum for both day and night periods and the recent results from the CC even rates at SNO [4]. As for atmospheric neutrinos we include in our analysis all the contained events from the latest 79.5 kton-yr Super-Kamiokande data set, as well as the upward-going neutrino-induced muon fluxes from both Super-Kamiokande and the MACRO detector [5]. The constraints arising from the relevant reactor (CHOOZ and Bugey) and accelerator (CDHSW) experiments [6] are also imposed.

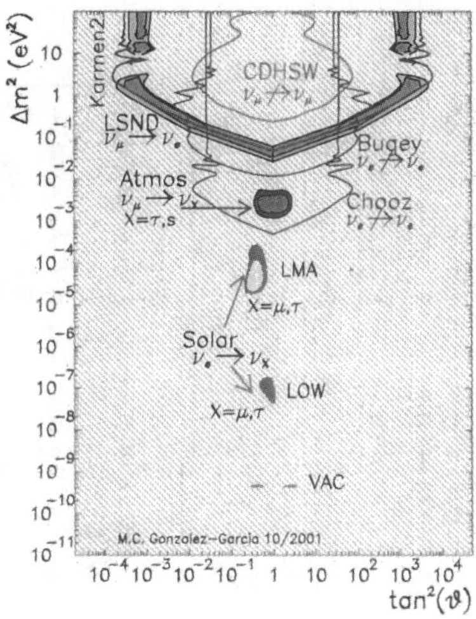

Figure 1: Summary of the present pieces of evidence for neutrino masses and mixing as obtained from 2-ν oscillation analyses. The allowed regions correspond to 90% and 99% CL for 2 d.o.f. We also show relevant bounds from laboratory experiments.

2 Three-Neutrino Mixing

Here we present a brief summary and update of the analysis performed in Ref. [2] and we refer to that publication for further details and references.

The combined description of both solar and atmospheric anomalies requires that all three known neutrinos take part in the oscillations. The mixing parameters are encoded in the 3×3 lepton mixing matrix U relating the flavour and the mass basis. In general U contains 3 mixing angles and 1 or 3 CP violating phases depending on whether the neutrinos are Dirac or Majorana. We will neglect the CP violating phases as they are not accessible by the existing experiments. In this case the mixing matrix can be conveniently chosen in the form $U = U_{23}U_{13}U_{12}$, where U_{ij} is a rotation matrix in the plane ij. With this the parameter set relevant for the joint study of solar and atmospheric conversions becomes five-dimensional:

$$\begin{array}{cc} \Delta m^2_\odot \equiv \Delta m^2_{21} \;, & \Delta m^2_{atm} \equiv \Delta m^2_{32}, \\ \theta_\odot \equiv \theta_{12} \;, & \theta_{atm} \equiv \theta_{23} \;,\; \theta_{reac} \equiv \theta_{13} \;, \end{array} \tag{1}$$

where all mixing angles are assumed to lie in the full range from $[0, \pi/2]$.

In general the transition probabilities will present an oscillatory behaviour

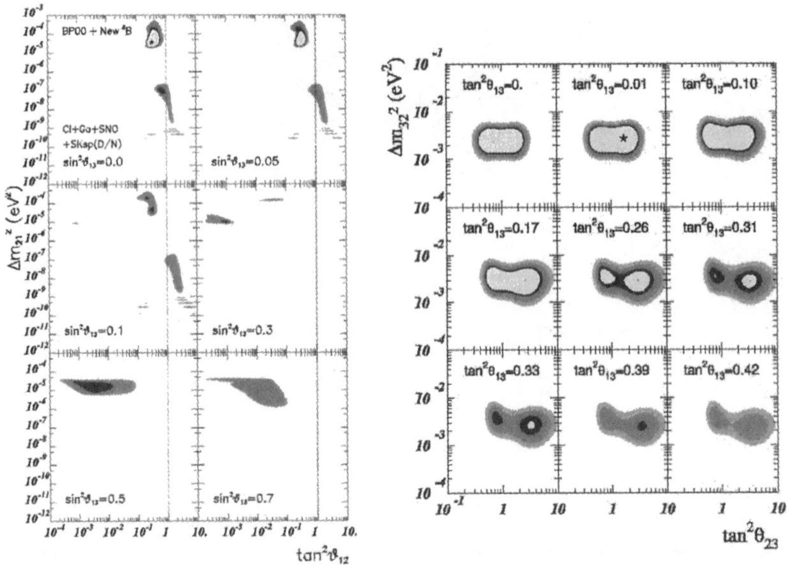

Figure 2: (Left) Allowed regions (at 90, 95, and 99% CL) in the $(\Delta m^2_{21}, \tan^2\theta_{12})$ plane from the global analysis of the solar neutrino data in the framework of three-neutrino oscillations for various values of $\sin^2\theta_{13}$. The global best fit point is marked by the star. (Right) Allowed regions (at 90, 95, 99, and 99.7% CL) in the $(\Delta m^2_{32}, \tan^2\theta_{23})$ plane from the global analysis of the atmospheric neutrino data in the framework of three-neutrino oscillations for various values of $\tan^2\theta_{13}$. The global best fit point is marked by the star.

with two oscillation lengths. However from the required hierarchy in the splittings $\Delta m^2_{atm} \gg \Delta m^2_{\odot}$ indicated by the solutions to the solar and atmospheric neutrino anomalies it follows that:

- For solar neutrinos the oscillations with the atmospheric oscillation length are averaged out and the survival probability takes the form: $P^{3\nu}_{ee,MSW} = \sin^4\theta_{13} + \cos^4\theta_{13}P^{2\nu}_{ee,MSW}$ where $P^{2\nu}_{ee,MSW}$ is obtained with the modified sun density $N_e \to \cos^2\theta_{13}N_e$. So the analyses of solar data constrain three of the five independent oscillation parameters: $\Delta m^2_{21}, \theta_{12}$ and θ_{13}.
- Conversely for atmospheric neutrinos, the solar wavelength is too long and the corresponding oscillating phase is negligible. As a consequence the atmospheric data analysis restricts Δm^2_{32}, θ_{23} and θ_{13}, the latter being the only parameter common to both solar and atmospheric neutrino oscillations and which may potentially allow for some mutual influence.

Therefore solar and atmospheric neutrino oscillations decouple in the limit

$\theta_{13} = 0$. In this case the values of allowed parameters can be obtained directly from the results of the analysis in terms of two–neutrino oscillations presented in Fig. 1. Deviations from the two–neutrino scenario are determined by the size of the mixing θ_{13}. This angle is constrained by the CHOOZ reactor experiment. To analyze the CHOOZ constraints we need to evaluate the survival probability for $\bar{\nu}_e$ of average energy $E \sim$ few MeV at a distance of $L \sim 1$ Km. For these values of energy and distance, one can safely neglect Earth matter effects. The survival probability takes the analytical form:

$$P_{ee}^{\text{CHOOZ}} = 1 - c_{13}^4 \sin^2 2\theta_{12} \sin^2\left(\frac{\Delta m_{21}^2 L}{4E_\nu}\right) - \sin^2 2\theta_{13}\left[c_{12}^2 \sin^2\left(\frac{\Delta m_{31}^2 L}{4E_\nu}\right)\right.$$
$$\left. + s_{12}^2 \sin^2\left(\frac{\Delta m_{32}^2 L}{4E_\nu}\right)\right] \simeq 1 - \sin^2 2\theta_{13} \sin^2\left(\frac{\Delta m_{32}^2 L}{4E_\nu}\right).$$

The second equality holds in the approximation $\Delta m_{21}^2 \ll E_\nu/L$ which can only be safely made for $\Delta m_{21}^2 \leq 3 \times 10^{-4}\text{eV}^2$. Thus in general the analysis of the CHOOZ data involves four oscillation parameters: Δm_{32}^2, θ_{13}, Δm_{21}^2, and θ_{12}.

The first question to answer is how the presence of θ_{13} affects the analysis of the solar and atmospheric neutrino data. In Fig. 2(left) we show the allowed regions for the oscillation parameters Δm_{21}^2 and $\tan^2\theta_{12}$ from the global analysis of the solar neutrino data in the framework of three-neutrino oscillations for different values of $\sin^2\theta_{13}$. The allowed regions for a given CL are defined as the set of points satisfying the condition $\chi^2(\Delta m_{12}^2, \tan^2\theta_{12}, \tan^2\theta_{13}) - \chi^2_{\text{min}} \leq \Delta\chi^2(\text{CL, 3 d.o.f})$ where, for instance, $\Delta\chi^2(\text{CL, 3 d.o.f})$=6.25, 7.83, and 11.36 for CL=90, 95, and 99%, respectively. The global minimum used in the construction of the regions lies in the LMA region and corresponds to $\tan^2\theta_{13} = 0$, that is, to the *decoupled* scenario. As seen in the figure, the modifications to the decoupled case are significant only for large θ_{13}. As $\sin^2\theta_{13}$ increases, all the allowed regions disappear, leading to an upper bound on $\sin^2\theta_{13}$ independent of the values taken by the other parameters in the three-neutrino mixing matrix. For instance, no region of parameter space is allowed (at 99% C.L. for 3 d.o.f) for $\sin^2\theta_{13} = |U_{e3}|^2 > 0.80$. This fact is also illustrated in Fig. 3 where we plot the shift in χ^2 as a function of $\sin^2\theta_{13}$ when the mass and mixing parameters Δm_{12}^2 and $\tan^2\theta_{12}$ are chosen to minimize χ^2.

In Fig. 2(right) we show the allowed regions for the oscillation parameters Δm_{32}^2 and $\tan^2\theta_{23}$ from the analysis of the atmospheric neutrino data in the framework of three-neutrino oscillations for different values of $\sin^2\theta_{13}$. In the upper-left panel $\tan^2\theta_{13} = 0$ which corresponds to pure $\nu_\mu \to \nu_\tau$ oscillations. Note the exact symmetry of the contour regions under the transformation $\theta_{23} \to \pi/4 - \theta_{23}$. This symmetry follows from the fact that in the pure $\nu_\mu \to \nu_\tau$ channel matter effects cancel out and the oscillation probability depends

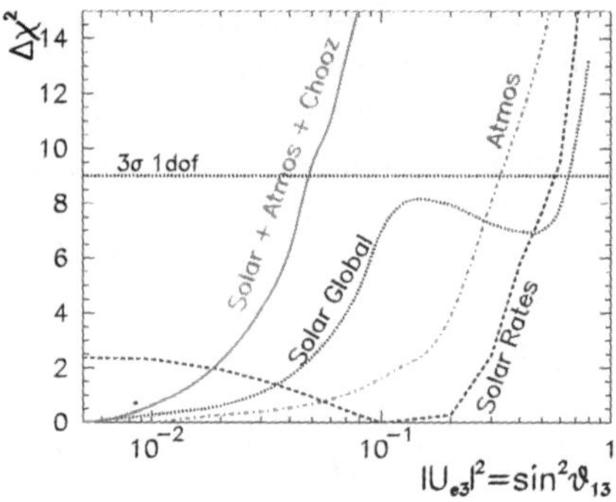

Figure 3: Dependence of $\Delta\chi^2$ on $\sin^2\theta_{13}$ in the analysis of the atmospheric, solar and CHOOZ neutrino data. The dotted horizontal line corresponds to the 3σ limit for a single parameter.

on θ_{23} only through the double-valued function $\sin^2 2\theta_{23}$. For $\theta_{13} \neq 0$ this symmetry breaks due to the three-neutrino mixing structure even if matter effects are neglected. The analysis of the full atmospheric neutrino data in the framework of three-neutrino oscillations clearly favours the $\nu_\mu \to \nu_\tau$ oscillation hypothesis. As a matter of fact, the best fit corresponds to a very small value, $\theta_{13} = 6°$, but it is statistically indistinguishable from the decoupled scenario, $\theta_{13} = 0°$. No region of parameter space is allowed (at 99% C.L. for 3 d.o.f) for $\sin^2\theta_{13} = |U_{e3}|^2 > 0.40$. The physics reason for this limit is clear: large values of θ_{13} imply a too large contribution of the $\nu_\mu \to \nu_e$ channel and would spoil the otherwise successful description of the angular distribution of contained events. This situation is illustrated also in Fig. 3 where we plot the shift of χ^2 for the analysis of atmospheric data in the framework of oscillations between three neutrinos as a function of $\sin^2\theta_{13}$ when the mass and mixing parameters Δm^2_{32} and $\tan^2\theta_{23}$ are chosen to minimize χ^2.

For any value of the mixing parameters, the mass-squared difference relevant for the atmospheric analysis is restricted to lie within the range of sensitivity of the CHOOZ experiment. Indeed, as illustrated in Fig. 3, the limit on $\sin^2\theta_{13}$ becomes stronger when the CHOOZ data are combined with the atmospheric and solar neutrino results.

One can finally perform a global analysis in the five dimensional parameter space combining the full set of solar, atmospheric and reactor data. Such analysis leads to the following allowed 3σ ranges for individual parameters (the other parameters have been chosen to minimize the global χ^2):

$$\begin{aligned} 1.9\times10^{-5} < \Delta m^2_{21}/\text{eV}^2 < 2.7\times10^{-4} \quad & \text{LMA}\,, \\ 4.3\times10^{-8} < \Delta m^2_{21}/\text{eV}^2 < 1.8\times10^{-7} \quad & \text{LOW}\,, \\ 1.4\times10^{-3} < \Delta m^2_{32}/\text{eV}^2 < 6.0\times10^{-3}, & \end{aligned} \tag{2}$$

and the moduli of the mixing matrix U:

$$|U| = \begin{pmatrix} 0.67-0.91 & 0.41-0.73 & <0.22 \\ 0.21-0.70 & 0.20-0.77 & 0.51-0.87 \\ 0.05-0.63 & 0.36-0.83 & 0.49-0.85 \end{pmatrix}. \tag{3}$$

In conclusion, we learn that at present the large mixing-type solutions provide the best fit. As concerns $|U_{e3}|$, both solar and atmospheric data favour small values and this trend is strengthened by the reactor data.

3 Four-Neutrino Mixing

There are six possible four-neutrino schemes that can accommodate the results from solar and atmospheric neutrino experiments as well as the LSND evidence. They can be divided in two classes: (3+1) and (2+2) In the (3+1) schemes there is a group of three neutrino masses separated from an isolated one by a gap of the order of 1 eV^2, which is responsible for the short-baseline oscillations observed in the LSND experiment. In (2+2) schemes there are two pairs of close masses separated by the LSND gap. In this talk we concentrate on the (2+2) schemes and present a brief summary of the analysis performed in Ref.[3] and we refer to that publication for further details and references.

The main difference between (3+1) and (2+2) schemes is that, if a (2+2)-spectrum is realized in nature, the transition into the sterile neutrino is a solution of either the solar or the atmospheric neutrino problem, or the sterile neutrino has to take part in both, whereas with a (3+1)-spectrum it could be only slightly mixed with the active neutrinos and mainly provide a description of the LSND result. Oscillations into sterile neutrinos are disfavoured for both atmospheric and solar neutrinos[4,5]. One expects then that these schemes will be also disfavoured. However when considered in the framework of (2+2) schemes, oscillations into pure active or pure sterile states are only limiting cases of the most general possibility of oscillations into an admixture of active and sterile neutrinos. One can wonder then, if some finite admixture of active-sterile oscillations is able to give a suitable description of both solar and atmospheric data.

For the phenomenology of neutrino oscillations in (2+2) schemes the mass spectrum can be taken as: $\Delta m^2_\odot = \Delta m^2_{21} \ll \Delta m^2_{\text{atm}} = \Delta m^2_{43} \ll \Delta m^2_{\text{LSND}} =$

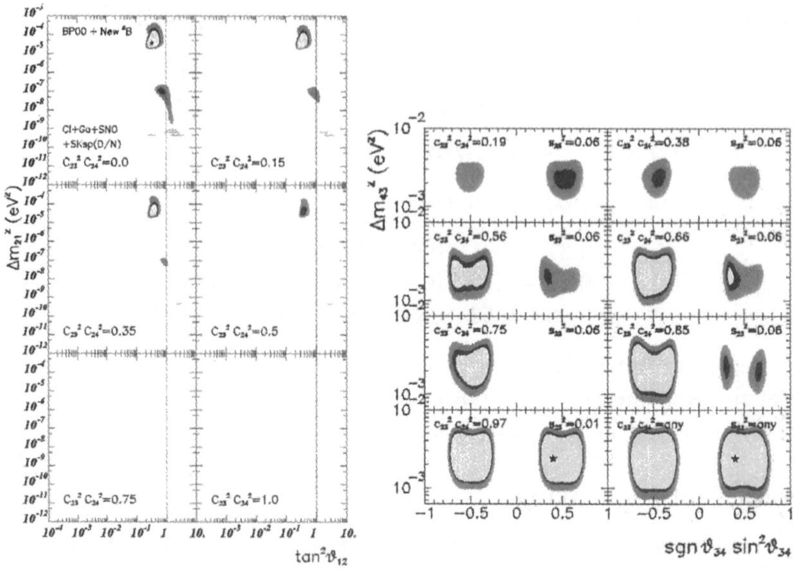

Figure 4: (Left) Allowed regions (at 90, 95, 99% CL) in the $(\Delta m^2_{21}, \tan^2\theta_{21})$ plane from the global analysis of the atmospheric neutrino data in the framework of four-neutrino oscillations for various values of $c^2_{23}c^2_{24}$. The global best fit point is marked by the star. (Right) Results of the analysis of atmospheric neutrino data for the allowed regions in Δm^2_{43} and θ_{34} for the 2+2 four-neutrino oscillations.

$\Delta m^2_{41} \simeq \Delta m^2_{42} \simeq \Delta m^2_{31} \simeq \Delta m^2_{32}$ and neglecting possible CP phases, the matrix U can be written as a product of six rotations, and can be conveniently written as $U = U_{24}\,U_{23}\,U_{14}\,U_{13}\,U_{34}\,U_{12}$, where U_{ij} represents a rotation of angle θ_{ij} in the ij plane. This general form can be further simplified by taking into account the negative results from the reactor experiments, in particular the Bugey experiment, which in the range of Δm^2_{41} relevant to the LSND experiment implies that $|U_{e3}|^2 + |U_{e4}|^2 = c^2_{14}s^2_{13} + s^2_{14} \leq 10^{-2}$ so that the two angles θ_{13} and θ_{14} give negligible contributions to solar and atmospheric neutrino transitions for which the U matrix takes the effective form $U = U_{24}\,U_{23}\,U_{34}\,U_{12}$. The full parameter space relevant to solar and atmospheric neutrino oscillation can be covered with three of the angles θ_{ij} in the first quadrant, $\theta_{ij} \in [0, \pi/2]$ while one (which we choose to be θ_{34}) is allow to vary in the range $-\frac{\pi}{2} \leq \theta_{34} \leq \frac{\pi}{2}$.

Thus solar neutrino oscillations occur with a mixing angle θ_{12} between the states $\nu_e \to \nu_\alpha$ with $\nu_\alpha = c_{23}c_{24}\;\nu_s + \sqrt{1 - c^2_{23}c^2_{24}}\;\nu_a$ where ν_a is a linear combination of ν_μ and ν_τ. We remind the reader that ν_μ and ν_τ cannot be

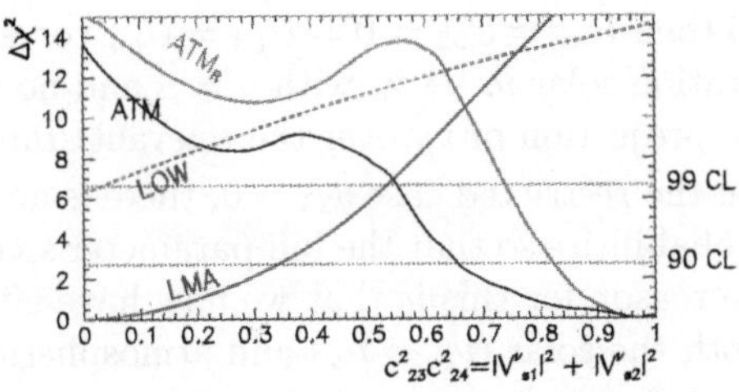

Figure 5: $\Delta\chi^2$ as a function of the active–sterile admixture $|U_{s1}|^2+|U_{s2}|^2=c^2_{23}c^2_{24}$ for the analysis of solar and atmospheric data in 2+2 four–neutrino schemes.

distinguished in solar neutrino experiments, because their matter potential and their interaction in the detectors are equal, due to only NC weak interactions. Thus solar neutrino oscillations cannot depend on the mixing angle θ_{34} and depend on θ_{23} and θ_{24} through the combination $c^2_{23}c^2_{24}$.

Atmospheric neutrino oscillations, *i.e.* oscillations with the mass difference Δm^2_{34} and mixing angle θ_{34}, occur between the states $\nu_\beta \to \nu_\gamma$ with $\nu_\beta = s_{23}c_{24}\nu_s + c_{23}\nu_\mu - s_{23}s_{24}\nu_\tau$ and $\nu_\gamma = s_{24}\nu_s + c_{24}\nu_\tau$.

Thus the mixing angles θ_{23} and θ_{24} determine two projections. First, the projection of the sterile neutrino onto the states in which the solar ν_e oscillates is given by

$$c^2_{23}c^2_{24} = 1-|U_{a1}|^2-|U_{a2}|^2 = |U_{s1}|^2+|U_{s2}|^2. \tag{4}$$

Second, the projection of the ν_μ over the solar neutrino oscillating states is given by

$$s^2_{23} = |U_{\mu1}|^2+|U_{\mu2}|^2 = 1-|U_{\mu3}|^2-|U_{\mu4}|^2 \tag{5}$$

One expects s_{23} to be small in order to explain the atmospheric neutrino deficit. We will see that this is indeed the case. Furthermore, the negative results from the CDHS and CCFR searches for ν_μ-disappearance also constrain such a projection to be smaller than 0.2 at 90% CL for $\Delta m^2_{\rm LSND} \geq 0.4$ eV^2.

We distinguish the following limiting cases:

(i) If $c_{23}=1$ then $U_{\mu1}=U_{\mu2}=0$. The atmospheric $\nu_\mu=\nu_\beta$ state oscillates into a state $\nu_\gamma = c_{24}\nu_\tau + s_{24}\nu_s$. We will denote this case as *restricted* (ATM$_{\rm R}$). In particular:

– If $c_{23}=c_{24}=1$, $U_{a1}=U_{a2}=0$ ($U_{s3}=U_{s4}=0$) and we have the limit of pure two-generation solar $\nu_e \to \nu_s$ transitions and atmospheric $\nu_\mu \to \nu_\tau$ transitions.

– If $c_{24}=0$ then $U_{s1}=U_{s2}=0$ and $U_{\tau3}=U_{\tau4}=0$, corresponding to the limit of pure two-generation solar $\nu_e \to \nu_\tau$ transitions and atmospheric $\nu_\mu \to \nu_s$ transitions.

(ii) If $c_{23} = 0$ then $U_{s1} = U_{s2} = 0 = U_{\mu 3} = U_{\mu 4}$ corresponding to the limit of pure two-generation solar $\nu_e \to \nu_a$ with $a = \mu$ and no atmospheric neutrino oscillations as the projection of ν_μ over the relevant states cancels out.

Notice that in the restricted case $\theta_{23} = 0$, there is an additional symmetry in the relevant probabilities so that the full parameter space can be spanned by $0 \leq \theta_{34} \leq \frac{\pi}{2}$. The reason for this is that we now have effectively two-neutrino oscillations for both the solar ($\nu_e \to \nu_\alpha$) and atmospheric ($\nu_\mu \to \nu_\gamma$) cases.

To summarize, solar neutrino oscillations depend on the new mixing angles only through the product $c_{23}c_{24}$ and therefore the analysis of the solar neutrino data in four-neutrino mixing schemes is equivalent to the two-neutrino analysis but taking into account that the parameter space is now three-dimensional $(\Delta m^2_{21}, \tan^2\theta_{12}, c^2_{23}c^2_{24})$. Atmospheric neutrino oscillations are affected independently by the angles θ_{23} and θ_{24}, and the analysis of the atmospheric neutrino data in the four-neutrino mixing schemes is equivalent to the two-neutrino analysis, but taking into account that the parameter space is now four-dimensional $(\Delta m^2_{43}, \theta_{34}, c^2_{23}, c^2_{24})$.

In Fig. 4(left) we show the allowed regions for the oscillation parameters Δm_{21} and $\tan^2(\theta_{12})$ from the global analysis of the solar neutrino data in the framework of four–neutrino oscillations for different values of $c^2_{23}c^2_{24}$. The global minimum used in the construction of the regions lies in the LMA region and for pure ν_e–active oscillations. As seen in the figure as the sterile component increases the regions became smaller till they totally disappear.

This behaviour is also illustrated in Fig. 5 where we show the shift on the χ^2 for the analysis of the solar data in the framework of oscillations between four neutrinos as a function of the active–sterile admixture $|U_{s1}|^2 + |U_{s2}|^2 = c^2_{23}c^2_{24}$. From the figures we conclude that the solar neutrino data favour pure $\nu_e \to \nu_a$ oscillations but sizeable active-sterile admixtures are still allowed. For instance at 99% CL, both the LMA and LOW-QVO regions are allowed for maximal active-sterile mixing, $c^2_{23}c^2_{24} = 0.5$.

For the analysis of the atmospheric data in Fig. 4(right) we plot the sections of such a volume in the plane for different values of the projections $|U_{\mu 1}|^2 + |U_{\mu 2}|^2$ and $|U_{s1}|^2 + |U_{s2}|^2$. The global minimum used in the construction of the regions lies almost in the pure atmospheric ν_μ–ν_τ oscillations. As shown in the figure the regions becomes considerably smaller for increasing values of the mixing angle θ_{23}, which determines the size of the projection of the ν_μ over the "atmospheric" neutrino oscillating states, and for increasing values of the mixing angle θ_{24}, which determines the active–sterile admixture in which the "almost–ν_μ" oscillates. Therefore from the analysis of the atmospheric neutrino data we obtain an upper bound on both mixings, which, in particular, implies a lower bound on the combination $c^2_{23}c^2_{24} = |U_{s1}|^2 + |U_{s2}|^2$ limited from above by the solar neutrino data. Fig. 5 we show the shift on the χ^2 for the analysis of the atmospheric data in the framework of oscillations between four neutrinos as a function of the active–sterile admixture $|U_{s1}|^2 + |U_{s2}|^2 = c_{23}^2 c_{24}^2$ for the general case (in which the analysis is optimized with respect to the parameter $s_{23}^2 = |U_{\mu 1}|^2 + |U_{\mu 2}|^2$) as well as the restricted case in which $s_{23}^2 = 0$.

The 90% (99%) CL (for 1 d.o.f) lower bounds on $c_{23}^2 c_{24}^2$ from the analysis of the atmospheric neutrino data are:

$$|U_{s1}|^2 + |U_{s2}|^2 > 0.64\ (0.52)\ \text{general}\ [0.83\ (0.74)\ \text{restricted}] \tag{6}$$

In summary, the analysis of the solar data favours the scenario with $\nu_e - \nu_a$ oscillations in the $1-2$ plane and gives an upper bound on the projection of ν_s on this plane. Conversely, the analysis of the atmospheric data prefers the oscillations of the $3-4$ states to occur between a close-to-pure ν_μ and an active (ν_τ) neutrino and gives an upper bound on the projection of the ν_s over the $3-4$ states or, equivalently, a lower bound on its projection over the $1-2$ states. Fig. 5 shows that the exclusion curves from the solar and atmospheric analyses are only marginally compatible at $\sim 99\%$ CL.

What is the best scenario for the active-sterile admixture once these two bounds are put together?. The combined analysis still favours close-to-pure active and sterile oscillations (which gives a bad fit to either the solar or the atmospheric data) and disfavours oscillations into a near-maximal active-sterile admixture which could, in principle, lead to a compromise between the fits to the two data samples. This can be interpreted as a hint of incompatibility between the present solar, atmospheric and LSND results even in the context of 2+2 four-neutrino mixing although, at present, these schemes cannot be ruled out from the pure statistical point of view.

References

1. For a recent review see M.C. Gonzalez-Garcia and Y. Nir, hep-ph/0202058.
2. M.C. Gonzalez-Garcia *et al.*. Phys. Rev. **D63** (2001) 033005.
3. M. C. Gonzalez-Garcia, M. Maltoni and C. Pena-Garay, Phys. Rev. D **64**, 093001 (2001); hep-ph/0108073.
4. See M. C. Gonzalez-Garcia and M. Smy in these proceedings.
5. See M. Messier in these proceedings.
6. CHOOZ Coll., M. Apollonio *et al.*. Phys. Lett. **B420**, 397 (1998); Bugey Coll.,B. Achkar *et al.*, Nucl. Phys. **B424**, 503 (1995); CDHSW Coll.,F. Dydak *et al.*, Phys. Lett. B **134**, 281 (1984).

THREE-FLAVOR ANALYSIS OF ATMOSPHERIC AND SOLAR NEUTRINOS

G.L. FOGLI, E. LISI, A. MARRONE,* AND A. PALAZZO

Dipartimento di Fisica & Sezione INFN, Via Amendola 173, 70126 Bari, Italy

The status of three-flavor oscillations of solar and atmospheric neutrinos is reviewed. Bounds on the mass-mixing neutrino parameters are shown. Effects beyond the usual hierarchical approximation are studied and illustrated.

1 Introduction

The most robust evidence for neutrino oscillations comes from atmospheric and solar ν experiments. The strong indication of atmospheric ν_μ disappearance observed in the Super-Kamiokande (SK) experiment [1] is corroborated by MACRO and Soudan-2 data.[2] There is also evidence for active ν_e oscillations from global analyses of solar neutrino experiments,[3] confirmed at a significance level $> 3\sigma$ by a model independent the comparison [4] of SK [5] and SNO data.[6] Oscillations of solar and atmospheric neutrinos are controlled by two squared mass differences, that we denote as δm^2 and m^2, respectively. They are separated at least by one order of magnitude. Therefore, it is reasonable to use the hierarchical approximation $\delta m^2 \ll m^2$ in combined solar+atmospheric analyses. However, given the increasing accuracy of experimental results, it is important to check the impact of non-hierarchical effects on the analysis.

2 Notation

In this work, we consider 3ν mixing between active neutrino flavor states ν_α $(\alpha = e, \mu, \tau)$ and mass eigenstates ν_i $(i = 1, 2, 3)$, with mass m_i. The elements of the mixing matrix U are functions of three mixing angles $(\omega, \phi, \psi) \equiv (\theta_{12}, \theta_{13}, \theta_{23})$, plus a possible CP violating phase (here neglected).

We define two squared mass differences as follows:

$$\delta m^2 = m_2^2 - m_1^2\,, \tag{1}$$

$$m^2 = \left| m_3^2 - \left(\frac{m_1^2 + m_1^2}{2} \right) \right| . \tag{2}$$

*SPEAKER

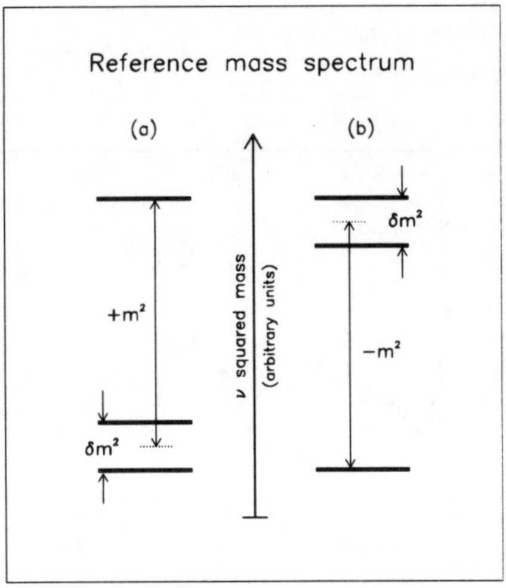

Figure 1. Neutrino mass spectrum.

The associated neutrino mass spectrum is shown in Fig. 1, for the two cases of direct ($+m^2$) and inverted ($-m^2$) hierarchy, corresponding to $m_3 > m_1, m_2$ or $m_3 < m_1, m_2$, respectively. Conventionally, we take $\delta m^2 > 0$. Usually, solar+atmospheric combined analyses assume the hierarchical approximation ($\delta m^2 \ll m^2$) at zeroth order in the small parameter $\delta m^2/m^2$. This corresponds to take $m^2 \to \infty$ for solar neutrinos, and $\delta m^2 \to 0$ for atmospheric neutrinos. Under this approximation the full five-dimensional parameter space $(\delta m^2, m^2, \omega, \phi, \psi)$ reduces to the two three-dimensional subspaces, $(\delta m^2, \omega, \phi)$ and (m^2, ϕ, ψ), for solar and atmospheric neutrinos, respectively.

3 Analysis of atmospheric neutrino data

3.1 Hierarchical approximation ($\delta m^2 = 0$).

Two-flavor analysis. The parameter space of two-flavor $\nu_\mu \leftrightarrow \nu_\tau$ mixing is (ψ, m^2), because ν_e is decoupled ($U^2_{e3} = \sin^2\phi{=}0$), and ω can be rotated away for $\delta m^2 = 0$. This 2ν scenario explains well the atmospheric ν anomaly. In particular, Fig. 2 shows the zenith distributions of sub-GeV (SG), multi GeV (MG), e-like and μ-like (e, μ) contained events and upward stopping (US) and through-going (UT) muons, together with the best-fit distributions for two flavor $\nu_\mu \leftrightarrow \nu_\tau$ oscillations.[7] Our recent analysis [8] of the first KEK-to-Kamioka

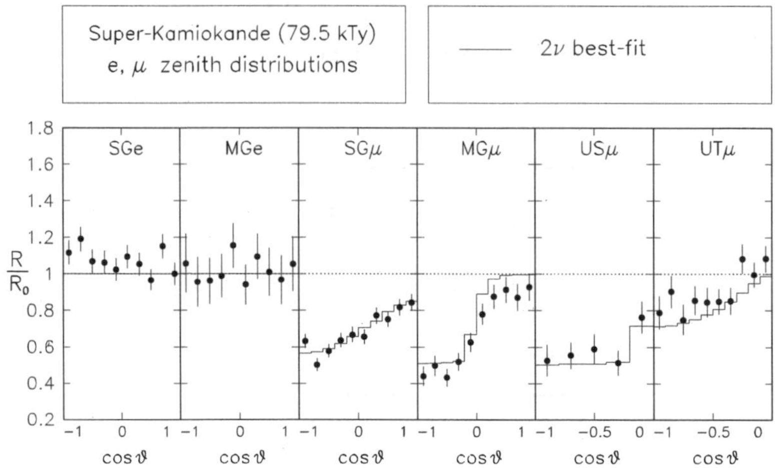

Figure 2. SK atmospheric zenith distributions of leptons, normalized to the standard no oscillation expectations. Dots with error bars: 79.5 kTy SK data. Histograms: 2ν best-fit. Dotted horizontal line: standard expectations.

(K2K) experiment results (44 observed events, 64 expected in absence of oscillations) is also consistent with (2ν) oscillations in the $\nu_\mu \leftrightarrow \nu_\tau$ channel . The measured number of events is in accordance with the expectations for $m^2 \sim 3 \times 10^{-3}$ eV^2 and maximal mixing ($\psi \sim \pi/4$), i.e. in the same region of the parameter space allowed by SK fit, so that it makes sense to combine K2K with SK, as shown in Fig 3(a). The upper left panel of Figure 3(a) shows the allowed region in the $(m^2, \tan^2\psi)$ plane from SK data only (the stars are the two degenerate best-fit points $(m^2, \tan^{\pm 2}\psi) = (3 \times 10^{-3}\,\mathrm{eV}^2, 0.76)$. The upper right panel corresponds to the K2K fit only. The limits set by K2K are weaker than those from SK, especially on the angle ψ. The lower left panel shows the result of the combination of K2K and SK data. The m^2 value at the best-fit point is only slightly lowered with respect to the case of SK data alone (2.9×10^{-3} eV^2), but the allowed m^2 range is reduced from $[1.3, 5.6] \to [1.5, 4.8]$ (in units of 10^{-3} eV^2). In the lower right panel the same combination is shown, but with K2K total error hypothetically reduced by a factor 2.

Three-flavor mixing. The 3ν case is realized for $\phi \neq 0$, corresponding to additional ν_e mixing. The results of the three-flavor analysis of SK+CHOOZ

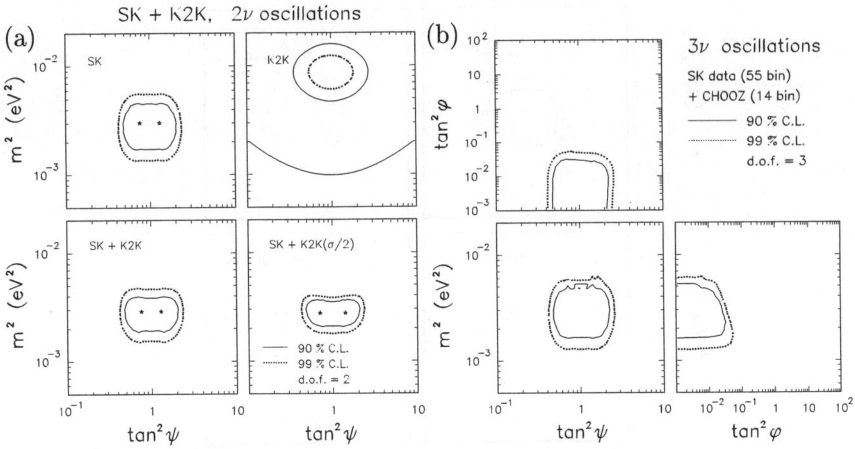

Figure 3. (a) K2K expected number of events as a function of m^2; (b) combined three-flavor analysis of SK+CHOOZ.

are shown in Fig. 3(b). The best-fit point to SK data only (not shown) is at $(m^2, \tan^2\psi, \tan^2\phi) = (3\times10^{-3}\,\mathrm{eV}^2, 0.9, 0.01)$, with $\chi^2_{min} = 38.1$ for $55-2$ degrees of freedom. The nonzero value of ϕ at best-fit is, however, not statistically significant, since the χ^2 difference with the pure 2ν case ($\phi = 0$) is very small ($\Delta\chi^2 \sim 1$). The upper limit on $\tan^2\phi$ is 0.35 at 90% C.L., so that a relatively large ν_e mixing is still allowed from SK data only, although not preferred. The inclusion of the CHOOZ experiment in the analysis severely constrains large ϕ values as shown in Fig. 3(b). The best-fit point ($\chi^2_{min} = 45.7$ for $69-2$ d.o.f) is reached at $(m^2, \tan^2\psi, \tan^2\phi) = (3\times10^{-3}\,\mathrm{eV}^2, 0.75, 0)$. The upper bound on $\tan^2\phi$ is decreased by about one order of magnitude, and the allowed m^2 range at 90% C.L., $[1.6, 5.3]\times10^{-3}\,\mathrm{eV}^2$, becomes approximately symmetric around the maximal mixing axis ($\psi = \pi/4$), implying no evidence for matter effects and for sign ($\pm m^2$) discrimination.

3.2 *Non-hierarchical case* ($\delta m^2 \neq 0$)

When the hierarchical approximation is relaxed ($\delta m^2 \neq 0$), it is very important to check if the solutions found for $\delta m^2 = 0$ are stable. In particular, it is very important to check wether the minimum allowed m^2 value does not became smaller (and thus dangerously close to δm^2).

We have checked [9] that the m^2 best-fit values in a full five parameter analysis is not significantly changed from its value in the hierarchical approx-

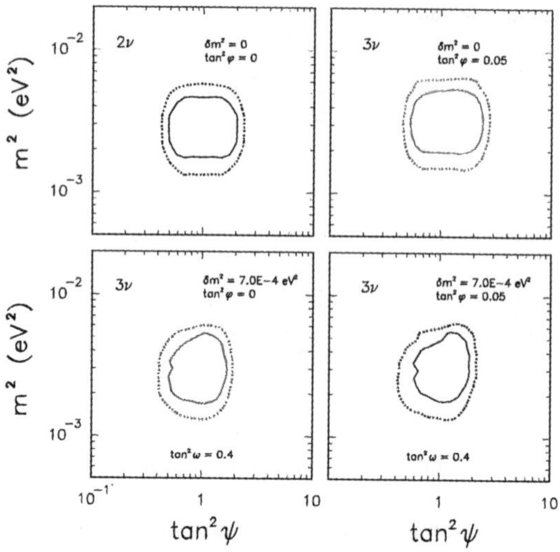

Figure 4. Allowed regions in the $(m^2, \tan^2\psi)$ plane, for fixed $(\delta m^2, \tan^2\phi, \tan^2\omega)$ values, in the case of normal hierarchy.

imation (not shown). Also the allowed regions are only mildly modified. In particular Fig. 4 shows the allowed projections in the $(\tan^2\psi, m^2)$ plane in the case of direct mass spectrum, for different values of δm^2 and $\tan^2\phi$ (the case of inverted hierarchy is very similar but not shown). The values $\delta m^2 = 7\times10^{-4}$ eV^2 and $\tan^2\phi = 0.05$ are at the limit of the region allowed by CHOOZ (see also Fig. 6), while ω is fixed at the LMA solution for solar neutrinos ($\tan^2\omega \cong 0.4$). Figure 4 shows that the lowest allowed m^2 value, at a given C.L., is never lower than the corresponding minimum value for the hierarchical case (upper left panel). Therefore, the hierarchical approximation is stable in m^2. Beyond the hierarchical approximation, the allowed region of the angle ψ is slightly modified, being shifted towards $\psi > \pi/4$ ($\psi < \pi/4$) when $\phi \neq 0$ ($\delta m^2 \neq 0$). Notice that, in the hierarchical approximation ($\delta m^2 = 0$), when $\phi \neq 0$ (upper right panel of Fig. 4) there is an excess of SG and MG e-like events proportional to $\sin^2\psi - 1/r$ ($r \sim 2$ being the ratio between the ν_μ and ν_e fluxes). Therefore, when ψ is in the second octant, there is an excess of e-like events that can help the fit.[10] The situation is reversed for $\phi = 0$ and $\delta m^2 \neq 0$ (lower left panels of Fig. 4), because the e-like event excess is proportional to $\cos^2\psi - 1/r$,[11] so that the first octant of ψ is preferred. The case with ϕ and δm^2 both different from zero is much more complicated, and there is no simple analytical approximation to understand the behavior of the allowed regions, although, from the lower right panel of Fig. 4, it can be seen that there is no dramatic change of the results of the fit.

4 Solar neutrinos

The model-independent analysis of SK and SNO data favors the oscillation hypothesis and confirms the validity of the standard solar model (SSM).[4] Assuming the SSM ν fluxes as input, the 2ν and 3ν oscillation scenarios can be studied through a χ^2 analysis of all available data, including the CHOOZ reactor data. The results are shown in the $(\delta m^2, \tan^2 \omega)$ plane in both Fig. 5(a) (2ν analysis, $\phi = 0$) and Fig. 5(b) (3ν, analysis $\phi \geq 0$). The solution preferred in the 2ν analyses is the so-called large mixing angle (LMA), while there is no allowed small mixing angle solution (SMA) at 3σ level, mostly due to nonobservation of spectral distortion in SK data. At lower δm^2, the so called

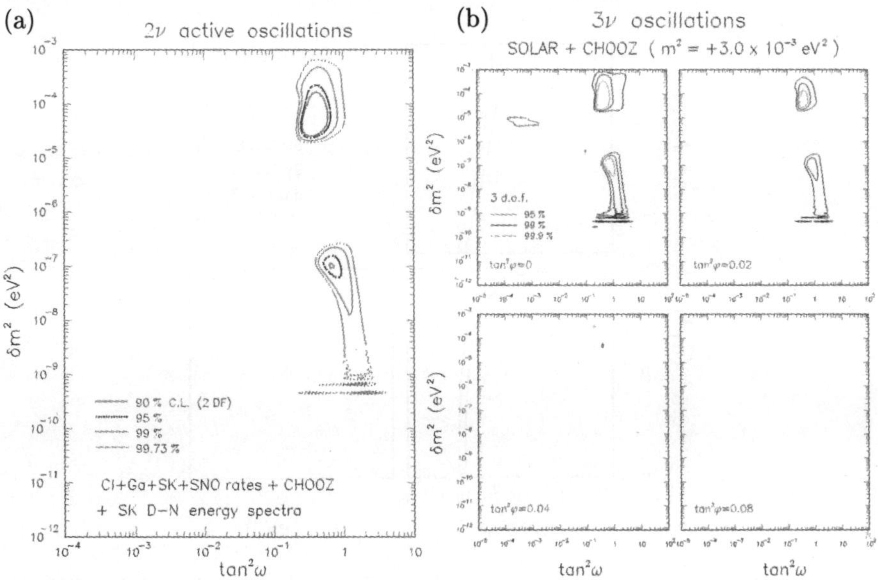

Figure 5. Solar neutrino analysis: a) 2ν oscillation analysis of total rates, SK day-night energy spectra, with CHOOZ data included; b) 3ν oscillation analysis with the same input data as in the 2ν case.

LOW ($\delta m^2 \sim 10^{-7}$ eV^2) and quasivacuum oscillation (QVO, $\delta m^2 \sim 10^{-8}$ eV^2) solutions still survive, although slightly disfavored with respect to the LMA one. The fit of SK data alone is practically insensitive to the value of m^2, in the range allowed by atmospheric neutrinos.[12] However, m^2 is relevant for the fit to CHOOZ. Note that the upper part of the LMA solution is cut by the negative CHOOZ result. Allowing nonzero value of ϕ, the 3ν results are shown in Fig. 5(b), for fixed $m^2 = 3 \times 10^{-3}$ eV^2. The upper left panel of Fig. 5(b) concides with the 2ν case, up to the different d.o.f.'s (3 in the 3ν case). With increasing ϕ, two features emerge: the maximum allowed value of δm^2 decreases and the LOW solution moves to larger mixing.

The dependence of δm^2_{max} on ϕ is connected with the CHOOZ analysis. Figure 6 shows the projections onto the $(m^2, \phi, \delta m^2)$ coordinate planes of the region allowed by the CHOOZ data. The lower left panel shows an anticorrelation between the maximum allowed values of δm^2 and ϕ. This result is important in constraining the upper part of the LMA solution in the 3ν solar analysis.

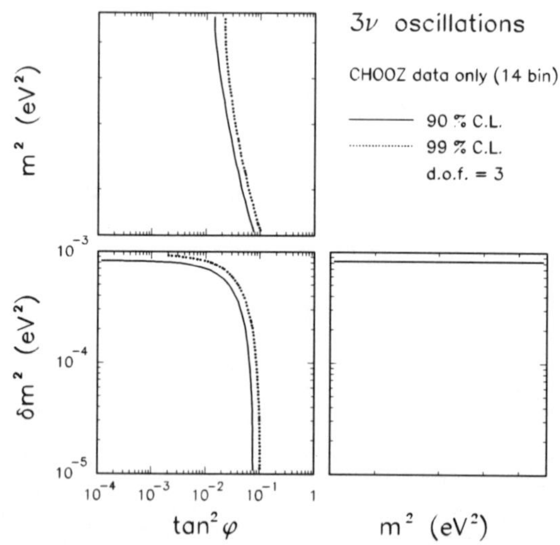

Figure 6. Projections onto the coordinate planes of the allowed region of the CHOOZ analysis.

5 Conclusions

The solar neutrino problem and the atmospheric neutrino anomaly can be well explained in a 3ν oscillation scenario. Solar neutrino data prefer the LMA solution. Atmospheric neutrino data are very well explained by almost pure $\nu_\mu \to \nu_\tau$ oscillations. Both solar and atmospheric neutrino analyses, and the combination with the negative results of the CHOOZ experiment, prefer small values of $\sin^2\phi = U_{e3}^2$, which is a nontrivial consistency check. If the standard hierarchical scenario is perturbed by considering the effect of $\delta m^2 \neq 0$ on atmospheric neutrinos and of $m^2 < \infty$ on solar neutrinos, the fit appears to be rather stable: the allowed m^2 and δm^2 regions are not significantly altered, while the effects on the mixing angles, although being also relatively small, are interesting.

References

1. C. McGrew in *Neutrino Telescopes 2001*, 9th International Workshop on Neutrino Telescopes, Venice, Italy, 2001, edited by M. Baldo Ceolin (Padua Univ. press, Italy, 2001), p. 93.
2. MACRO Collaboration, *Phys. Lett.* B **517**, 59 (2001); Soudan-2 Collaboration, *Phys. Lett.* B **449**, 137 (1999).
3. M. C. Gonzalez-Garcia, these Proceedings.
4. G.L. Fogli, E. Lisi, D. Montanino, and A. Palazzo, *Phys. Rev.* D **64**, 093007 (2001).
5. SK Collaboration, *Phys. Rev. Lett.* **86**, 5651 (2001); *Phys. Rev. Lett.* **86**, 5656 (2001).
6. SNO Collaboration, *Phys. Rev. Lett.* **87**, 071301 (2001).
7. G.L. Fogli, E. Lisi, and A. Marrone, *Phys. Rev.* D **64**, 093005 (2001).
8. G.L. Fogli, E. Lisi, and A. Marrone, *Phys. Rev.* D **65**, 073028 (2002).
9. G.L. Fogli, E. Lisi, A. Marrone, and A. Palazzo, work in progress.
10. E.K. Akhmedov, A. Dighe, P. Lipari, and A.Y. Smirnov, *Nucl. Phys.* B **542**, 3 (1999).
11. O.L.G. Peres and A.Yu. Smirnov, *Phys. Lett.* B **456**, 204 (1999).
12. G.L. Fogli, E. Lisi, and A. Marrone, D. Montanino, and A. Palazzo, in *Neutrino Telescopes 2001*, [1], p. 105.

IMPLICATIONS OF THE RECENT RESULTS OF SOLAR NEUTRINO EXPERIMENTS *

M. MARIS

Osservatorio Astronomico di Trieste, I-34113 Trieste, Italy

S.T. PETCOV

SISSA/INFN - Sezione di Trieste, I-34014 Trieste, Italy †

Detailed predictions for the D-N asymmetry for the Super-Kamiokande and SNO experiments, as well as for the ratio of the CC and NC event rates measured by SNO, in the cases of the LMA MSW and of the LOW solutions of the solar neutrino problem, are presented. The possibilities to use the forthcoming SNO data on these two observables to discriminate between the LMA and LOW solutions and/or to further constrain the regions of the two solutions are also discussed.

1 Introduction

With the publication of the first results of the SNO experiment [1] further strong evidences for oscillations/transitions of the solar ν_e on the way to the Earth have been obtained. The SNO experiment measured the rate of the charged current (CC) reaction $\nu_e + D \rightarrow e^- + p + p$ for $T_e \geq 6.75$ MeV, T_e being the kinetic energy of the final state electron [1]. The reaction is due to the flux of solar ^{8}B ν_e, having energy of $E \gtrsim 8.2$ MeV. Assuming that the ^{8}B neutrino energy spectrum is not substantially modified by the oscillations/transitions, the SNO collaboration obtained the following value of the solar ν_e flux:

$$\Phi^{CC}(\nu_e) = (1.75 \pm 0.15) \times 10^6 \text{ cm}^{-2}\text{s}^{-1}, \tag{1}$$

where the statistical and systematic errors and the estimated theoretical uncertainty (due to the uncertainty in the CC reaction cross section) given in [1], are added in quadrature. Combined with the data from the Super-Kamiokande experiment [2], the SNO result (1) clearly demonstrates [1] (see also [3]) the presence of $\nu_{\mu,\tau}$ and/or $\bar{\nu}_{\mu,\tau}$ component in the flux of solar neutrinos reaching the Earth and measured by the Super-Kamiokande collaboration:

$$\Phi(\nu_{\mu,\tau}) + \Phi(\bar{\nu}_{\mu,\tau}) = (3.69 \pm 1.13) \times 10^6 \text{ cm}^{-2}\text{s}^{-1}. \tag{2}$$

This flux is different from zero at more than 3 s. d., which represents a compelling evidence for oscillations and/or transitions of the solar ν_e.

---

*PRESENTED BY S. T. PETCOV.

†Also at: Institute of Nuclear Research and Nuclear Energy, Bulgarian Academy of Sciences, BG–1784 Sofia, Bulgaria.

Global analyses of the solar neutrino data [2,4], including the Super-Kamiokande data on the e^-−spectrum and day-night (D-N) asymmetry, show that the data favor the large mixing angle (LMA) MSW, the LOW and the quasi-vacuum oscillation (QVO) solutions of the solar neutrino problem with transitions into active neutrinos [3,5,6,7,8]. In the case of the LMA solution, the range of values of the neutrino mass-squared difference $\Delta m^2 > 0$, characterizing the two-neutrino solar $\nu_e \rightarrow \nu_{\mu(\tau)}$ transitions, found, e.g., in [3] reads:

$$\text{LMA MSW}: \quad 2.0\times 10^{-5}\ \text{eV}^2 \lesssim \Delta m^2 \lesssim 5.0\times 10^{-4}\ \text{eV}^2 \quad (99\%\ \text{C.L.}). \tag{3}$$

The best fit values of Δm^2 obtained in [3,5] are grouped in the narrow interval $(\Delta m^2)_{BFV} = (4.3 - 4.9) \times 10^{-5}$ eV^2. A smaller value was found in [7], $(\Delta m^2)_{BFV} = (3.3 - 3.7) \times 10^{-5}$ eV^2, while larger values were obtained, e.g., in [6,8]: $(\Delta m^2)_{BFV} = (6.0-6.3) \times 10^{-5}$ eV^2. For the mixing parameter $\sin^2 2\theta$, which controls the oscillations of the solar ν_e, it was found, e.g., in [3]:

$$\text{LMA MSW}: \quad 0.60 \lesssim \sin^2 2\theta \lesssim 0.99 \quad (99\%\ \text{C.L.}). \tag{4}$$

The best fit values of $\sin^2 2\theta$ obtained by the different authors are all confined to the interval $(\sin^2 2\theta)_{BFV} = (0.75 - 0.82)$.

Detailed results were obtained for the LOW solution as well [3,5,6,7,8]. The 95% C.L. allowed intervals of values of Δm^2 and $\sin^2 2\theta$ found, e.g., in [3] read:

$$\text{LOW}: \quad 6.0 \times 10^{-8}\ \text{eV}^2 \lesssim \Delta m^2 \lesssim 1.8 \times 10^{-7}\ \text{eV}^2\ , \quad 0.94 \lesssim \sin^2 2\theta \lesssim 1.0. \tag{5}$$

The best fit values of Δm^2 and $\sin^2 2\theta$ for the LOW solution, derived, e.g., in [3,5,7] are all approximately given by $(\Delta m^2)_{BFV} \cong 10^{-7}$ eV^2 and $(\sin^2 2\theta)_{BFV} \cong (0.94 - 0.97)$. A substantially different value of $(\Delta m^2)_{BFV}$ was found in [8]: $(\Delta m^2)_{BFV} \cong 5.5 \times 10^{-8}$ eV^2 and $(\sin^2 2\theta)_{BFV} \cong 0.99$.

The analyses [3,5,6,8] were based, in particular, on the standard solar model (SSM) predictions of ref. [9] (BP2000) for the different components of the solar neutrino flux (*pp*, *pep*, ^{7}Be, ^{8}B, CNO, *hep*, ^{17}F). In [3,5,6,7] the published Super-Kamiokande data on the D - N asymmetry [2] were used as input in the analyses, while in [8] the latest (preliminary) results on the D-N asymmetry, obtained from the analysis of *all currently available Super-Kamiokande solar neutrino data* were utilized (see further). The authors of ref. [7] have used in their analysis a new larger value of the ^{8}B neutrino flux [a] which is suggested

[a] The ^{8}B neutrino flux predicted in [9] reads $\Phi(B)_{BP2000} = 5.05 \times (1^{+0.20}_{-0.16}) \times 10^6$ cm^{-2}s^{-1}, while the flux utilized in the analysis performed in [7] is $\Phi(B)_{NEW} = 5.93 \times (1^{+0.14}_{-0.13}) \times 10^6$ cm^{-2}s^{-1}. Let us note also that, e.g., in the first and the last articles quoted in [5] and in [7,8] results obtained by treating the ^{8}B neutrino flux as a free parameter in the analysis were also reported.

by the results of the latest (and more precise) experimental measurement [10] of the cross section of the reaction $p+{}^7\mathrm{Be} \to {}^8\mathrm{B} + \gamma$.

We have updated recently [11] our earlier predictions [12,13,14,15] for the D-N asymmetry for the Super-Kamiokande and SNO experiments, taking into account the recent progress in the studies of solar neutrinos. We derived also [11] predictions for another important observable - the ratio of the event rates of the CC reaction, $R_{SNO}(CC)$, and of the neutral current (NC) reaction $\nu + D \to \nu + n + p$, $R_{SNO}(NC)$, induced by the solar neutrinos in SNO,

$$R^{SNO}_{CC/NC} \equiv \frac{R_{SNO}(CC)}{R_{SNO}(NC)} / \frac{R^0_{SNO}(CC)}{R^0_{SNO}(NC)} , \tag{6}$$

where $R^0_{SNO}(CC)/R^0_{SNO}(NC)$ is the value of the ratio in the absence of oscillations of solar neutrinos. The present article contains a concise review of the results obtained in [11]. We focus on the predictions for the D-N asymmetry and the ratio $R^{SNO}_{CC/NC}$ for the LMA and the LOW solutions of the solar neutrino problem, which are favored by the current solar neutrino data.

The experimental observation of a non-zero D-N asymmetry

$$A^N_{D-N} \equiv \frac{R_N - R_D}{(R_N + R_D)/2}, \tag{7}$$

where R_N and R_D are, e.g., the one year averaged event rates in a given detector respectively during the night and the day, would be a very strong evidence in favor of an MSW solution of the solar neutrino problem (see, e.g., [16,17] and [12,13,14,15] and the references quoted therein). The current Super-Kamiokande data [2] do not contain evidence for a substantial D-N asymmetry: the latest published result on A^N_{D-N} (from 1258 days of data-taking) reads [2]

$$A^N_{D-N}(SK) = 0.033 \pm 0.022 \; (stat.) \; ^{+0.013}_{-0.012} \; (syst.), \tag{8}$$

while the result of the latest analysis of *all currently available Super-Kamiokande solar neutrino data* (1496 days of measurements) gives [8]

$$A^N_{D-N}(SK) = 0.021 \pm 0.022 \; (stat.) \; ^{+0.013}_{-0.012} \; (syst.). \tag{9}$$

Adding the errors in (8) and (9) in quadrature, one finds that at 1.5 (2.0) s.d., $A^N_{D-N}(SK) < 0.072$ (0.085) and $A^N_{D-N}(SK) < 0.060$ (0.073), respectively.

First results on the D-N asymmetry and on the CC to NC event rate ratio $R^{SNO}_{CC/NC}$ are expected to be published in the near future by the SNO collaboration. We discuss also the possibilities to further constrain the regions of the LMA MSW and LOW-QVO solutions of the solar neutrino problem by using the forthcoming SNO data on the D-N asymmetry A^N_{D-N} and on the CC to NC event rate ratio $R^{SNO}_{CC/NC}$.

2 The D-N Asymmetry in the LMA and LOW Solution Regions

Our detailed predictions [11] for the (*Full Night*) D-N asymmetry in the regions of the LMA MSW and LOW solutions of the solar neutrino problem for the Super-Kamiokande and SNO experiments, $A^N_{D-N}(SK)$ and $A^N_{D-N}(SNO)$, are shown in Figs. 1 and 2, respectively [b]. In Fig. 2 we show contours of constant $A^N_{D-N}(SNO)$ in the $\Delta m^2 - \sin^2 2\theta$ plane for two values of the threshold kinetic energy of the final state electron, $T_{\rm e,th} = 6.75$ MeV (upper panel) and 5.0 MeV (lower panel). The published SNO data were obtained using the first value [1], while the second one is the threshold energy planned to be reached at a later stage of the experiment. A comparison of the upper and lower panels in Fig. 2 shows that the D-N asymmetry $A^N_{D-N}(SNO)$ *decreases* somewhat in the LMA MSW solution region - approximately by $\sim (8-10)\%$, when $T_{\rm e,th}$ is decreased from 6.75 MeV to 5.0 MeV. The change in the LOW solution region is opposite and larger in magnitude than in the LMA MSW solution region: $A^N_{D-N}(SNO)$ *increases* by about $\sim (15-20)\%$. The above results imply that for given $\sin^2 2\theta$, the same values of the asymmetry at $T_{\rm e,th} = 5.0$ MeV occur in both the LMA MSW and LOW solution regions at smaller values of Δm^2 than for $T_{\rm e,th} = 6.75$ MeV.

As Figs. 1 - 2 show, at $\Delta m^2 \gtrsim 1.5 \times 10^{-4}$ eV^2 in the LMA solution region, both $A^N_{D-N}(SK)$ and $A^N_{D-N}(SNO)$ are smaller than 1%. For given $\Delta m^2 \lesssim 10^{-4}$ eV^2 and $\sin^2 2\theta$ from the LMA region we have [15] $A^N_{D-N}(SNO) \cong (1.5-2.0)A^N_{D-N}(SK)$. The difference between $A^N_{D-N}(SK)$ and $A^N_{D-N}(SNO)$ is due to i) the contribution of the NC $\nu_{\mu(\tau)} - e^-$ elastic scattering reaction (in addition to that due to the $\nu_e - e^-$ elastic scattering) to the solar neutrino event rate measured by the Super-Kamiokande experiment (there is no similar contribution to the SNO CC event rate), and ii) to the relatively small value of the solar ν_e survival probability in the Sun, $\bar{P} \sim 0.3$. Thus, in the case of the LMA solution, the D-N asymmetry measured in SNO can be considerably larger than the D-N asymmetry measured in the Super-Kamiokande experiment [15]. The 2 s.d. upper limit on the D-N asymmetry, $A^N_{D-N}(SK) < 8.5\%$ (7.3%), following from the Super-Kamiokande data, eq. (7) (eq. (8)), for instance, does not exclude a D-N asymmetry in the SNO CC event rate as large as $\sim (10-15)\%$. As Fig. 2 shows, $A^N_{D-N}(SNO)$ can reach a value of $\sim 20\%$ in the 99% C.L. region of the LMA solution, eq. (3). In the

[b] The calculations of the D-N effect for the Super-Kamiokande and SNO detectors performed in [11] are based on the methods developed for our earlier studies of the D-N effect for these detectors, which are described in detail in [12,13,15]. We used in [11] the BP2000 SSM [9] predictions for the electron number density distribution in the Sun.

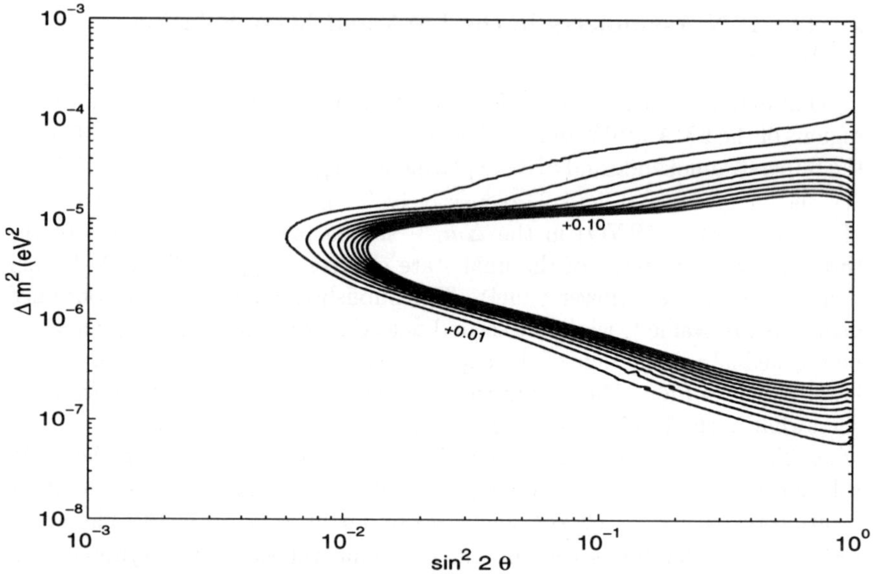

Figure 1. Iso-(D-N) asymmetry contour plot for the Super-Kamiokande experiment for $T_{\rm e,th} = 5$ MeV, $T_{\rm e,th}$ being the detected e^- kinetic energy threshold. The contours correspond to values of the *Full Night* asymmetry $A^N_{D-N}(SK) = 0.01,\ 0.02,\ 0.03,...,0.09,0.10$.

best fit point of the LMA solution, found in [3,5] and in [7], we get for $T_{\rm e,th} =$ 6.75 MeV (5.0 MeV), $(A^N_{D-N}(SNO))^{LMA}_{BF1} \cong 7.1\ (6.6)\%$ and $10.1\ (9.3)\%$, respectively. One finds, however, a considerably smaller value of $A^N_{D-N}(SNO)$ in the best fit point derived in [8]: $(A^N_{D-N}(SNO))^{LMA}_{BF3} \cong 5.0\ (4.6)\%$. The D-N asymmetry in the Super-Kamiokande detector in the indicated three different best fit points reads, respectively, for $T_{\rm e,th} = 5.0$ MeV: $(A^N_{D-N}(SK))^{LMA}_{BF} \cong$ 3.6%; 5.0%; 2.5%.

Actually, as it is not difficult to show, the following approximate relation between $A^N_{D-N}(SK)$ and $A^N_{D-N}(SNO)$ holds for fixed Δm^2 and $\sin^2 2\theta$ from the region of the LMA solution:

$$A^N_{D-N}(SNO) \cong A^N_{D-N}(SK) \left[1 + \frac{r}{(1-r)\bar{P}}\right] , \tag{10}$$

where $r \equiv \sigma(\nu_{\mu(\tau)}e^-)/\sigma(\nu_e e^-)$, $\sigma(\nu_l e^-)$ being the $\nu_l - e^-$ elastic scattering cross section, $l = e, \mu, \tau$, and $\bar{P}$ is the average probability of solar ν_e survival

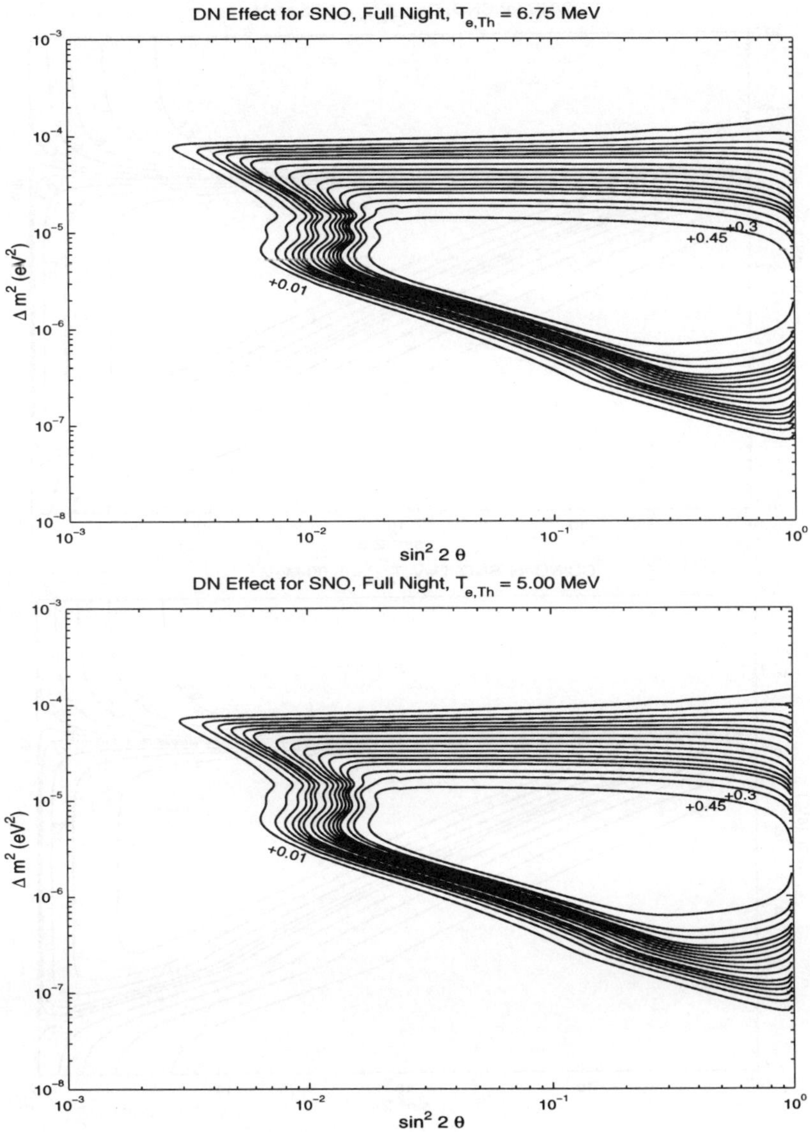

Figure 2. Iso-(D-N) asymmetry ($A^N_{D-N}(SNO)$) contour plot for the SNO experiment for $T_{\rm e,th} = 6.75$ MeV (upper panel) and $T_{\rm e,th} = 5.00$ MeV (lower panel)., The contours correspond to values of the *Full Night* asymmetry in the CC event rate $A^N_{D-N}(SNO) = 0.01$, 0.02, 0.03, 0.04, 0.05, 0.06, 0.08, 0.10, 0.12, 0.14, 0.16, 0.18, 0.20, 0.25, 0.30, 0.45.

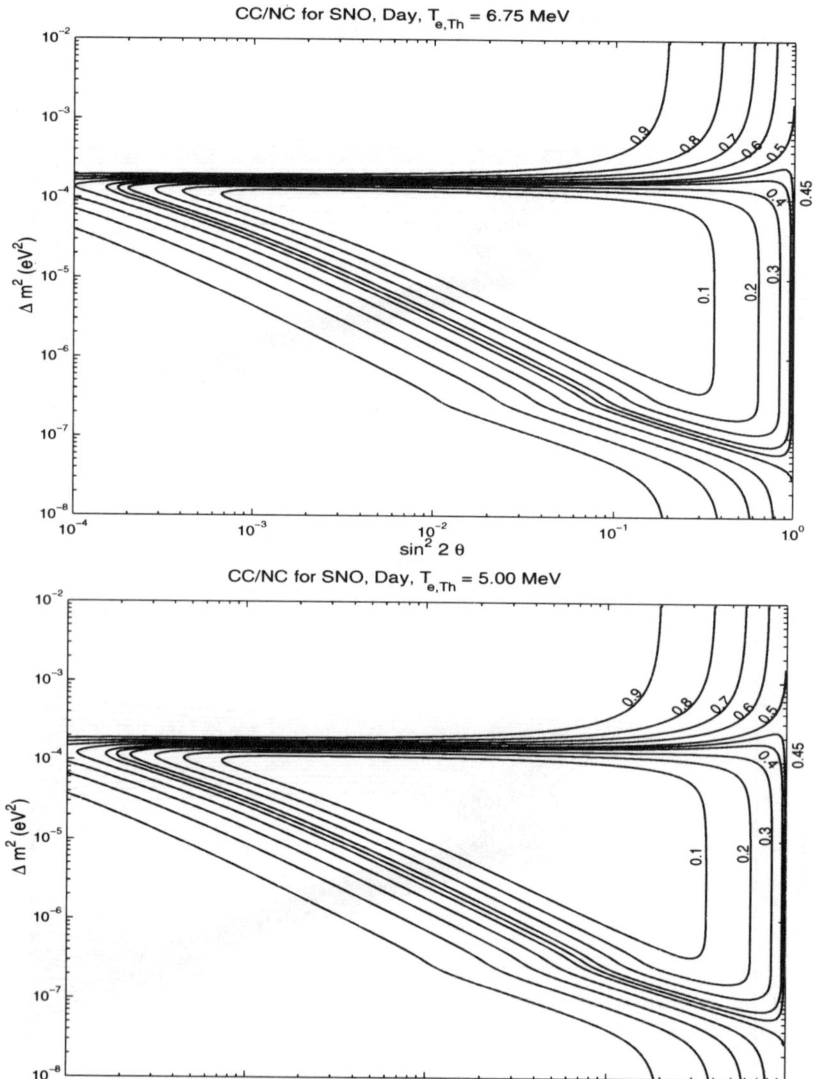

Figure 3. Iso$-R^{SNO}_{CC/NC}(D)$ contour plots for $T_{\rm e,th} = 6.75$ MeV (upper panel) and $T_{\rm e,th} = 5.00$ MeV (lower panel): the contours correspond to fixed values of the *Day* ratio of CC and NC event rates measured in SNO. In the absence of solar ν_e oscillations $R^{SNO}_{CC/NC}(D) = 1$. The values of $R^{SNO}_{CC/NC}(D)$ are indicated on the contours.

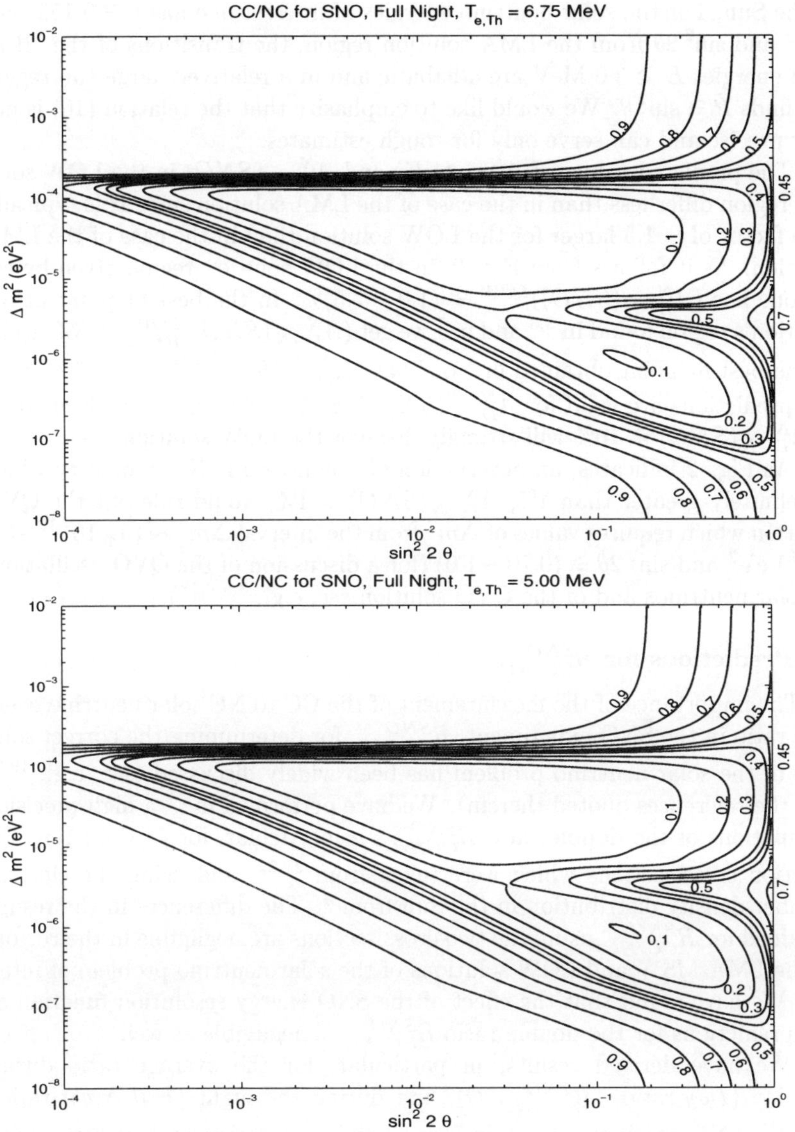

Figure 4. The same as in Fig. 4 for the *Night* ratio of the CC and NC solar neutrino event rates, $R^{SNO}_{CC/NC}(N)$, measured in the SNO experiment. In the case of absence of oscillations of solar neutrinos $R^{SNO}_{CC/NC}(N) = 1$. The contours shown are for $R^{SNO}_{CC/NC}(N) = 0.10$, 0.20, 0.30, 0.40, 0.45, 0.50, 0.60, 0.70, 0.80, 0.90.

in the Sun. For the solar neutrino energies of interest one has $r \cong 0.155$. For Δm^2 and $\sin^2 2\theta$ from the LMA solution region, the transitions of the ^{8}B ν_e with energies $E \gtrsim 5.0$ MeV are adiabatic and in a relatively large sub-region one finds $\bar{P} \cong \sin^2\theta$. We would like to emphasize that the relation (10) is not very precise and can serve only for rough estimates.

The predicted values of $A^N_{D-N}(SK)$ and $A^N_{D-N}(SNO)$ in the LOW solution region differ less than in the case of the LMA solution since $\bar{P}$ is typically by a factor of ~ 1.5 larger for the LOW solution than in the case of the LMA solution. As it follows from Fig. 2, in the LOW solution region given by eq. (5) one has $(A^N_{D-N}(SNO))^{LOW} \cong (1.0 - 7.5)\%$. In the best fit point of the solution's region found in [3,5] and in [7] we get $(A^N_{D-N}(SNO))^{LOW}_{BF1,2} \cong 3\%$, while in the best fit point obtained in [8] one has $(A^{N,C}_{D-N}(SNO))^{LOW}_{BF3} < 0.5\%$. Similar predictions are valid for $A^N_{D-N}(SK)$ (Fig. 1). Obviously, an observation of $A^N_{D-N}(SNO) \gtrsim 10\%$ will strongly disfavor the LOW solution.

As Fig. 2 indicates, an observation of a non-zero D-N asymmetry which is definitely greater than 1%, $A^N_{D-N}(SNO) > 1\%$, would rule out the QVO solution which requires values of Δm^2 from the interval $\Delta m^2 \sim (5\times10^{-10}-5\times10^{-8})$ eV^2 and $\sin^2 2\theta \cong (0.70 - 1.0)$ (for a discussion of the QVO oscillations of solar neutrinos and of the QVO solution see, e.g., [3,5,6,18]).

3 Predictions for $R^{SNO}_{CC/NC}$

The importance of the measurement of the CC to NC solar neutrino event rate ratio in the SNO experiment, $R^{SNO}_{CC/NC}$, for determining the correct solution of the solar neutrino problem has been widely discussed (see, e.g., [19,20] and the references quoted therein). We have performed in [11] a high precision calculations of the double ratio $R^{SNO}_{CC/NC}$, in particular, for two different CC reaction cross-sections which were taken from [21,22], and using the electron number density distribution in the Sun from [9]. The differences in the results obtained for $R^{SNO}_{CC/NC}$ using the two cross sections are negligible in the regions of the LMA MSW and LOW solutions of the solar neutrino problem of interest. We have found that the effect of the SNO energy resolution function on the predictions for the double ratio $R^{SNO}_{CC/NC}$ is negligible as well.

We have derived results, in particular, for the average ratio during the day (*Day* ratio), $R^{SNO}_{CC/NC}(D)$, and during the night (*Full Night* ratio), $R^{SNO}_{CC/NC}(N)$, which are shown in Figs. 3 and 4. A comparison of Figs. 3 and 4 shows that CC to NC ratio increases substantially during the night for values of Δm^2 and $\sin^2 2\theta$ from the region 2×10^{-7} eV$^2 \lesssim \Delta m^2 \lesssim 2\times10^{-5}$ eV^2, $10^{-2} \lesssim \sin^2 2\theta \lesssim 0.98$, which, however, is not favored by the current solar neutrino data. The increase is due the Earth matter effect; the Earth mantle-core interference effect [23] also contributes to it. At $\Delta m^2 \gtrsim 8 \times 10^{-5}$ eV^2 in the LMA region, and at $\Delta m^2 \lesssim 10^{-7}$ eV^2 in the LOW-QVO region, the *Day* and the *Night* ratios practically coincide, $R^{SNO}_{CC/NC}(D) \cong R^{SNO}_{CC/NC}(N)$.

As the results exhibited in Figs. 3 and 4 indicate (see also Fig. 6 in [11]), when $T_{\rm e,th}$ is decreased from 6.75 MeV to 5.0 MeV, the ratios $R^{SNO}_{CC/NC}(X)$, $X = D, N$, change little and only in relatively small sub-regions of the LMA MSW and of the LOW solution regions (compare the upper and the lower panels in each of Figs. 3 and 4).

In the 99% C.L. LMA MSW solution region, eqs. (3) and (4), we find that each of the *Day* and *Night* double ratios $R^{SNO}_{CC/NC}(X)$, $X = D, N$, can take values in the interval $R^{SNO}_{CC/NC}(X) \cong (0.20 - 0.65)$ for both $T_{\rm e,th} = 6.75$ MeV and $T_{\rm e,th} = 5.00$ MeV. If $\Delta m^2 \lesssim 2 \times 10^{-4}$ eV^2, which corresponds to the 95% C.L. solution's region of ref. [3], we have $R^{SNO}_{CC/NC}(X) \cong (0.20 - 0.45)$, $X = D, N$. In the best fit points in the LMA MSW solution region, obtained in [3,5], [7] and [8], we get, respectively, $R^{SNO}_{CC/NC}(D) \cong 0.29;\ 0.28;\ 0.27$, $R^{SNO}_{CC/NC}(N) \cong 0.31;\ 0.29;\ 0.29$, for both values of $T_{\rm e,th}$, $T_{\rm e,th} = 6.75$ MeV; 5.00 MeV. The "best fit" ratios are very sensitive to the best fit value of $\sin^2 2\theta$.

In the case of the LOW solution, the interval of possible values of $R^{SNO}_{CC/NC}(X)$, $X = D, N$, is much narrower if Δm^2 and $\sin^2 2\theta$ lie within the region given by eq. (5): $R^{SNO}_{CC/NC}(X) \cong (0.38 - 0.45)$. Somewhat larger values of $R^{SNO}_{CC/NC}(X)$ - up to ~ 0.55, are possible in the 99.73% C.L. LOW solution regions derived in refs. [5,8]. In the LOW solution best fit points found in [3,5,7] and in [8], we obtain for $T_{\rm e,th} = 6.75$ MeV (5.00 MeV), respectively, $R^{SNO}_{CC/NC}(D) \cong 0.44\ (0.43)$ and 0.49, $R^{SNO}_{CC/NC}(N) \cong 0.45\ (0.44)$ and 0.49.

If the average probability of survival of the solar (^{8}B) ν_e with energy 8.2 MeV $\lesssim E \lesssim$ 14.0 MeV (the flux of which was measured by the SNO experiment [1]) does not exhibit i) a strong dependence on the neutrino energy and ii) a large day-night variation, we have, as it is not difficult to show,

$$R^{SNO}_{CC/NC} \cong \frac{\Phi^{CC}(\nu_e)}{\Phi^{CC}(\nu_e) + \Phi(\nu_{\mu,\tau})} \cong 0.32 \pm 0.07, \tag{11}$$

where $R^{SNO}_{CC/NC}$ is the averaged ratio over the period of SNO data-taking [1], and we have used eqs. (1) and (2). Taking into account an uncertainty corresponding to " 1 standard deviation" and to "2 standard deviations", we find from eq. (11), respectively, $0.25 \leq R^{SNO}_{CC/NC} \leq 0.39$ and $0.18 \leq R^{SNO}_{CC/NC} \leq 0.46$. As Figs. 3 and indicate, an upper limit $R^{SNO}_{CC/NC} \leq 0.45$ would imply that in the cases of the LMA MSW and of the LOW solutions one has $\Delta m^2 \lesssim 2 \times 10^{-4}$ eV^2 and $\Delta m^2 \gtrsim 6 \times 10^{-8}$ eV^2, respectively.

4 Constraining the LMA and LOW Solution Regions

It should be clear from the discussions in Sections 2 and 3 that a measured value of $A^N_{D-N}(SNO) > 1.0\%$ and/or of $R^{SNO}_{CC/NC} \leq 0.50$ in the SNO experiment can strongly diminish the regions of the allowed values of Δm^2 and $\sin^2 2\theta$ of the LMA and of the LOW solutions of the solar neutrino problem. As it follows from the results shown graphically in Figs. 2, 3 and 4, an experimental upper limit on $A^N_{D-N}(SNO)$ in the case of the LMA (LOW) solution would imply a lower (upper) limit on Δm^2; an experimental upper limit on $R^{SNO}_{CC/NC}$ would lead to an upper (lower) limit on Δm^2. Thus, even upper limits on $A^N_{D-N}(SNO)$ of the order of 10% and on $R^{SNO}_{CC/NC}$ of the order of 0.50 can significantly reduce the LMA and LOW solution regions.

5 Conclusions

The Super-Kamiokande solar neutrino data combined with the data from the SNO experiment provide a remarkable evidence for oscillations/transitions of the solar neutrinos into $\nu_{\mu,\tau}$ and/or $\bar{\nu}_{\mu,\tau}$. Global analyses of the solar neutrino data in terms of the hypothesis of $\nu_e \to \nu_{\mu,\tau}$ transitions of solar neutrinos favor the LMA MSW and the LOW-QVO solutions. The measurement of the D-N asymmetry $A^N_{D-N}(SNO)$ in the CC event rate, and of the ratio of the CC and NC event rates, $R^{SNO}_{CC/NC}$, in the SNO experiment can allow, in particular, to discriminate between these solutions. A value of $A^N_{D-N}(SNO) \gtrsim 10\%$ would favor the LMA solution, would strongly disfavor the LOW solution, while an observation of $A^N_{D-N}(SNO) > 1\%$ would rule out the QVO solution. For the LMA solution, the D-N asymmetry measured in SNO can be considerably larger than that measured in the Super-Kamiokande experiment [15]: one can have $A^N_{D-N}(SNO) \cong (1.5 - 2.0) A^N_{D-N}(SK)$. We find [11] further that in the LMA region at $\Delta m^2 \lesssim 2 \times 10^{-4}$ eV^2 one has $R^{SNO}_{CC/NC}(X) \cong (0.20 - 0.45)$. In the LOW solution region given by eq. (5) we obtain $R^{SNO}_{CC/NC}(X) \cong (0.38 - 0.45)$.

Finally, a measured value of $A^N_{D-N}(SNO) > 1.0\%$ and/or of $R^{SNO}_{CC/NC} \leq 0.50$ in the SNO experiment can strongly diminish the $\Delta m^2 - \sin^2 2\theta$ regions of the LMA and of the LOW-QVO solutions. Even upper limits on $A^N_{D-N}(SNO)$ of the order of 10% and on $R^{SNO}_{CC/NC}$ of the order of 0.50 can significantly reduce the LMA MSW and the LOW-QVO solution regions.

6 Acknowledgements

S.T.P. would like to thank the organizers of the International Workshop on Neutrino Oscillations and their Origin 2001, and especially Prof. Y. Suzuki, for the scientifically stimulating program of the workshop and kind hospitality.

References

1. SNO Collaboration, Q.R. Ahmad et al., nucl-ex/0106015.
2. Super-Kamiokande Collaboration, Y. Fukuda *et al.*, Phys. Rev. Lett. **86** (2001) 5651 and 5656.
3. G.F. Fogli et al., Phys. Rev. D **64** (2001) 093007.
4. B.T. Cleveland et al., Astrophys. J. **496** (1998) 505; Y. Fukuda *et al.*, Phys. Rev. Lett. **77** (1996) 1683; V. Gavrin, Nucl. Phys. Proc. Suppl. **91** (2001) 36; W. Hampel *et al.*, Phys. Lett. B **447** (1999) 127; M. Altmann *et al.*, Phys. Lett. B **490** (2000) 16.
5. J.N. Bahcall, M.C. Gonzalez-Garcia and C. Peña-Garay, JHEP **08** (2001) 014; P. Creminelli, G. Signorelli and A. Strumia, hep-ph/0102234; P.I. Krastev and A.Yu. Smirnov, hep-ph/0108177.
6. M.V. Garzelli and C. Giunti, hep-ph/0108191 and hep-ph/0111254.
7. J.N. Bahcall, M.C. Gonzalez-Garcia and C. Peña-Garay, hep-ph/0111150.
8. M. Smy, hep-ex/0202020.
9. J.N. Bahcall, M.H. Pinsonneault and S. Basu, Astrophys. J. **555** (2001) 990.
10. A.R. Junghans et al., submitted to Phys. Rev. Lett.
11. M. Maris and S.T. Petcov, hep-ph/0201087.
12. Q.Y. Liu, M. Maris and S.T. Petcov, Phys. Rev. D**56** (1997) 5991.
13. M. Maris and S.T. Petcov, Phys. Rev. D**56** (1997) 7444.
14. M. Maris and S.T. Petcov, Phys. Rev. D **58**, 113008 (1998).
15. M. Maris and S.T. Petcov, Phys. Rev. D **62** (2000) 093006.
16. N. Hata and P. Langacker, Phys. Rev. D**50** (1994) 632; A.J. Baltz and J. Weneser, Phys. Rev. D**D50** (1994) 5971.
17. V. Barger et al., hep-ph/0104166.
18. S.T. Petcov, Phys. Lett. B**214** (1988) 139; S.T. Petcov and J. Rich, Phys. Lett. B**224** (1989) 426; E. Lisi et al., Phys. Rev. D**63** (2001) 093002.
19. The SNO homepage: http://ewiserver.npl.washington.edu/sno.
20. J.N. Bahcall, P.I. Krastev and A. Yu. Smirnov, Phys. Rev. D**63** (2001) 053012; V. Barger, D. Marfatia and K. Whisnant, Phys. Lett. B**509** (2001) 19; G.L. Fogli, E. Lisi and D. Montanino, Phys. Lett. B**434** (1998) 333.
21. K. Kubodera, S. Nozawa, Int. J. Mod. Phys. E**3** (1994) 101.
22. S. Ying, W.C. Haxton, E.M. Henley, Phys. Rev. C**45** (1992) 1982.
23. S.T. Petcov, Phys. Lett. B**434** (1998) 321 (see also hep-ph/9809587 and hep-ph/9811205); M.V. Chizhov and S. T. Petcov, Phys. Rev. Lett. **83** (1999) 1906 and Phys. Rev. D**63** (2001) 073003.

NEUTRINOLESS DOUBLE BETA DECAY AND NEUTRINO OSCILLATIONS

H.V. KLAPDOR-KLEINGROTHAUS

Max-Planck-Institut für Kernphysik,
P.O. Box 10 39 80, D-69029 Heidelberg, Germany
Spokesman of HEIDELBERG-MOSCOW and GENIUS Collaborations
E-mail: klapdor@gustav.mpi-hd.mpg,
Home-page: http://www.mpi-hd.mpg.de.non_acc/

Double beta decay is indispensable to solve the question of the neutrino mass matrix together with ν oscillation experiments. Recent analysis of the most sensitive experiment since eight years - the HEIDELBERG-MOSCOW experiment in Gran-Sasso - yields an evidence for the neutrinoless decay mode. This result is the first evidence for lepton number violation and proves the neutrino to be a Majorana particle. We give the present status of the analysis in these Proceedings. It excludes several of the neutrino mass scenarios allowed from present neutrino oscillation experiments - only degenerate and inverse hierarchy mass scenarios survive. This result allows neutrinos to still play an important role as dark matter in the Universe. To improve the present accuracy, considerably enlarged experiments are required, such as GENIUS. A GENIUS Test Facility has just been funded and will come into operation by end of 2002.

1 Introduction

Recently atmospheric and solar neutrino oscillation experiments have shown that neutrinos are massive. This was the first indication of beyond standard model physics. The absolute neutrino mass scale was, however, still unknown, and only neutrino oscillations and neutrinoless double beta decay *together* can solve this problem (see, e.g. [3,4,5]). Another question of fundamental importance is, that whether the neutrino is a Dirac or Majorana particle. Only double beta decay can solve this problem. Finally double beta decay has a unique sensitivity to probe lepton number conservation.

In this paper we will discuss the status of double beta decay search. We shall, in section 2, discuss the expectations for the observable of neutrinoless double beta decay, the effective neutrino mass $\langle m_\nu \rangle$, from recent ν oscillation experiments. In section 3 we shall discuss the recent evidence for the neutrinoless decay mode, and discuss the consequences for the neutrino mass scenarios which could be realized in nature. In section 4 we discuss the possible future potential of $0\nu\beta\beta$ experiments, which could improve the present accuracy.

2 Allowed Ranges of $\langle m \rangle$ by ν Oscillation Experiments

The observable of double beta decay

$$\langle m \rangle = |\sum U_{ei}^2 m_i| = |m_{ee}^{(1)}| + e^{i\phi_2}|m_{ee}^{(2)}| + e^{i\phi_3}|m_{ee}^{(3)}|,$$

with U_{ei} denoting elements of the neutrino mixing matrix, m_i neutrino mass eigenstates, and ϕ_i relative Majorana CP phases, can be written in terms of oscillation parameters [4,5]

$$|m_{ee}^{(1)}| = |U_{e1}|^2 m_1, \tag{1}$$

$$|m_{ee}^{(2)}| = |U_{e2}|^2 \sqrt{\Delta m_{21}^2 + m_1^2}, \tag{2}$$

$$|m_{ee}^{(3)}| = |U_{e3}|^2 \sqrt{\Delta m_{32}^2 + \Delta m_{21}^2 + m_1^2}. \tag{3}$$

The effective mass $\langle m \rangle$ is related with the half-life for $0\nu\beta\beta$ decay via

$$\left(T_{1/2}^{0\nu}\right)^{-1} \sim \langle m_\nu \rangle^2, \tag{4}$$

and for the limit on $T_{1/2}^{0\nu}$ deducible in an experiment we have

$$T_{1/2}^{0\nu} \sim a\sqrt{\frac{Mt}{\Delta E B}}. \tag{5}$$

Here a is the isotopical abundance of the $\beta\beta$ emitter; M is the active detector mass; t is the measuring time; ΔE is the energy resolution; B is the background count rate.

Neutrino oscillation experiments fix or restrict some of the parameters in (1)–(3), e.g. in the case of normal hierarchy solar neutrino experiments yield Δm_{21}^2, $|U_{e1}|^2 = \cos^2\theta_\odot$ and $|U_{e2}|^2 = \sin^2\theta_\odot$. Atmospheric neutrinos fix Δm_{32}^2, and experiments like CHOOZ, looking for ν_e disappearance restrict $|U_{e3}|^2$. The phases ϕ_i and the mass of the lightest neutrino, m_1 are free parameters. Double beta decay can fix the parameter m_1 and thus the absolute mass scale. The expectations for $\langle m \rangle$ from oscillation experiments in different neutrino mass scenarios have been analyzed in [3,4,5] and are shown in Fig. 7.

3 Indications for the Neutrinoless Decay Mode

The status of present double beta experiments is shown in Fig. 1,2 and is extensively discussed in [6]. The HEIDELBERG-MOSCOW experiment using 11 kg of enriched ^{76}Ge in form of five HP Ge-detectors is running since August 1990 in the Gran-Sasso underground laboratory [6,2,8,32,26], and is the most sensitive one since more than 8 years (running since 1990).

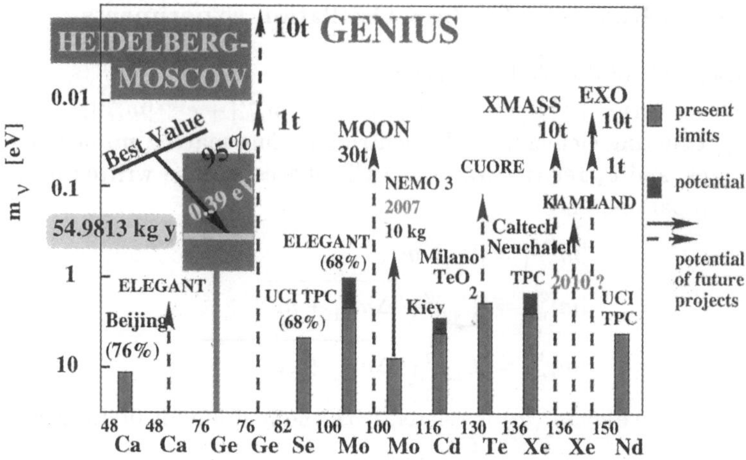

Figure 1. Present sensitivity, and expectation for the future, of the most sensitive $\beta\beta$ experiments. Given are limits for $\langle m \rangle$, except for the HEIDELBERG-MOSCOW experiment where the recently observed *value* is given (95% c.l. range), and the best value [1]. Framed parts of the bars: present status; not framed parts: expectation for running experiments; solid and dashed lines: experiments under construction or proposed, respectively. For references see [6,8,35,59].

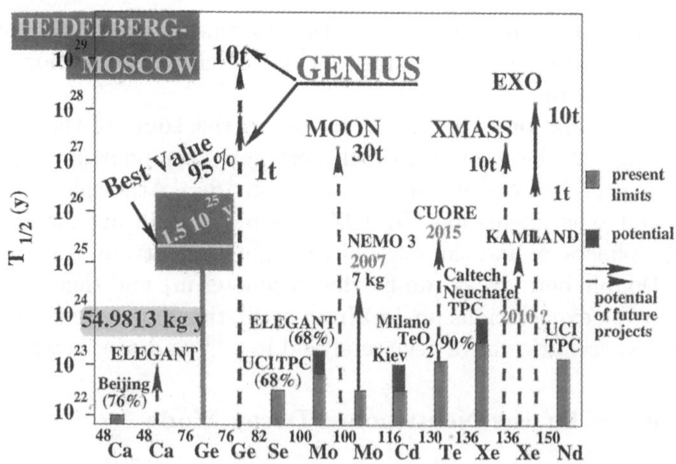

Figure 2. Present evidence for $0\nu\beta\beta$decay from data of the HEIDELBERG-MOSCOW experiment and the potential of present and future $\beta\beta$ experiments. Vertical axis - half life limits (only HEIDELBERG-MOSCOW gives a value) in years, horizontal - isotopes used in the various experiments.

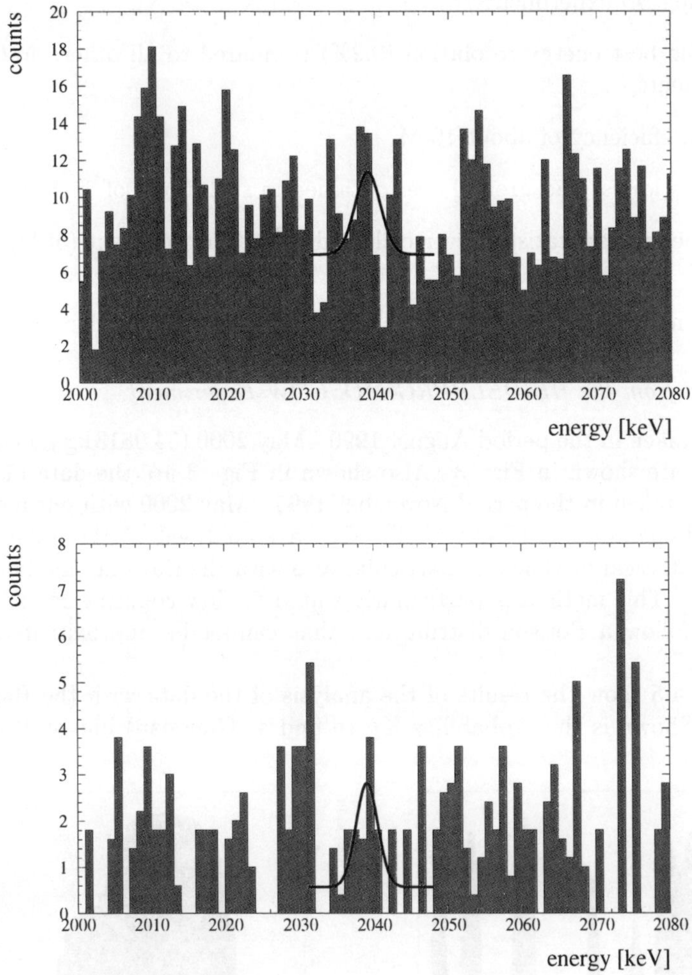

Figure 3. HEIDELBERG-MOSCOW experiment. Upper part: energy spectrum in the range between 2000 keV and 2080 keV, after 54.9813 kg y. $Q_{\beta\beta}$ = 2039.006(50) keV according to [18]. Lower part: Spectrum of single site events (SSE) after 28.053 kg y, corrected for the efficiency of SSE identification. The accepted events have been identified as SSE events by all our three methods of pulse shape analysis [9,10,11]. Shown are also the lines at $Q_{\beta\beta}$ identified by the Bayesian method (see text).

It has

- (a) the largest source strength (11.0 kg of ^{76}Ge enriched to 86%) of all running $\beta\beta$ experiments,
- (b) the best energy resolution (0.2%) compared to all other, no-Ge experiments,
- (c) an efficiency of about 100%,
- (d) the lowest background ever obtained in such types of experiment,
- (e) the largest statistics ever collected in a $\beta\beta$ experiment (55 kg y until 2000).

We communicate here the status of the analysis of November 2001.

3.1 Data from the HEIDELBERG-MOSCOW Experiment

The data taken in the period August 1990 - May 2000 (54.9813 kg y, or 723.44 mol-years are shown in Fig. 3. Also shown in Fig. 3 are the data of single site events taken in the period November 1995 - May 2000 with our methods of pulse shape analysis (PSA) [9,11,10]. We have analysed [1,8] those data with various statistical methods, in particular also with the Bayesian method (see, e.g. [13,14]). This method is particularly suited for low counting rates, where the data follow a Poisson distribution, that cannot be approximated by a Gaussian.

Figs. 4,5 show the results of the analysis of the data with the Bayesian method. Shown is the probability K_E to find a (Gaussian) line with known

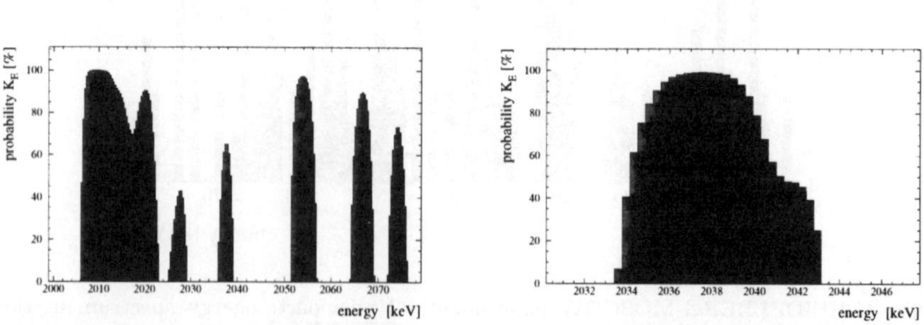

Figure 4. Left: Probability K_E that a line exists at a given energy in the range of 2000-2080 keV derived via Bayesian inference from the spectrum shown in Fig. 3 (left part). Right: Result of a Bayesian scan for lines as in the left part of this figure, but in the energy range of interest for double beta decay.

standard deviation of $\sigma = 1.70$ (1.59) keV corresponding to 4.00 (3.74) keV FWHM, as function of energy. This information was obtained by two-parametric Bayesian inference. We have asked for the intensity I of a line at a given energy E and for the background level. The background was assumed to be constant over the spectral range of 2000-2080 keV. For details of the procedure see [1,8]. The probability K_E is the ordinate in Figs. 4,5.

This procedure reproduces γ-lines at the position of known weak lines from the decay of ^{214}Bi at 2010.7, 2016.7, 2021.8 and 2052.9 keV [17]. The lines at 2010.7 and 2052.9 keV are observed at a confidence level of 3.7 and 2.6σ, respectively. The observed intensities are consistent with the expectations from the strong Bi lines in our spectrum, and the branching ratios given in [17], within the 2σ experimental error see [8]. The expectations here are calculated including pile-up effects, by Monte-Carlo simulation of our set-up. *Only* in this way the strong dependence of the relative intensities on the location of the impurities in the set-up can be properly taken into account.

In addition, a line centered at 2039 keV shows up. This is compatible with the Q-value [18,19] of the double beta decay process. We emphasize, that at this energy no γ-line is expected according to the Monte Carlo simulations of an the set-up, and the compilations in [17]. We have at present no convincing identification of the lines around 2070 keV indicated by the peak identification procedure.

It may be important to note, that essentially the same lines as found by the peak scan procedure in Figs. 4,5, are found [8] when doing the same kind of analysis with the best existing *natural* Ge experiment of D. Caldwell et al. [41], which has a by a factor of ~3 better statistics *of the background.* This experiment does, however, not see the line at 2039 keV - which is consistent

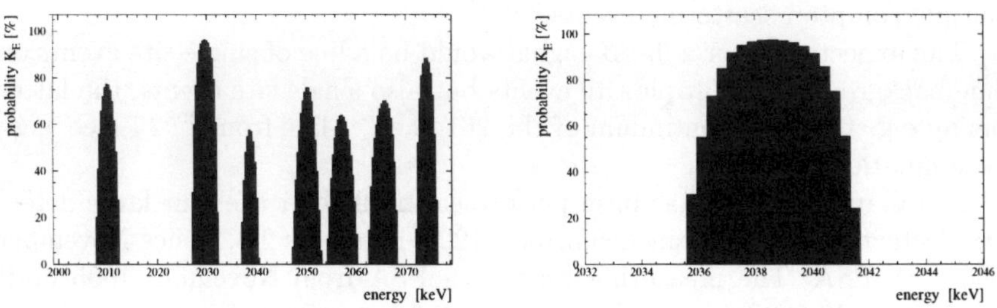

Figure 5. Scan for lines in the single site event spectrum taken from 1995-2000 with detectors Nr. 2,3,5, (Fig. 3, right part), with the Bayesian method. Left: Energy range 2000 -2080 keV. Right: Energy range of interest for double beta decay.

with the expectation of 0.6 events in that experiment for a $0\nu\beta\beta$ signal from the half-life given in Table 1.

On the left-hand side of Figs. 4,5, the background intensity (1-η) identified by the Bayesian procedure is too high because the procedure averages the background over all the spectrum (including lines) except for the line it is trying to single out. Inclusion of the known lines into the fit naturally improves the background. For example with a simultaneous fit of the range 2000 - 2060 keV of the spectrum shown in Fig. 3 (upper part) (assuming lines at 2010.78, 2016.70, 2021.60, 2052.94, 2034.76, 2039.0, 2041.16 keV), the probability for a line at 2039 keV is 86% (see Ref. [8]).

Finally, on the right-hand side of Fig. 4, (and also Fig. 5) the peak detection procedure is carried out within an energy interval that seems to not contain (according to the left-hand side) lines other than the one at $Q_{\beta\beta}$, and is broad enough (about $\pm$ 5 standard deviations of the Gaussian line) for a meaningful analysis. We find, with the Bayesian method, the probability K_E = 96.5% that there is a line at 2039.0 keV in the spectrum shown in Fig. 3. This is a confidence level of 2.1 σ in the usual language. The number of events is found to be 0.8 to 32.9 (7.6 to 25.2) with 95% (68%) c.l., with best value of 16.2 events. For the spectrum shown in Fig. 2 in [1], we find a probability for a line at 2039.0 keV of 97.4% (2.2 σ). In this case the number of events is found to be 1.2 to 29.4 with 95% c.l.. It is 7.3 to 22.6 events with 68.3% c.l.. The most probable number of events (best value) is 14.8 events.

We also applied the method recommended by the Particle Data Group [14]. This method (which does not use the information that the line is Gaussian) finds a line at 2039 keV on a confidence level of 3.1 σ (99.8% c.l.).

The spectrum of single site events selects events confined to a few mm region in the detector corresponding to the track length of the emitted electrons - such as double beta events, and rejects multiple-site events - such as multiple compton scattering events [9,10,11].

The expectation for a $0\nu\beta\beta$ signal would be a line of single site events on some background of multiple site events but also single site events, the latter coming e.g. from the continuum of the 2614 keV γ-line from ^{208}Tl (see, e.g., the simulation in [9]).

Installation of PSA has been performed in 1995 for the four large detectors. Detector Nr.5 runs since February 1995, detectors 2,3,4 since November 1995 with PSA. The measuring time with PSA from November 1995 until May 2000 is 36.532 kg years, for detectors 2,3,5 it is 28.053 kg y.

Fig. 3 (lower part) shows the SSE spectrum obtained under the restriction that the signal simultaneously fulfills the criteria of *all three* methods for a single site event. Fig. 6 shows typical SSE and MSE events from our spectrum.

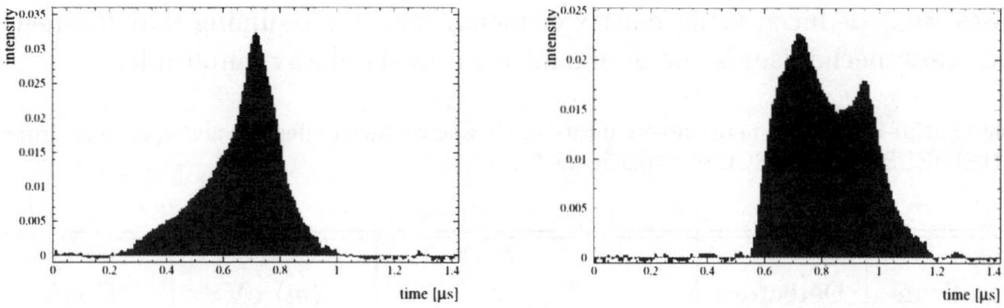

Figure 6. Left: Shape of one candidate for $0\nu\beta\beta$ decay (energy 2038.61 keV) classified as SSE by all three methods of pulse shape discrimination. Right: Shape of one candidate (energy 2038.97 keV) classified as MSE by all three methods.

In total, we find 9 SSE events in the region 2034.1 - 2044.9 keV ($\pm\ 3\sigma$ around $Q_{\beta\beta}$).

Analysis with the Bayesian method of the single site event spectrum shows again evidence for a line at the energy of $Q_{\beta\beta}$ (Fig. 5). Analyzing the range of 2032 - 2046 keV, we find the probability of 96.8% that there is a line at 2039.0 keV. We thus see a signal of single site events, as expected for neutrinoless double beta decay, precisely at the $Q_{\beta\beta}$ value [18]. The analysis of the line at 2039.0 keV before correction for the efficiency yields 4.6 events (best value) or (0.3 - 8.0) events within 95% c.l. ((2.1 - 6.8) events within 68.3% c.l.). Corrected for the efficiency to identify an SSE signal by successive application of all three PSA methods, which is 0.55 ± 0.10, we obtain a $0\nu\beta\beta$ signal with 92.4% c.l.. The signal is (3.6 - 12.5) events with 68.3% c.l. (best value 8.3 events). Thus, with proper normalization concerning the running times (kg y) of the full and the SSE spectra, we see that more than 90% of the full signal remain after the single site cut (Bayes analysis, best value), while the ^{214}Bi lines (best values) are considerably reduced.

The PDG method gives a signal at 2039.0 keV of 2.8 σ (99.4%). The possibility, that the single site signal is the double escape line corresponding to a (much more intense!) full energy peak of a γ-line at 2039+1022=3021 keV is excluded from the high-energy part of our spectrum.

3.2 Half-Life and Effective Neutrino Mass

We have shown that to our present knowledge the signal at $Q_{\beta\beta}$ does not originate from a background γ-line. On this basis we translate the observed number of events into half-lifes. In Table 1 we give conservatively the values obtained with the Bayesian method. Also given are the effective neutrino masses $\langle m_\nu \rangle$ deduced using matrix elements from [15], assuming that the neutrino mass mechanism is the dominant term in the decay amplitude.

Table 1. Half-life for the neutrinoless decay mode and deduced effective neutrino mass from the HEIDELBERG-MOSCOW experiment [8].

Significance $[kg\,y]$	Detectors	$T^{0\nu}_{1/2}$ y	$\langle m \rangle$ eV	Conf. level
54.9813	1,2,3,4,5	$(0.80 - 35.07) \times 10^{25}$	(0.08 - 0.54)	95% *c.l.*
54.9813	1,2,3,4,5	$(1.04 - 3.46) \times 10^{25}$	(0.26 - 0.47)	68% *c.l.*
54.9813	1,2,3,4,5	1.61×10^{25}	0.38	Best Value
46.502	1,2,3,5	$(0.75 - 18.33) \times 10^{25}$	(0.11 - 0.56)	95% *c.l.*
46.502	1,2,3,5	$(0.98 - 3.05) \times 10^{25}$	(0.28 - 0.49)	68% *c.l.*
46.502	1,2,3,5	1.50×10^{25}	0.39	Best Value
28.053	2,3,5 SSE	$(0.88 - 22.38) \times 10^{25}$	(0.10 - 0.51)	90% *c.l.*
28.053	2,3,5 SSE	$(1.07 - 3.69) \times 10^{25}$	(0.25 - 0.47)	68% *c.l.*
28.053	2,3,5 SSE	1.61×10^{25}	0.38	Best Value

We derive from the data taken with 46.502 kg y the half-life $T^{0\nu}_{1/2} = (0.8 - 18.3) \times 10^{25}$ y (95% c.l.). The analysis of the other data sets, shown in Table 1, and in particular of the single site events data, confirm this result.

The result obtained is consistent with the limits given earlier by the HEIDELBERG-MOSCOW experiment [2]. It is also consistent with all other double beta experiments - which still reach less sensitivity. A second Ge-experiment, [55] which has stopped operation in 1999 after reaching a significance of 8.8 kg y, yields (if one believes their method of 'visual inspection' in their data analysis), in a conservative analysis, a limit of $T^{0\nu}_{1/2} > 0.55 \times 10^{25}$ y (90% c.l.). The ^{128}Te geochemical experiment [42] yields $\langle m_\nu \rangle < 1.1$ eV (68 % c.l.), the ^{130}Te cryogenic experiment [50,48] yields $\langle m_\nu \rangle < 1.8$ eV and the $CdWO_4$ experiment [44] $\langle m_\nu \rangle < 2.6$ eV, all derived with the matrix elements of [15] to make the results comparable to the present value.

Concluding we obtain, with about 95% probability, a first indication for the neutrinoless double beta decay mode. As a consequence, at this confidence level, lepton number is not conserved. Further the neutrino is a Majorana particle. The effective mass $\langle m \rangle$ is deduced to be $\langle m \rangle = (0.11 - 0.56)$ eV (95% c.l.), with best value of 0.39 eV. Allowing conservatively for an uncertainty of the nuclear matrix elements of ± 50% (for detailed discussions of the status

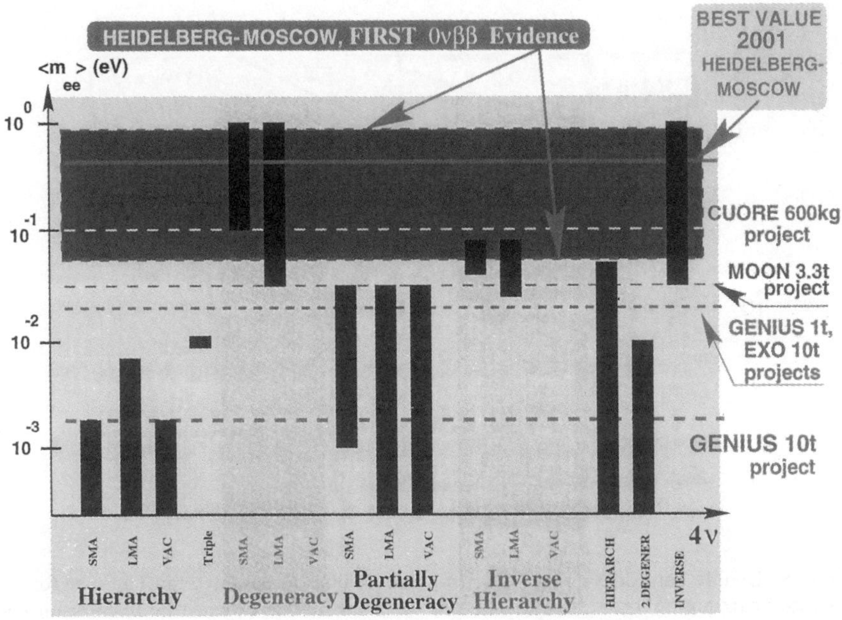

Figure 7. The impact of the evidence obtained for neutrinoless double beta decay in this paper (best value of the effective neutrino mass $\langle m \rangle = 0.39\,\mathrm{eV}$, 95% confidence range (0.05 - 0.84) eV - allowing already for an uncertainty of the nuclear matrix element of $\pm$ 50%) on possible neutrino mass schemes. The bars denote allowed ranges of $\langle m \rangle$ in different neutrino mass scenarios, still allowed by neutrino oscillation experiments (see [3]). Hierarchical models are excluded by the new $0\nu\beta\beta$ decay result [1]. Also shown are the expected sensitivities for the future potential double beta experiments CUORE, MOON, EXO and the 1 ton and 10 ton project of GENIUS [6,34,61].

of nuclear matrix elements we refer to [6,16] and references therein) this range may widen to $\langle m \rangle = (0.05 - 0.84)\,\mathrm{eV}$ (95% c.l.).

In this conclusion, it is assumed that contributions to $0\nu\beta\beta$ decay from processes other than the exchange of a Majorana neutrino (see, e.g. [6] and references therein) are negligible.

With the value deduced for the effective neutrino mass, the HEIDELBERG-MOSCOW experiment excludes several of the neutrino mass scenarios allowed from present neutrino oscillation experiments (see Fig. 7) - allowing only for degenerate mass scenarios and an inverse hierarchy 3ν - scenario. In particular hierarchical mass schemes are excluded at the present level of accuracy.

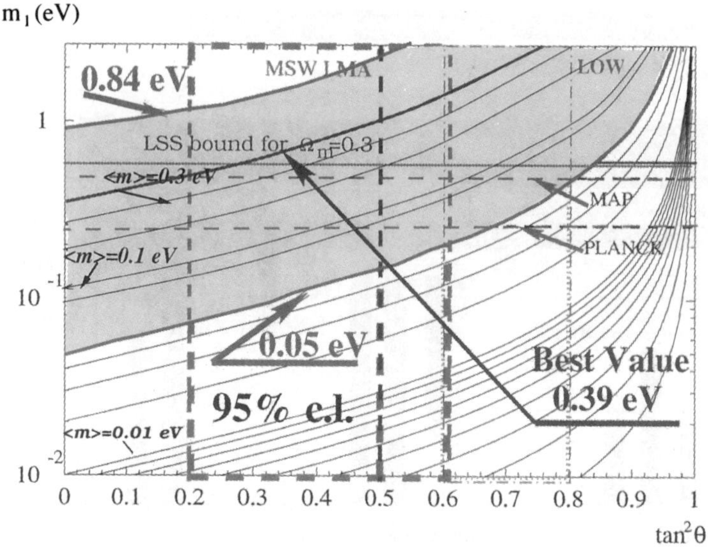

Figure 8. Double beta decay observable $\langle m \rangle$ and oscillation parameters: The case for degenerate neutrinos. Plotted on the axes are the overall scale of neutrino masses m_0 and mixing $\tan^2\theta_{12}$. Also shown is a cosmological bound deduced from a fit of CMB and large scale structure [23] and the expected sensitivity of the satellite experiments MAP and PLANCK. The present limit from tritium β decay of 2.2 eV [29] would lie near the top of the figure. The range of $\langle m \rangle$ fixed by the HEIDELBERG-MOSCOW experiment [1,8] is, in the case of small solar neutrino mixing, already in the range to be explored by MAP and PLANCK [23].

Assuming the degenerate scenarios to be realized in nature we fix - according to the formulae derived in [4] - the common mass eigenvalue of the degenerate neutrinos to m = (0.05 - 3.4) eV. Part of the upper range is already excluded by tritium experiments, which give a limit of m $<$ 2.2 eV (95% c.l.) [29]. The full range can only partly (down to $\sim$ 0.5 eV) be checked by future tritium decay experiments, but could be checked by some future $\beta\beta$ experiments (see, e.g. next section) The deduced best value for the mass is consistent with expectations from experimental $\mu \rightarrow e\gamma$ branching limits in models assuming the generating mechanism for the neutrino mass to be also responsible for the recent indication for as anomalous magnetic moment of the muon [68]. It lies in a range of interest also for Z-burst models recently discussed as explanation for super-high energy cosmic ray events beyond the GKZ-cutoff [67]. The range of $\langle m \rangle$ fixed in this work is, already now, in the range to be explored by the satellite experiments MAP and PLANCK [23] (see

Fig. 8).

The neutrino mass deduced allows neutrinos to still play an important role as hot dark matter in the Universe.

4 Future of $\beta\beta$ Experiments

To improve the present sensitivity for the effective neutrino mass considerably, and to fix this quantity more accurately, requires new experimental approaches, as discussed extensively in [6,59,28,20,21,22,28]. Some of them are indicated in Figs. 1,2,7.

Of present generation experiments probably no one has a potential to probe $\langle m \rangle$ below the present HEIDELBERG-MOSCOW level (see Fig. 1,2).

The Milano cryogenic experiment using TeO_2 bolometers improved their values for the $\langle m \rangle$ from $\beta\beta$ decay of ^{130}Te, from 5.3 eV in 1994 to 1.8 eV in 2000 [48]. NEMO-III, originally aiming at a sensitivity of 0.1 eV, reduced their goals recently to $0.3 \div 0.7$ eV (see [52]) (which is more consistent with estimates given by [51]), to be reached in some years after starting of running, foreseen for the year 2002.

4.1 GENIUS and other Proposed Future Double Beta Experiments

With the era of the HEIDELBERG-MOSCOW experiment the time of the small smart experiments is over.

To reach significantly larger sensitivity, $\beta\beta$ experiments have according to eq. (5) to become large. On the other hand source strengths of up to 10 tons of enriched material touch the world production limits. This means that the background has to be reduced by the order a factor of 1000 and more compared to that of the HEIDELBERG-MOSCOW experiment.

Table 2 lists some key numbers for GENIUS, [20,21,28] which was the first proposal for a third generation double beta experiment, and of some other proposals made after the GENIUS proposal. The potential of some of them is shown also in Fig. 7, and it is seen that not all of them will lead to large improvements in sensitivity (see also Table 3). Since it was realized in the HEIDELBERG-MOSCOW experiment, that the remaining small background is coming from the material close to the detector (holder, copper cap, ...), elimination of *any* material close to the detector will be decisive. Experiments which do not take this into account, like, e.g. CUORE [48], and MAJORANA [56], will allow only rather limited steps in sensitivity. The latter do not really apply a striking new strategy for background reduction, particularly also after it was found that the projected segmentation of detectors for MAJORANA may not work.

Table 2. Some key numbers of future double beta decay experiments (and of the HEIDELBERG-MOSCOW experiment). Explanations: ∇ - assuming the background of the present pilot project. $**$ - with matrix element from [15], [16], [38], [39], [40] (see Table II in [27] and *including* an assumed uncertainty of $\pm 50\%$ of the nuclear matrix element). $\triangle$ - this case shown to demonstrate **the ultimate limit** of such experiments. For details see [6].

$\beta\beta$– *Isotope*	*Name*	*Status*	Mass (*tonnes*)	Assumed backgr. † events/ kg y keV, ‡ events/kg y FWHM, $*$ events /yFWHM	*Running* Time (tonn. years)	Results limit for $0\nu\beta\beta$ half-life (years)	$<m_\nu>$ (eV)
^{76}**Ge**	**HEIDEL-BERG MOSCOW** [1,2,34,8]	**running since 1990**	0.011 (enriched)	† 0.06 ‡ 0.24 $*$ 2	**54.98 kg y**	**(0.8 – 18.3) x 10^{25}y 95% c.l. NOW !!**	**0.05-0.84 eV)** 95% c.l. NOW !!**
^{100}**Mo**	NEMO III [52]	*under constr. 2002*	~0.01 (enri-ched)	† **0.0005** ‡ 0.2 $*$ 2	50 kg y	10^{24}	0.3-0.7
^{130}**Te**	CUORE$^{\nabla}$ [49]	*idea since 1998*	0.75 (natural)	† 0.5 ‡ 4.5/$*$ 1000	5	$9 \cdot 10^{24}$	0.2-0.5
^{130}**Te**	CUORE [49,54]	*idea since 1998*	0.75 (natural)	† 0.005 ‡ 0.045/ $*$ 45	5	$9 \cdot 10^{25}$	0.07-0.2
^{100}**Mo**	MOON [43,57]	*idea since 1999*	10 (enrich.) 100(nat.)	?	30 300	?	0.03
^{116}**Cd**	CAMEOII CAMEOIII[58]	*idea since 2000*	0.65 1(enr.)	$*$ 3. ?	5-8 5-8	10^{26} 10^{27}	0.06 0.02
^{136}**Xe**	EXO [45,46]	Proposal since 1999	1 10	$*$ 0.4 $*$ 0.6	5 10	$8.3 \cdot 10^{26}$ $1.3 \cdot 10^{28}$	0.05-0.14 0.01-0.04
^{76}**Ge**	**GENIUS - TF** [36,37]	*under constr. end 2002?*	11 kg (enr.)	† $\mathbf{6 \cdot 10^{-3}}$	3	$\mathbf{1.6 \cdot 10^{26}}$	**0.15**
^{76}**Ge**	**GENIUS** [20,21]	Proposal since 1997	1 (enrich.) 1	† $\mathbf{0.04 \cdot 10^{-3}}$ ‡ $\mathbf{0.15 \cdot 10^{-3}}$ $*$ **0.15** $*$ **1.5**	1 10	$\mathbf{5.8 \cdot 10^{27}}$ $\mathbf{2 \cdot 10^{28}}$	**0.02-0.05** **0.01-0.028**
^{76}**Ge**	**GENIUS** [20,21]	Proposal since 1997	10 (enrich.)	‡ $\mathbf{0.15 \cdot 10^{-3}}$ $\mathbf{0^{\triangle}}$	10 10	$\mathbf{6 \cdot 10^{28}}$ $\mathbf{5.7 \cdot 10^{29}}$	**0.006 - 0.016** **0.002 - 0.0056**

EXO [45] needs still very extensive research and development to probe the applicability of the proposed detection method. In particular if it would be confirmed that tracks will be too short to be identified, it would act essentially only as a highly complicated calorimeter. Another crucial point is - eq. (5) - the energy resolution, which can be optimized *only* in experiments using Germanium detectors or bolometers. It will be difficult to probe evidence for this rare decay mode in experiments, which have to work - as result of their limited resolution - with energy windows around $Q_{\beta\beta}$ of several hundreds of keV, such as NEMO III [53], EXO [46]. E.g. for EXO, its limited energy resolution of 130 keV [47,46] would have the consequence, that half of the final mass of 1 ton would already be needed, only to compensate this bad energy resolution, if comparing with the sensitivity of the HEIDELBERG-MOSCOW experiment.

In the GENIUS project a reduction by a factor of more than 1000 down to a background level of 0.1 events/tonne y keV in the range of $0\nu\beta\beta$ decay is planned to be reached by removing all material close to the detectors, and by using naked Germanium detectors in a large tank of liquid nitrogen. It has been shown that the detectors show excellent performance under such conditions [21,31]. For technical questions and extensive Monte Carlo simulations of the GENIUS project for its application in double beta decay we refer to [21,31].

Table 3. Some of the new projects under discussion for future double beta decay experiments (see ref.[6]).

NEW PROJECTS				
	BACKGROUND REDUCTION	MASS INCREASE	POTENTIAL FOR DARK MATTER	POTENTIAL FOR SOLAR ν s
GENIUS	+	+	+	+ *)
CUORE	(+)	+	−	−
MOON	(+)	+	−	+
EXO	+	+	−	−
MAJORANA	−	+	−	−
*) real time measurement of pp neutrinos with threshold of 20 keV (!!)				

4.2 GENIUS and Other Beyond Standard Model Physics

GENIUS will allow besides the large increase in sensitivity for double beta decay described above, the access to a broad range of other beyond SM physics topics in the multi-TeV range. Already now $\beta\beta$ decay probes the TeV scale on which new physics should manifest itself (see, e.g. [20,33,34]). Basing to a large extent on the theoretical work of the Heidelberg group in the last six years, the HEIDELBERG-MOSCOW experiment yields results for SUSY models (R-parity breaking, neutrino mass), leptoquarks (leptoquarks-Higgs coupling), compositeness, right-handed W mass, nonconservation of Lorentz invariance and equivalence principle, mass of a heavy left or righthanded neutrino, competitive to corresponding results from high-energy accelerators like TEVATRON and HERA. The potential of GENIUS extends into the multi-TeV region for these fields and its sensitivity would correspond to that of LHC or NLC and beyond (for details see [6,33,34]).

4.3 GENIUS-Test Facility

Construction of a test facility for GENIUS – GENIUS-TF – consisting of ~ 40 kg of HP Ge detectors suspended in a liquid nitrogen box has been started. Up to summer of 2001, six detectors each of ~ 2.5 kg and with a threshold of as low as ~ 500 eV have been produced.

Besides test of various parameters of the GENIUS project, the test facility would allow, with the projected background of 2-4 events/(kg y keV) in the low-energy range, to probe the DAMA evidence for dark matter by the seasonal modulation signature, (see [36,37]).

5 Conclusion

The status of present double beta decay search has been discussed, and recent evidence for a non-vanishing Majorana neutrino mass obtained by the HEIDELBERG-MOSCOW experiment has been presented. Future projects to improve the present accuracy of the effective neutrino mass have been discussed. The most sensitive of them and perhaps at the same time most realistic one, is the GENIUS project. GENIUS is the only of the new projects (see also Table 3) which simultaneously has a huge potential for cold dark matter search, and for real-time detection of low-energy neutrinos (see [21,28,30,34,57,59]).

References

1. H.V. Klapdor-Kleingrothaus, A. Dietz, H.V. Harney and I.V. Krivosheina, *Modern Physics Letters* **A 16, No. 37** (2001) 2409-2420.
2. H.V. Klapdor-Kleingrothaus et al.,(HEIDELBERG-MOSCOW Collaboration), *Eur. Phys.* **J. A 12** (2001) 147 and hep-ph/0103062 and in Proc. of "Third International Conference on Dark Matter in Astro- and Particle Physics", DARK2000 H.V. Klapdor-Kleingrothaus (Editor), Springer-Verlag Heidelberg, (2001) 520 - 533 and *http://www.mpi-hd.mpg.de/non_acc/*.

3. H.V. Klapdor-Kleingrothaus and U. Sarkar, Modern Physics Letters **A 16, No. 38** (2001) 2469-2482.
4. H.V. Klapdor-Kleingrothaus, H. Päs and A.Yu. Smirnov, *Phys. Rev.* **D 63** (2001) 073005 and *hep-ph/***0003219**.
5. H.V. Klapdor-Kleingrothaus, H. Päs and A.Yu. Smirnov, in Proc. of DARK'2000, Heidelberg, 10-15 July, 2000, Germany, ed. H.V. Klapdor-Kleingrothaus, *Springer, Heidelberg* (2001) 420 - 434.
6. H.V. Klapdor-Kleingrothaus, "60 Years of Double Beta Decay - From Nuclear Physics to Beyond the Standard Model", *World Scientific, Singapore* (2001) 1281 p.
7. H.V. Klapdor-Kleingrothaus, to be Publ. in *Part. Nucl. Lett.* (2002), Proc. NANP2001, ed. V. Bednjakov, et al.; and H.V. Klapdor-Kleingrothaus, to be publ. in Proc. of TAUP'01 (2002), Gran-Sasso, Italy, September 7-13, 2001, eds A. Bettini et al.
8. H.V. Klapdor-Kleingrothaus et al., to be publ.
9. J. Hellmig and H.V. Klapdor-Kleingrothaus, *Nucl. Instrum. Meth.* **A 455** (2000) 638 - 644.
10. J. Hellmig, F. Petry and H.V. Klapdor-Kleingrothaus, *Patent* **DE19721323A**.
11. B. Majorovits and H.V. Klapdor-Kleingrothaus. *Eur. Phys. J.* **A 6** (1999) 463.
12. B. Maier, Dissertation, November 1995, MPI-Heidelberg; F. Petry, Dissertation, November 1995, MPI-Heidelberg; J. Hellmig, Dissertation, November 1996, MPI-Heidelberg; B. Majorovits, Dissertation, December 2000, MPI-Heidelberg; A. Dietz, Dipl. Thesis, Univ. Heidelberg, 2000 (unpublished), and Dissertation, in preparation.
13. G. D'Agostini, hep-ex/0002055, W. von der Linden and V. Dose, Phys. Rev. **E 59** 6527 (1999), and F.H. Fröhner, JEFF Report **18** NEA OECD (2000) and Nucl. Sci. a. Engineering **126** (1997) 1, K. Weise and W. Wöger, "Messunsicherheit und Messdatenauswertung", Wiley-VCH, Weinheim 1999, and P.M. Lee, "Bayesian Statistics: An Introduction", Second edition, Arnold, London 1997, and A. O'Hagan, "Bayesian Inference", Volume 2B of "Kendall's Advanced Theory of Statistics", Arnold, London, 1997.
14. Particle Data Group, R.M. Barnett et al., *Phys. Rev.* **D 54** (1996) 155.
15. A. Staudt, K. Muto and H.V. Klapdor-Kleingrothaus, *Europh. Lett.* **13** (1990) 31.
16. T. Tomoda, *Rept. Prog. Phys.* **54** (1991) 53 - 126.
17. R.B. Firestone and V.S. Shirley, Table of Isotopes, 8-th Edition, *John Wiley and Sons*, Incorp., N.Y. (1998) and S.Y.F. Chu, L.P. Ekström and R.B. Firestone The Lund/LBNL Nuclear Data Search, Vers.2.0, Febr.1999; and http://nucleardata.nuclear.lu.se/nucleardata/toi/radSearch.asp
18. G. Douysset et al., *Phys. Rev. Lett.* **86** (2001) 4259 - 4262.
19. J.G. Hykawy et al., *Phys. Rev. Lett.* **67** (1991) 1708.

20. H.V. Klapdor-Kleingrothaus in Proceedings of BEYOND'97, First International Conference on Particle Physics Beyond the Standard Model, Castle Ringberg, Germany, 8-14 June 1997, edited by H.V. Klapdor-Kleingrothaus and H. Päs, *IOP Bristol* (1998) 485 - 531 and *Int. J. Mod. Phys.* **A 13** (1998) 3953 .
21. H.V. Klapdor-Kleingrothaus et al. **MPI-Report MPI-H-V26-1999** and *Preprint: hep-ph/***9910205** and in Proceedings of the 2nd Int. Conf. on Particle Physics Beyond the Standard Model BEYOND'99, Castle Ringberg, Germany, 6-12 June 1999, edited by H.V. Klapdor-Kleingrothaus and I.V. Krivosheina, *IOP Bristol* (2000) 915 - 1014.
22. H.V. Klapdor-Kleingrothaus, in Proc. of 18th Int. Conf. on NEUTRINO 98, Takayama, Japan, 4-9 Jun 1998, (eds) Y. Suzuki et al. *Nucl. Phys. Proc. Suppl.* **77** (1999) 357 - 368.
23. R.E. Lopez, *astro-ph/***9909414**; J.R. Primack and M.A.K. Gross, *astro-ph/***0007165**; J.R. Primack, *astro-ph/***0007187**; J. Einasto, in Proc. of DARK2000, Heidelberg, Germany, July 10-15, 2000, Ed. H.V. Klapdor-Kleingrothaus, *Springer, Heidelberg,* (2001).
24. Y. Suzuki in Proc. of NEUTRINO2000, ed. J. Law et al. (2001), Nucl. Phys. **B 91** (2001) 29 - 35.
25. M.C. Gonzalez-Garcia et al., *hep-ph/***0009350** and *Phys. Rev.* **D 63** (2001) 033005.
26. HEIDELBERG-MOSCOW Collaboration (M. Günther et al.), Phys. Rev. **D 55** (1997) 54.
27. HEIDELBERG-MOSCOW Coll., *Phys. Rev. Lett.* **83** (1999) 41 - 44.
28. H.V. Klapdor-Kleingrothaus, in Proc. of Int. Conference NOW2000 - "Origins of Neutrino Oscillations", ed. G. Fogli, *Nucl. Phys.* **B 100** (2001) 309 - 313.
29. J. Bonn et al., in Proc. of NEUTRINO2000, (2000), ed. J. Law et al. *Nucl. Phys.* **B 91** (2001) 273 - 279.
30. V.A. Bednyakov and H.V. Klapdor-Kleingrothaus, *hep-ph/***0011233** and *Phys. Rev.* **D 63** (2001) 095005.
31. H.V. Klapdor-Kleingrothaus, J. Hellmig and M. Hirsch, *J. Phys.* **G 24** (1998) 483.
32. H.V. Klapdor-Kleingrothaus, in Proc. of the Int. Symposium on Advances in Nuclear Physics, eds.: D. Poenaru and S. Stoica, *World Scientific, Singapore* (2000) 123 - 129.
33. H.V. Klapdor-Kleingrothaus, in Proc. of Int. Symposium on Lepton and Baryon Number Violation, Trento, Italy, 20-25 April, 1998, ed. H.V. Klapdor-Kleingrothaus and I.V. Krivosheina, *IOP, Bristol,* (1999) 251-301 and *Preprint: hep-ex/***9901021**.
34. H.V. Klapdor-Kleingrothaus, *Springer Tracts in Modern Physics,* **163** (2000) 69-104, *Springer-Verlag, Heidelberg, Germany* (2000).
35. H.V. Klapdor-Kleingrothaus, in Proc. Int. Workshop on Low Energy Solar Neutrinos, LowNu2, Dec.4-5. ed: Y. Suzuki et al. *WS, Singapore* (2001).
36. H.V. Klapdor-Kleingrothaus et al., Internal Report **MPI-H-V32-2000**.

37. H.V. Klapdor-Kleingrothaus et al., *Nucl. Instr. Meth.* **A 481** (2002) 149-159.
38. W.C. Haxton and G.J. Stephenson, *Prog. Part.Nucl.Phys.* **12** (1984) 409-479.
39. X.R. Wu, A. Staudt, H.V. Klapdor-Kleingrothaus, Cheng-Rui Ching and Tso-Hsiu Ho, *Phys. Lett.* **B 272** (1991) 169 - 172.
40. X.R. Wu, A. Staudt, T.T.S. Kuo and H.V. Klapdor-Kleingrothaus, *Phys. Lett.* **B 276** (1992) 274 - 278.
41. D. Caldwell, J. *Phys. G* **17** (1991) S137-S144.
42. T. Bernatowicz, R. Cowsik, C. Hohenberg and F. Podosek, Phys. Rev. Lett. **69** (1992) 2341 - 2344 and Phys. Rev. **C 47** (1993) 806 - 825
43. H. Ejiri et al., *Phys. Rev. Lett.* **85** (2000) 2917-2920 and *Preprint: nucl-ex/***9911008**.
44. F. A. Danevich et al., Phys. Rev. **C 62** (2000) 045501.
45. M. Danilov et al., *Phys. Lett.* **B 480** (2000) 12 - 18.
46. G. Gratta in Proc. of LowNu2, Dec. 4-5 (2000) Tokyo, Japan, ed: Y. Suzuki, World Scientific (2001).
47. G. Gratta, talk given on ApPEC, Astroparticle Physics European Coordination, Paris, France 21.01.2002.
48. A. Alessandrello et al., *Phys. Lett.* **B 486** (2000) 13-21 and S. Pirro et al., *Nicl. Instr. Methods* **A 444** (2000) 71-76.
49. A. Giuliani (CUORE Collaboration), in Proc. "Lepton-Baryon'98", eds. H.V. Klapdor-Kleingrothaus and I.V. Krivosheina, Trento, Italy, April 20 - 25, 1998, *IOP, Bristol* (1999) 302 - 308.
50. (CUORE Coll.), talk given on ApPEC, Astroparticle Physics European Coordination, Paris, France 21.01.2002.
51. V.I. Tretyak and Yu.G. Zdesenko, *At. Data Nucl. Data Tables* **61** (1995) 43-62.
52. NEMO Collaboration, Contr. paper for XIX Int. Conf. NEUTRINO2000, Sudbury, Canada, June 16 - 21, 2000 LAL 00-31 (2000) 1 - 10 and NEMO-III Collaboration in Proc. Intern. Conf. of NANPino2000, Dubna, Russia, July 2000, ed. V. Bednjakov et al. (2001).
53. X. Sarazin (for the NEMO Collaboration), talk given on ApPEC, Astroparticle Physics European Coordination, Paris, France 21.01.2002; (NEMO Collaboration), Contr. paper for XIX Int. Conf. NEUTRINO2000, Sudbury, Canada, June 16 - 21, 2000 LAL 00-31 (2000) 1 - 10 and (NEMO-III Collaboration) in Proc. of NANPino2000, Dubna, Russia, July 2000, ed. V. Bednjakov, *Part. and Nucl. Lett.* **3** (2001) 62.
54. E. Fiorini in Proc. of Int. Conf. NEUTRINO2000, ed. J. Law et al. *Nucl. Phys.* **B 91** (2001) 262 - 269.
55. C.E. Aalseth et al. (IGEX Collaboration), *Yad. Fiz.* **63, No 7** (2000) 1299 - 1302.
56. L. DeBraekeleer, talk at Worksh. on the Next Generation U.S. Underground Science Facility, WIPP, June 12-14, 2000, Carlsbad, New Mexico, USA; C.E. Aalseth et al. (Majorana Collaboration), in Proc. of TAUP'2001, Gran Sasso, Italy, September 2001, ed. A. Bettini et al. (2002).

57. Proc. Int. Workshop on Low Energy Solar Neutrinos, LowNu2, December 4 and 5 (2000) Tokyo, Japan, ed: Y. Suzuki, World Scientific, Singapore (2001).
58. G. Bellini et al. *Phys. Lett.* **B 493** 216 - 228 (2000).
59. H.V. Klapdor-Kleingrothaus, in Proc. of NANPino-2000, Dubna, July 19-22, 2000, *Particles and Nuclei, Letters* iss. 1/2 (2001) and *hep-ph/* **0102319**.
60. H. Minakata and H. Nunokawa, *Phys. Lett.* **B 504** (2001) 301 - 308 and *hep-ph*/0010240.
61. H.V. Klapdor-Kleingrothaus, hep-ph/0103074 and in Proc. of NOON 2000, Tokyo, Japan, 6-18 Dec. 2000, ed: Y. Suzuki et al., *World Scientific* (2001).
62. H.V. Klapdor-Kleingrothaus, Prog. Part. Nucl. Phys. **48, issue 1** (2002).
63. C. Weinheimer, in Proc. of 3rd International Conference on Dark Matter in Astro and Particle Physics (Dark 2000), Heidelberg, Germany, 10-16 July, 2000, ed. H.V. Klapdor-Kleingrothaus, Springer, Heidelberg (2001) 513 - 519.
64. X. Wang, M. Tegmark and M. Zaldarriaga, *astro-ph/* **0105091**.
65. M. Tegmark et al., hep-ph/0008145.
66. V. Barger, S.L. Glashow, D. Marfatia and K. Whisnant, *hep-ph/***0201262**.
67. T.J. Weiler, in Proc. Beyond the Desert 1999, 'Accelerator, Non-Accelerator and Space Approaches', Ringberg Castle, Tegernsee, Germany, 6-12 Juni 1999, edited by H.V. Klapdor-Kleingrothaus and I.V. Krivosheina, IOP Bristol (2000) 1085 - 1106; H. Päs and T.J. Weiler, *Phys. Rev.* **D 63** (2001) 113015.
68. E. Ma and M. Raidal, *Phys. Rev. Lett.* **87** (2001) 011802; Erratum-ibid. **87** (2001) 159901.

ABSOLUTE NEUTRINO MASSES: PHYSICS BEYOND SM, DOUBLE BETA DECAY AND COSMIC RAYS

HEINRICH PÄS

Institut für Theoretische Physik und Astrophysik, Universität Würzburg, Am Hubland, 97074 Würzburg, Germany
E-mail: paes@physik.uni-wuerzburg.de

THOMAS J. WEILER

Department of Physics and Astronomy, Vanderbilt University, Nashville, TN 37235, USA
E-mail: tom.weiler@vanderbilt.edu

Absolute neutrino masses provide a key to physics beyond the standard model. We discuss the impact of absolute neutrinos masses on physics beyond the standard model, the experimental possibilities to determine absolute neutrinos masses, and the intriguing connection with the Z-burst model for extreme-energy cosmic rays.

1 Introduction

Solar and atmospheric neutrino oscillations have established solid evidence for non-vanishing neutrino masses and determined mass squared differences δm^2 and the mixing matrix U with increasing accuracy. The puzzle of the absolute mass scale for neutrinos, though, is still unsolved [1]. In fact it is a true experimental challenge to determine an absolute neutrino mass below 1 eV. Three approaches have the potential to accomplish the task, id sunt larger versions of the tritium end-point distortion measurements, limits from the evaluation of the large scale structure in the universe, and next-generation neutrinoless double beta decay ($0\nu\beta\beta$) experiments. In addition there is a fourth possibility: the extreme-energy cosmic-ray experiments in the context of the recently emphasized Z-burst model.

This article is organized as follows: in section 2 the specific role of the neutrino among the elementary fermions of the SM is reviewed, as are the two most popular mechanisms for neutrino mass generation. Also, the link of absolute neutrino masses to the physics underlying the standard model is discussed. Section 3 deals with direct determinations of the absolute neutrinos mass via tritium beta decay and cosmology. In section 4 $0\nu\beta\beta$ is discussed, which may test very small values of neutrino masses, when information obtained from oscillation studies is input. Section 5 finally deals with the connection of the sub-eV neutrino mass scale and the ZeV energy scale of extreme energy cosmic rays in the Z-burst model.

2 Neutrino masses and physics beyond the standard model

The specific role of the neutrino among the elementary fermions of the standard model (SM) is twofold: It is the only neutral particle, and its mass is much smaller than the masses of the charged fermions. Thus it is self-evident that these properties may be related in a deeper theoretical framework underlying the standard model – usually via Majorana mass-generating mechanisms. In the following we comment on the two most popular mechanisms to generate small neutrino masses, namely the see-saw mechanism and radiative neutrino mass generation.

2.1 The see-saw mechanism

The see-saw mechanism is based on the observation, that in order to generate Dirac neutrino masses

$$m^D \overline{\nu_L} \nu_R \tag{1}$$

analogous to the mass terms of the charged leptons, the introduction of right-handed neutrinos is required. However a lepton-number violating Majorana mass term for right-handed neutrinos

$$\overline{\nu_R} M^R (\nu_R)^c \tag{2}$$

is not prohibited by any gauge symmetry of the standard model. Thus by buying a Dirac neutrino mass term m^D, one inevitably invites mixing with a Majorana mass $M^R \gg m^D$ which may live a priori at the mass scale of the underlying unified theory. The diagonalization of the general mass matrix yields mass eigenvalues

$$m_\nu \simeq (m^D)^2/M^R \ll m^D \tag{3}$$

$$M \simeq M^R, \tag{4}$$

explaining the smallness of the light mass. Though the fundamental scale M^R is unaccessible for any kind of direct experimental testing, it is obvious from eq. 3 that with information on the low-energy observables m_ν and m^D the "beyond the SM" mass scale of M^R can be reconstructed. While it turns out to be unrealistic to determine m^D in the standard model, this option exists indeed in supersymmetry. In the supersymmetric version of the see-saw mechanism, lepton flavor violation (LFV) in the neutrino mass matrix

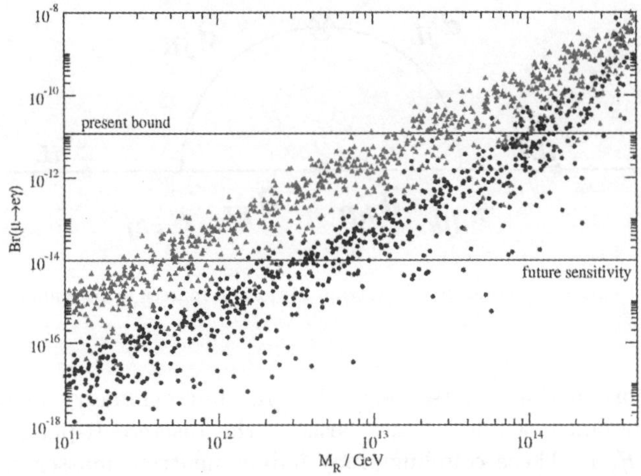

Figure 1. Branching ratio $\mu \to e\gamma$ as a function of the right-handed Majorana mass M_R. Shown are large (lower curve) and small (upper curve) neutrino mass scales, for a specific mSUGRA scenario. The scatter points correspond to estimated uncertainties in neutrino parameters after planned neutrino experiments have been performed.

(as required by neutrino oscillations) generates also LFV soft terms in the slepton mass matrix proportional to the Dirac neutrino Yukawa couplings [2],

$$\delta \tilde{m}_L^2 \propto Y_\nu Y_\nu^\dagger . \tag{5}$$

These LFV soft terms induce large branching ratios for SUSY mediated loop-decays such as $\mu \to e\gamma$,

$$\Gamma(\mu \to e\gamma) \propto \alpha^3 \frac{|(\delta \tilde{m}_L)^2_{ij}|^2}{m_S^8} \tan^2 \beta . \tag{6}$$

Here m_S denotes the slepton mass scale in the loop. Thus in the supersymmteric framework it is possible to probe the heavy mass scale M_R by determining the (light) neutrino mass scale m_ν and the (Dirac) Yukawa couplings. This fact is illustrated in fig. 1, where the branching ratio $\mu \to e\gamma$ is shown as a function of M_R for a specific mSUGRA scenario, both for a large (lower curve) and small (upper curve) neutrino mass scale [3].

2.2 Radiative neutrino masses

An alternative mechanism generates neutrino masses via loop graphs at the SUSY scale, in contrast to the tree level generation of charged lepton masses

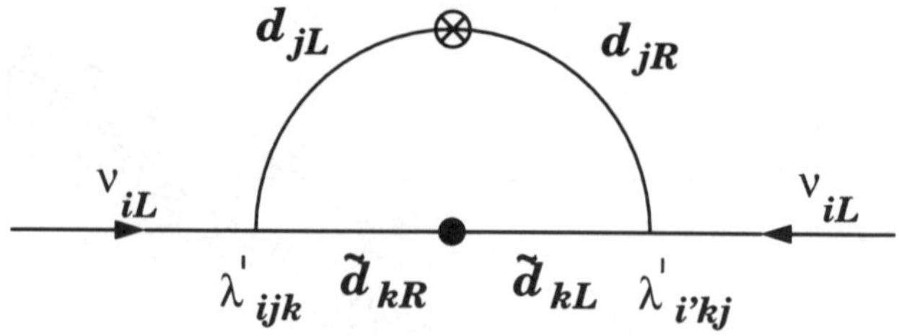

Figure 2. Radiative generation of neutrino Majorana masses in $R\!\!\!/_p$-violating SUSY.

via the Higgs mechanism (see e.g. [4]). In supersymmetry lepton-number violating couplings λ and λ' may arise if the discrete R-parity symmetry is broken ($R\!\!\!/_p$). These couplings may induce neutrino masses via one loop self-energy graphs, see fig. 2. The entries in the neutrino mass matrix, given by

$$m_{\nu_{ii'}} \simeq \frac{N_c \lambda'_{ijk}\lambda'_{i'kj}}{16\pi^2} m_{d_j} m_{d_k} \left[\frac{f(m^2_{d_j}/m^2_{\tilde{d}_k})}{m_{\tilde{d}_k}} + \frac{f(m^2_{d_k}/m^2_{\tilde{d}_j})}{m_{\tilde{d}_j}} \right], \tag{7}$$

are proportional to the products of $R\!\!\!/_p$-couplings and depending on the values of superpartner masses. A determination of the absolute neutrinos mass scale would allow one to constrain all entries in the mass matrix, using the smallness of atmospheric and solar δm^2's and unitarity of the neutrino mixing matrix U. In fact, recent bounds on absolute neutrino masses improve previous bounds on $R\!\!\!/_p$ -couplings by up to 4 orders of magnitude [5]. Thus determining the neutrino mass probes physics at heavy mass scales beyond the SM also in the case of radiative generated neutrino masses.

3 Absolute neutrino masses: direct determinations

3.1 *Tritium beta decay*

In tritium decay, the larger the mass states comprising $\bar{\nu}_e$, the smaller is the Q-value of the decay. The manifestation of neutrino mass is a reduction of phase space for the produced electron at the high energy end of its spectrum. An expansion of the decay rate formula about m_{ν_e} leads to the end point sensitive factor

$$m^2_{\nu_e} \equiv \sum_j |U_{ej}|^2 \, m^2_j \,, \tag{8}$$

where the sum is over mass states which can kinematically alter the end-point spectrum. If the neutrino masses are nearly degenerate, then unitarity of U leads immediately to a bound on $\sqrt{m_{\nu_e}^2} = m_3$. The design of a larger tritium decay experiment (KATRIN) to reduce the present 2.2 eV m_{ν_e} bound is under discussion; direct mass limits as low as 0.4 eV, or even 0.2 eV, may be possible in this type of experiment [6].

3.2 CMB/LSS cosmological limits

According to Big Bang cosmology, the masses of nonrelativistic neutrinos are related to the neutrino fraction of closure density by $\sum_j m_j = 40\,\Omega_\nu\,h_{65}^2$ eV, where h_{65} is the present Hubble parameter in units of 65 km/s/Mpc. As knowledge of large-scale structure (LSS) formation has evolved, so have the theoretically preferred values for the hot dark matter (HDM) component, Ω_ν. In fact, the values have declined. In the once popular HDM cosmology, one had $\Omega_\nu \sim 1$ and $m_\nu \sim 10$ eV for each of the mass-degenerate neutrinos. In the cold-hot CHDM cosmology, the cold matter was dominant and one had $\Omega_\nu \sim 0.3$ and $m_\nu \sim 4$ eV for each neutrino mass. In the currently favored ΛDM cosmology, there is scant room left for the neutrino component. The power spectrum of early-Universe density perturbations is processed by gravitational instabilities. However, the free-streaming relativistic neutrinos suppress the growth of fluctuations on scales below the horizon (approximately the Hubble size $c/H(z)$) until they become nonrelativistic at $z \sim m_j/3T_0 \sim 1000\,(m_j/\text{eV})$.

A recent limit [7] derived from the 2dF Galaxy Redshift Survey power spectrum constrains the fractional contribution of massive neutrinos to the total mass density to be less than 0.13, translating into a bound on the sum of neutrino mass eigenvalues, $\sum_j m_j < 1.8$ eV (for a total matter mass density $0.1 < \Omega_m < 0.5$ and a scalar spectral index $n = 1$). Previous cosmological bounds come from the data of galaxy cluster abundances, the Lyman α forest, data compilations of the cosmic microwave background (CMB), and peculiar velocities of large scale structure, and give upper bounds on the sum of neutrino masses in the range 3-6 eV. A discussion of possible limits from future supernova neutrino detection is given in [8].

Some caution is warranted in the cosmological approach to neutrino mass, in that the many cosmological parameters may conspire in various combinations to yield nearly identical CMB and LSS data. An assortment of very detailed data may be needed to resolve the possible "cosmic ambiguities".

4 Neutrinoless double beta decay

$0\nu\beta\beta$ [9] corresponds to two single beta decays occurring simultaneously in one nucleus and converts a nucleus (Z,A) into a nucleus (Z+2,A). While even the standard model (SM) allowed process emitting two antineutrinos

$$^A_ZX \rightarrow ^A_{Z+2} X + 2e^- + 2\overline{\nu}_e \tag{9}$$

is one of the rarest processes in nature with half lives in the region of 10^{21-24} years, more interesting is the search for the neutrinoless mode,

$$^A_ZX \rightarrow ^A_{Z+2} X + 2e^- \tag{10}$$

which violates lepton number by two units and thus implies physics beyond the SM.

The $0\nu\beta\beta$ rate is a sensitive tool for the measurement of the absolute mass-scale for Majorana neutrinos [10]. The observable measured in the amplitude of $0\nu\beta\beta$ is the ee element of the neutrino mass-matrix in the flavor basis (see fig. 3). Expressed in terms of the mass eigenvalues and neutrino mixing-matrix elements, it is

$$m_{ee} = |\sum_i U_{ei}^2 m_i| \,. \tag{11}$$

A reach as low as $m_{ee} \sim 0.01$ eV seems possible with proposed double beta decay projects such as GENIUSI, MAJORANA, EXO, XMASS or MOON. This provides a substantial improvement over the current bound, $m_{ee} < 0.6$ eV. A recent claim [11] by the Heidelberg-Moscow experiment reports a best fit value of $m_{ee} = 0.36$ eV, but is subject to possible systematic uncertainties. In the far future, another order of magnitude in reach is available to the 10 ton version of GENIUS, should it be funded and commissioned.

For masses in the interesting range $\gtrsim 0.01$ eV, the two light mass eigenstates are nearly degenerate and so the approximation $m_1 = m_2$ is justified. Due to the restrictive CHOOZ bound, $|U_{e3}|^2 < 0.025$, the contribution of the third mass eigenstate is nearly decoupled from m_{ee} and so $U_{e3}^2\, m_3$ may be neglected in the $0\nu\beta\beta$ formula. We label by ϕ_{12} the relative phase between $U_{e1}^2\, m_1$ and $U_{e2}^2\, m_2$. Then, employing the above approximations, we arrive at a very simplified expression for m_{ee}:

$$m_{ee}^2 = \left[1 - \sin^2(2\theta_{\rm sun})\, \sin^2\left(\frac{\phi_{12}}{2}\right)\right]\, m_1^2 \,. \tag{12}$$

The two CP-conserving values of ϕ_{12} are 0 and π. These same two values give maximal constructive and destructive interference of the two dominant terms

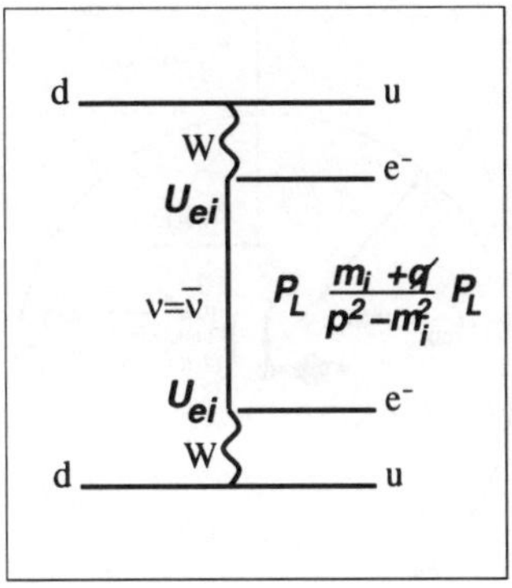

Figure 3. Diagram for neutrinoless double beta decay.

in eq. (11), which leads to upper and lower bounds for the observable m_{ee} in terms of a fixed value of m_1:

$$\cos(2\theta_{\rm sun})\, m_1 \le m_{ee} \le m_1\,, \quad \text{for fixed } m_1\,. \tag{13}$$

The upper bound becomes an equality, $m_{ee} = m_1$, if $\phi_{12} = 0$. The lower bound depends on Nature's value of the mixing angle in the LMA solution. A consequence of eq. (13) is that for a given measurement of m_{ee}, the corresponding inference of m_1 is uncertain over the range $[m_{ee},\, m_{ee}\, \cos(2\theta_{\rm sun})]$ due to the unknown phase difference ϕ_{12}, with $\cos(2\theta_{\rm sun}) \gtrsim 0.1$ weakly bounded even for the LMA solution [12]. This uncertainty disfavors $0\nu\beta\beta$ in comparison to direct measurements if a specific value of m_1 has to be determined, while $0\nu\beta\beta$ is more sensitive as long as bounds on m_1 are aimed at. Knowing the value of $\theta_{\rm sun}$ better will improve the estimate of the inherent uncertainty in m_1. For the LMA solar solution, the forthcoming Kamland experiment should reduce the error in the mixing angle $\sin^2 2\theta_{\rm sun}$ to ± 0.1 [13].

5 Extreme energy cosmic rays in the Z-burst model

It was expected that the EECR primaries would be protons from outside the galaxy, produced in Nature's most extreme environments such as the tori or radio hot spots of active galactic nuclei (AGN). Indeed, cosmic ray data show a spectral flattening just below 10^{19} eV which can be interpreted as a new extragalactic component overtaking the lower energy galactic component; the energy of the break correlates well with the onset of a Larmor radius for protons too large to be contained by the Galactic magnetic field. It was

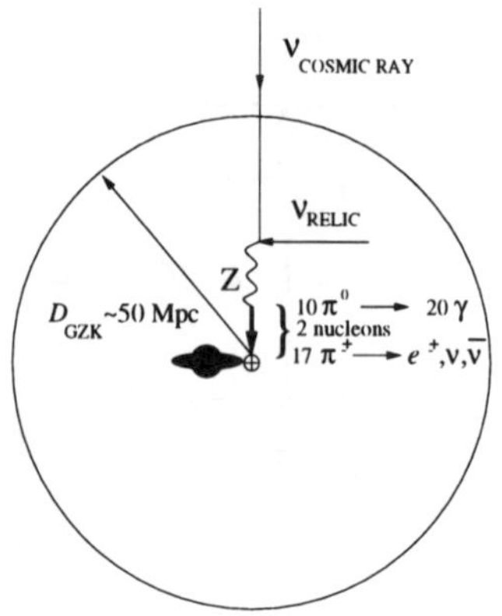

Figure 4. Schematic diagram showing the production of a Z-burst resulting from the resonant annihilation of a cosmic ray neutrino on a relic (anti)neutrino. If the Z-burst occurs within the GZK zone ($\sim$ 50 to 100 Mpc) and is directed towards the earth, then photons and nucleons with energy above the GZK cutoff may arrive at earth and initiate super-GZK air-showers.

further expected that the extragalactic spectrum would reveal an end at the Greisen-Kuzmin-Zatsepin (GZK) cutoff energy of $E_{\rm GZK} \sim 5 \times 10^{19}$ eV. The origin of the GZK cutoff is the degradation of nucleon energy by the resonant scattering process $N + \gamma_{2.7K} \rightarrow \Delta^* \rightarrow N + \pi$ when the nucleon is above the resonant threshold $E_{\rm GZK}$. The concomitant energy-loss factor is $\sim (0.8)^{D/6{\rm Mpc}}$ for a nucleon traversing a distance D. Since no AGN-like sources are known to exist within 100 Mpc of earth, the energy requirement for a proton arriving at earth with a super-GZK energy is unrealistically high. Nevertheless, to date more than twenty events with energies at and above 10^{20} eV have been observed (for recent reviews see [14]).

Several solutions have been proposed for the origin of these EECRs, ranging from unseen Zevatron accelerators (1 ZeV $= 10^{21}$ eV) and decaying supermassive particles and topological defects in the Galactic vicinity, to exotic primaries, exotic new interactions, and even exotic breakdown of conventional physical laws. A rather conservative and economical scenario involves cosmic ray neutrinos scattering resonantly on the cosmic neutrino background (CNB) predicted by Standard Cosmology, to produce Z-bosons [15]. These Z-bosons in turn decay to produce a highly boosted "Z-burst", containing on average twenty photons and two nucleons above $E_{\rm GZK}$ (see Fig. 4). The photons and nucleons from Z-bursts produced within 50 to 100 Mpc of earth can reach earth with enough energy to initiate the air-showers observed at $\sim 10^{20}$ eV.

The energy of the neutrino annihilating at the peak of the Z-pole is

$$E^R_{\nu_j} = \frac{M_Z^2}{2m_j} = 4\,(\frac{\text{eV}}{m_j})\,\text{ZeV}. \tag{14}$$

Even allowing for energy fluctuations about mean values, it is clear that in the Z-burst model the relevant neutrino mass cannot exceed ~ 1 eV. On the other hand, the neutrino mass cannot be too light or the predicted primary energies will exceed the observed event energies, and the primary neutrino flux will be pushed to unattractively higher energies. In this way, one obtains a rough lower limit on the neutrino mass of ~ 0.1 eV for the Z-burst model (with allowance made for an order of magnitude energy-loss for those secondaries traversing 50 to 100 Mpc). A detailed comparison of the predicted proton spectrum with the observed EECR spectrum in [16] yields a value of $m_\nu = 2.34^{+1.29}_{-0.84}$ eV for galactic halo and $m_\nu = 0.26^{+0.20}_{-0.14}$ eV for extragalactic halo origin of the power-like part of the spectrum.

A necessary condition for the viability of this model is a sufficient flux of neutrinos at $\gtrsim 10^{21}$ eV. Since this condition seems challenging, any increase of the Z-burst rate that ameliorates slightly the formidable flux requirement is helpful. If the neutrinos are mass degenerate, then a further consequence is that the Z-burst rate at E_R is three times what it would be without degeneracy. If the neutrino is a Majorana particle, a factor of two more is gained in the Z-burst rate relative to the Dirac neutrino case since the two active helicity states of the relativistic CNB depolarize upon cooling to populate all spin states.

Moreover the viability of the Z-burst model is enhanced if the CNB neutrinos cluster in our matter-rich vicinity of the universe. For smaller scales, the Pauli blocking of identical neutrinos sets a limit on density enhancement. As a crude estimate of Pauli blocking, one may use the zero temperature Fermi gas as a model of the gravitationally bound neutrinos. Requiring that the Fermi momentum of the neutrinos does not exceed mass times the virial velocity $\sigma \sim \sqrt{MG/L}$ within the cluster of mass M and size L, one gets the Tremaine-Gunn bound

$$\frac{n_{\nu_j}}{54\,\text{cm}^{-3}} \lesssim 10^3 \left(\frac{m_j}{\text{eV}}\right)^3 \left(\frac{\sigma}{200\text{km/s}}\right)^3 . \tag{15}$$

With a virial velocity within our Galactic halo of a couple hundred km/sec it appears that Pauli blocking allows significant clustering on the scale of our Galactic halo only if $m_j \gtrsim 0.5$ eV. Free-streaming (not considered here) also works against HDM clustering.

Thus, if the Z-burst model turns out to be the correct explanation of EECRs, then it is probable that neutrinos possess one or more masses in the range $m_\nu \sim (0.1 - 1)$ eV. Mass-degenerate neutrino models are then likely. Some consequences are:

- A value of $m_{ee} > 0.01$ eV, and thus a signal of $0\nu\beta\beta$ in next generation experiments, assuming the neutrinos are Majorana particles.
- Neutrino mass sufficiently large to affect the CMB/LSS power spectrum.

6 Conclusions

The absolute neutrino mass scale is a crucial parameter to learn about the theoretical structures underlying the SM. Information about absolute neutrino masses can be obtained from direct determinations via tritium beta decay or cosmology. More sensitive in giving limits but less valuable for determining the mass scale is neutrinoless double beta decay. The puzzle of EECRs above the GZK cutoff can be solved conservatively with the Z-burst model, connecting the ZeV scale of EECRs to the sub-eV scale of neutrino masses. If the Z-burst model turns out to be correct, neutrino masses in the region of 0.1-1 eV are predicted and degenerate scenarios are favored. In this case positive signals for future tritium beta decay experiments, CMB/LSS studies, and $0\nu\beta\beta$ can be expected.

Acknowledgements

HP would like to thank the organizers of NOON'01 for the kind invitation to this inspiring meeting. We also thank F. Deppisch for providing us with fig. 1. This work was supported by the DOE grant no. DE-FG05-85ER40226 and the Bundesministerium für Bildung und Forschung (BMBF, Bonn, Germany) under the contract number 05HT1WWA2.

References

1. H. Päs, T.J. Weiler, *Phys. Rev.* D **63**, 113015 (2001)
2. J. Hisano and D. Nomura, *Phys. Rev.* D **59**, 116005 (1999) J. Hisano, these proceedings.
3. F. Deppisch, H. Päs, A. Redelbach, R. Rückl, Y. Shimizu, Preprint WUE-ITP-2002-009.
4. M.A. Diaz, J.C. Romao, J.W.F. Valle, *Nucl. Phys.* B **524**, 23 (1998).

5. G. Bhattacharyya, H.V. Klapdor-Kleingrothaus, H. Päs, *Phys. Lett.* B **463**, 77 (1999).
6. C. Weinheimer, talk at the *Neutrino2000* conference at Sudbury/Canada; A. Osipowicz *et al.*, (KATRIN Collab.), hep-ex/0101091.
7. Ø. Elgarøy *et al.* (2dF Collab.) astro-ph/0204152.
8. J. Beacom, R. Boyd, and A. Mezzacappa, *Phys. Rev. Lett.* **85**, 3568 (2000), and J. Beacom, these proceedings.
9. Steven R. Elliott, Petr Vogel, hep-ph/0202264; H.V. Klapdor-Kleingrothaus, H. Päs, in: Proc. of the *International Workshop on Particle Physics and the Early Universe*, Trieste, Italy, September 27 - October 2, 1999, (Eds. U. Cotti, R. Jeannerot, G. Senjanovic. A. Smirnov), World Scientific, Singapore.
10. H.V. Klapdor-Kleingrothaus, H. Päs, A.Yu. Smirnov, *Phys. Rev.* D **63**, 073005 (2001).
11. H.V. Klapdor-Kleingrothaus, A. Dietz, H.L. Harney, I.V. Krivosheina, *Mod. Phys. Lett.* A **16** 2409 (2002); compare however C.E. Aalseth *et al.*, hep-ex/0202018 and F. Feruglio, A. Strumia, F. Vissani, hep-ph/0201291.
12. M.C. Gonzalez-Garcia, these proceedings; M. Smy, these proceedings.
13. V. Barger, D. Marfatia, B.P. Wood, *Phys. Lett.* B **498**, 53 (2001); H. Murayama and A. Pierce, *Phys. Rev.* D **65**, 013012 (2002).
14. S. Sarkar, hep-ph/0202013; T.J. Weiler, hep-ph/0103023.
15. T.J. Weiler, *Phys. Rev. Lett.* **49**, 234 (1982), and *Astroparticle Phys.* **11** 303 (1999); D. Fargion, B. Mele, A. Salis, *Astrophys. J.* **517** 725 (1999); T.J. Weiler, in "Beyond the Desert 99, Ringberg Castle", Tegernsee, Germany, June 6-12, 1999 [hep-ph/9910316].
16. Z. Fodor, S.D. Katz, A. Ringwald, *Phys. Rev. Lett.* **88**, 171101 (2002), and hep-ph/0203198; see also O. Kalashev et al., hep-ph/0112351 (2001); G. Gelmini and G. Varieschi, hep-ph/0201273 (2002).

RECENT RESULTS FROM THE K2K LONG–BASELINE NEUTRINO OSCILLATION EXPERIMENT

JAMES E. HILL

ICRR, Kashiwa, Japan, & SUNY Stony Brook, Stony Brook, NY, USA

The K2K Long–Baseline neutrino oscillation experiment has been aquiring data since mid-1999 and has analysed those up to July of 2001. Fifty-six fully contained events are observed in the fiducial volume of the far detector where $80.6^{+7.3}_{-8.0}$ are expected based partly on measurements near the beam production point. There is virtually no background for the contained event search. The methods established in this experiment are crucial for operation of future similar experiments to probe the nature of mixing in the neutral lepton sector, a necessary step in understanding the nature of family structure and of mass itself. A brief history and a few notes about the future and direction of the field precede the description of the experiment and its results.

1 Motivation for long–baseline experimentation

K2K [1] is the world's first experiment designed specifically to search for oscillations in the neutral lepton sector using the technique of long–baseline experimentation. The main motivation for long–baseline accelerator based neutrino experiments lies in the results provided by the analysis of atmospheric neutrino data [2]. Other parts of this proceedings describe those analyses in great detail.

1.1 Historic Results Implying Neutrino Oscillation

The first observation of atmospheric neutrinos [3] came in the 1960's from underground experiments searching for proton decay. Atmospheric neutrino studies began in earnest in the 1980's when proton decay experimenters realized the need to understand their backgrounds better. The first indications of the so-called "atmospheric neutrino anomaly" from the water Čerenkov detectors Kamiokande and IMB [4,5] were that the observed flavor ratio of atmospheric neutrinos did not match the expectation. Originally, the Soudan, NUSEX [6], and Frejus [7] experiments, which used iron calorimeters, appeared to disagree with the conclusion from the water target experiments, although these lacked sufficient statistics to refute any other claim. Later analyses [8] of of Soudan's full data set data appeared consistent with the previous water experiments and the new one, Super–Kamiokande, although still with smaller statistics.

The first generation of large water Čerenkov detectors began analyses of the directional distribution of neutrino events, but lacked the statistical power to fully exploit the information. The latest results from Super–Kamiokande are presented elsewhere in these proceedings, so they will not be covered here. The evidence for oscillations in atmospheric neutrinos is compelling, which makes confirmation (or not) imperative.

1.2 The Next Steps

Super–Kamiokande proves rather conclusively that the atmospheric neutrino results demonstrate some lack of understanding of the underlying physics. The simplest amendment to the Standard Model of electro-weak interactions which would explain this is neutrino oscillations. With this strong evidence and a firm testable hypothesis, the two tasks at hand for particle physicists are to verify or refute the effect in a controlled experiment, and to probe the values of relevant parameters of the theory, possibly with a whole series of experiments each drawing on the knowledge and experience gained in the previous.

K2K is designed to explore that first step. Through the 1990's, several similar experiments were proposed; a few of those proposals are in this generation of experiments, but only K2K has started data taking. The experiments which will follow in the next few years, MINOS [9], the CNGS projects [10], and some longer time scale projects generically referred to as "superbeam" or "neutrino factory" experiments represent a program of measurements to measure the full neutrino mixing matrix with precision competitive with (or in some estimations better than) the precision of corresponding measurements in the quark sector.

1.3 The Generic Strategy of Long–Baseline Experimentation

The atmospheric neutrino results –our main motivating measurement– have as their source mostly neutrinos in the GeV momentum range which travel geographical distances. Thus, a controlled experiment to check its validity might use accelerator produced neutrinos and a detector at a well known distance some hundreds of kilometers away. To ensure full understanding of the beam at production, a near detector is useful; to reduce uncertainties of the neutrino interaction model (on which detection relies) and detector based uncertainties, the near and far detectors should preferably use the same target materials and technology for detection.

There are several aspects of the technique for long–baseline experimentation that are generic to this whole class of experiments. Two crucial aspects of this technique are fine timing and precise beam aiming to a remote site too far away to be directly measured by conventional survey. The large distance scale drives us to rely strongly on the Global Positioning System (GPS) [13] for timing synchronization. The consequent small solid angle subtended by even a large far detector motivate a spatially wide beam with redundant systems in place to check its aiming stability and a well surveyed near neutrino detector [14] capable of measuring the beam direction precisely.

2 Design and overview of K2K

The K2K neutrino beam[15] is a horn-focused wide band ν_μ beam, with expected spectrum peaked at about 1 GeV. The primary beam for K2K is 12 GeV kinetic energy protons from the KEK proton-synchrotron [16]. Every 2.2 s, approximately 6×10^{12} protons in nine bunches are fast-extracted in a single turn, making a 1.1 μs beam spill. These protons are focused onto a 30 mm diameter, 66 cm long aluminum target which is a current carrying element in the first of a pair of horn magnets operating at 250 kA.

The near detector for K2K is actually a suite of different detectors with varying capabilities situated in a hall 300 m downstream of the pion production target. The main flux measurement is based on a one-kiloton water Čerenkov detector (1kt) which uses the same photo-multipliers as Super–Kamiokande, the far detector. The design is essentially the same except for scale and the analysis for the two detectors uses the same reconstruction algorithms.

A fine grained detector sits downstream of the 1kt. The first main component of it is a scintillating fiber tracker (SciFi) [17] with water targets enclosed between layers of tracking material. The next major component is a stack of iron plates interleaved with drift tubes to serve as a muon range detector (MRD) [18]. Both the SciFi and MRD have transverse pairs of tracking planes so tracks can be reconstructed in three dimensions. The SciFi has planes of plastic scintillator upstream and downstream of it to tag incoming and outgoing particles and to record fine timing for events in common with the MRD.

The fine grained detector as a whole is excellent for event classification studies since it can readily distinguish quasi-elastic and inelastic events. The MRD is has sufficient target mass to provide a rate of neutrino interactions sufficient for beam stability monitoring. In particular, its large transverse area makes it ideal for beam profile monitoring.

The far detector for K2K is the Super–Kamiokande detector, probably the best known water Čerenkov detector in the world. It is located in a mine in Kamioka town, Gifu prefecture, Japan, 250 km from the beamline at KEK.

3 Results From K2K

Any discussion of results from K2K naturally concentrates on the main goal of the experiment, the search for oscillations by comparing the near and far detector measurements. Here, we present results along this line by first detailing the near detector measurements before summarizing the data reduction and analysis at the far detector. There is not enough room here to include any deep discussion of the other results from measurements only at the near detector, so first we will list a few items currently being analyzed.

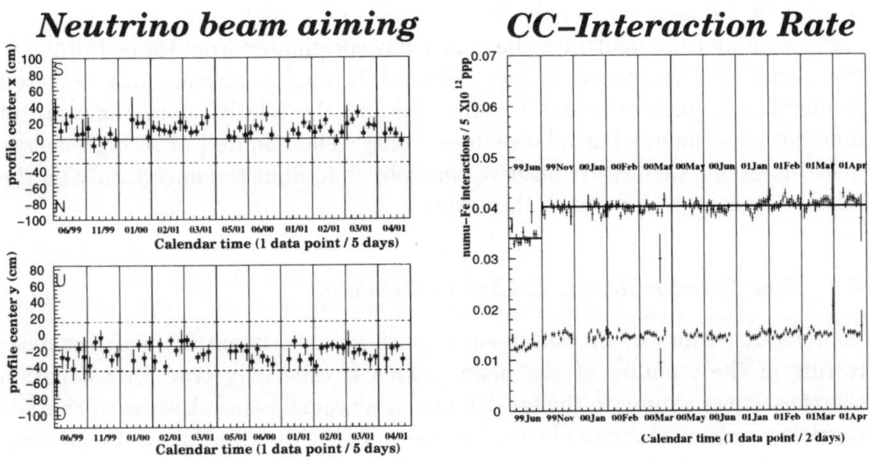

Figure 1. Beam stability as measured by the muon range detector. Left is the centroid of reconstructed vertices each five days through all the data analyzed. Right is the event rate measured each two days. The rate in June of 1999 is expected to be lower than later runs since the target area configuration was slightly different.

3.1 Topics Other than Neutrino Oscillations at K2K

The 1kt detector can tag electromagnetic showers with good efficiency in the momentum range of this beam. Since the beam contamination of ν_e is low ($\sim$ 1%), a sample of events with π^0 are straightforward to extract. Most π^0 events with no other visible ring are the result of neutral current interactions ($\nu N \rightarrow \nu N \pi^0$). Therefor, this detector can be used to measure well the cross section for this process compared to the relatively well known quasi-elastic cross section in this momentum range. Such a measurement will aid atmospheric neutrino analyses distinguishing $\nu_\mu \to \nu_s$ from $\nu_\mu \to \nu_\tau$ by the presence or absence of neutral current events.

Events with a π^0 and a non-showering ring are likely to be from the charged current process $\nu_\mu n \to \mu p \pi^0$, the dominant background for an important mode of proton decay predicted by some super–symmetric theories[19]. Precise measurement of this interaction cross section is crucial to understanding the backgrounds for future even larger proton decay experiments.

While the K2K neutrino beam consists predominantly of neutrinos from charged pion decay, the highest energy part of the beam ($>\sim$ 4GeV/c) comes almost exclusively from kaon decays. While this accounts for only about a percent of the neutrinos in the beam, there are enough such events to easily tag as coming from neutrinos that must have originated from kaons (although the geometrical acceptance is low). Most of the nuclear physics data on meson production by proton beams is quite old and the relative normalizations of different experiments has a large uncertainty. The hundred or so high energy neutrino events we already observe can help us to limit the uncertainty on the relative meson production in the target.

3.2 Near Detector Results Related to Oscillation

A prerequisite for any measurement in such an experiment is the direct monitoring of the stability of the beam. This is especially true for oscillation searches where aiming to the far detector is a crucial issue. A series of charged particle monitors along the beam-line measure the distribution of the progenitors and "sisters" of our neutrino beam. Many of these give crucial information about the pulse to pulse stability of the beam. More direct information about the neutrino beam can be gleaned from the near detector measurements, particularly with the large mass and area of the MRD. Figure 1 shows neutrino beam data from the MRD relating to the aiming and flux of neutrinos as a function of calendar time.

For a correct spectral measurement of the neutrino beam, background of non-quasi-elastic interactions needs to be well understood. The fine grained detector with scintillating fiber planes between water targets using the MRD as a muon calorimeter is well suited for this study since it can reconstruct two-track kinematics with great precision. Figure 2 gives some idea of how such a measurement can be done.

With the beam stability assured and the background measured, we can set about measuring a spectrum at the near detectors. The ideal detector for the

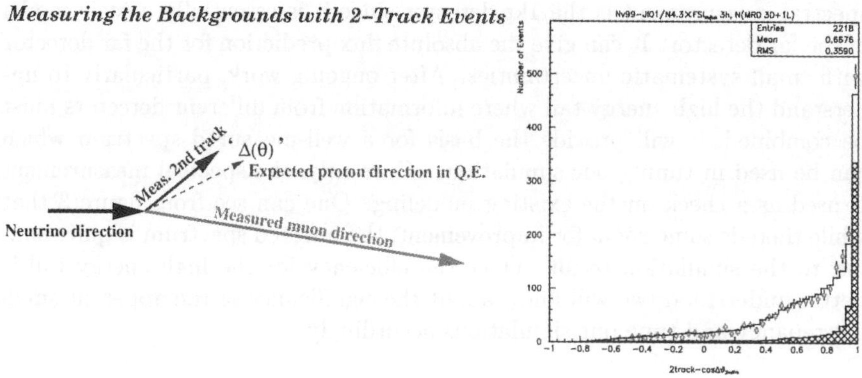

Figure 2. Background measurement with 2-track events from the fine-grained detector. The abscissa is the cosine of the angle between the second track reconstructed direction and the direction calculated for a proton based on the assumption that the other track was a muon and 2-body kinematics as is shown in the inset cartoon. Fitting this distribution to a pair of amplitudes times the histogram shapes from simulations can give a solid measurement of the non-quasi-elastic background in this data set.

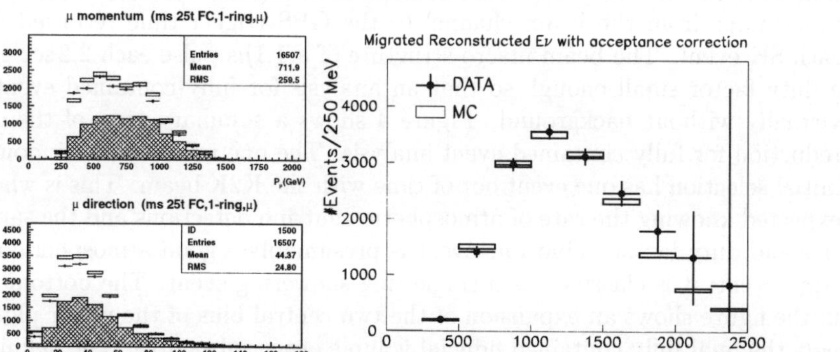

Figure 3. Neutrino momentum spectrum reconstructed from data from only the 1kt detector. Left are the basic distributions for single ring μ-like that go into calculating a spectrum. Points are data; the boxes are the total prediction for simulation; the filled histogram is the non-quasi-elastic contribution to the simulated sample. Only statistical error is shown. Right is a neutrino spectrum after background subtraction and corrections for efficiencies and acceptance. The inclusion of information from the other detectors will raise the efficiency for high energy events.

spectral measurement is the 1kt detector since it is essentially a scale model of the far detector. It can give the absolute flux prediction for the far detector with small systematic uncertainties. After ongoing work, particularly to understand the high energy tail where information from different detectors must be combined, it will provide the basis for a well-measured spectrum which can be used in tuning our simulations. Currently, the spectral measurement is used as a check on the existing modeling. One can see from figure 3 that while there is some room for improvement, the inferred spectrum is quite similar to the simulation result. Once the efficiency for the high energy tail is better understood, we will know about the significance of the apparent small discrepancy and tune our simulations accordingly.

3.3 Far Detector Results

The far detector, Super–Kamiokande, aquires data constantly with over 90% live time. Physics events are all self-triggered. Timing analysis for correlation with the arrival of the K2K beam is done off-line by comparing files of timestamps from the beam channel to the GPS trigger time recorded with each SK event. The beam macro-structure of a $1.1\mu s$ pulse each 2.2 sec gives a duty factor small enough so that an analysis for fully contained events is virtually without background. Figure 4 shows a summary view of the data reduction for fully contained event analysis. The one millisecond window for initial selection has one event out of time with the K2K beam. This is what is expected knowing the rate of atmospheric neutrino detections and the sample size and duty factor. That one event is presumably a good atmospheric neutrino event; it is classified as a single ring showering event. The bottom plot in the figure shows an expansion of the two central bins of the upper plot for only the final fully contained fiducial volume event subsample. The definition of zero time is the expected time of arrival of the K2K beam, so assuming a syncronization random error of less than 200 ns, we expect the good events to be included in a search window from -0.2 to $1.3\mu s$. The factors leading to the 200 ns error are really an upper limit, and the fact that these data show no tail beyond the expected beam window seems to suggest the synchronization is actually somewhat better than that. Table 1 shows a breakdown of the contained, in fiducial volume events into subclasses.

Analysis of outside fiducial volume and outer detector events gives a total sample with which to check timing of 104 events. The outer detector search has an expected background of order one.

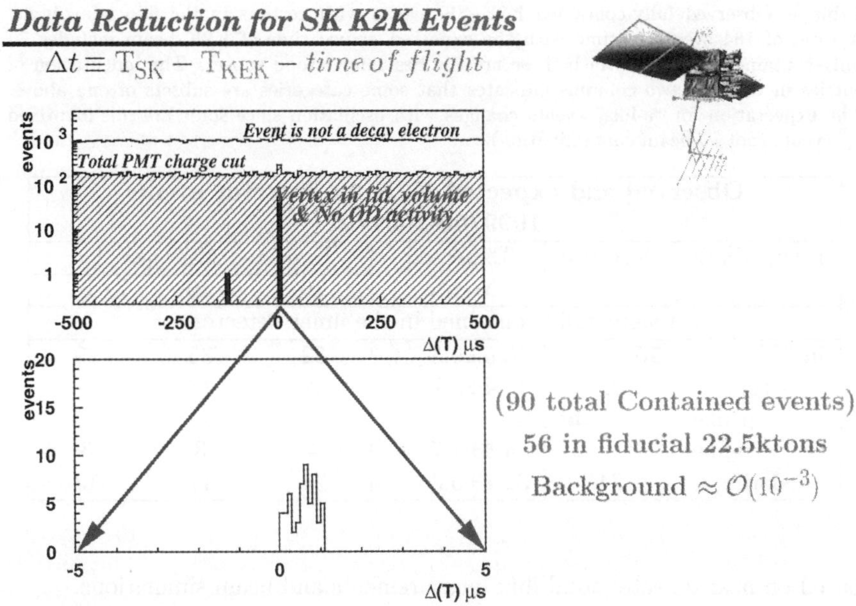

Figure 4. (A simplified view of) Data reduction for SK-K2K fully-contained events.

3.4 A Note About the Arrival Times of Events

A question which arises often regards the calendar time distribution of the events observed at the far detector. The rate of these should be proportional to the rate observed at the near detector at any time. To a very good approximation, this is proportional to the rate of protons reaching the target, so the cumulative observation at the far detector plotted against the accumulated protons on target should fall roughly along a straight line. The right plot in figure 5 shows that distribution. The diagonal line in the plot is drawn from (0,0) to the last point to show the average rate of arrival.

The first approximately 8% of K2K data were taken with a different beamline configuration. This is the one point left of the first solid vertical line. The contribution of this period to the whole result (or even to the total observed deficit) is too small to change any qualitative conclusion about it. The far detector expectation is calculated using the flux information from the near detector; the rate reduction is obvious in figure 1. For reference, the overall flux measured by the far detector is approximately 70% of what is expected

Table 1. Observed fully-contained K2K–SK events with vertices in the fiducial volume. A total of 104 events in time with the expected arrival time of K2K beam includes 70 fully-contained events of which these are the best understood subset. The indentation of entries in the first two columns indicates that some categories are subsets of one above. The expectation for "e-like" events changes with oscillation since some are mis-identified ν_μ events (not ν_e beam contamination).

Observed and expected K2K far detector events 1999/06-2001/07					
Event Class	Observed	Expected No osc.	Expected $\Delta(m^2)[eV^2] =$ $3 \cdot 10^{-3}$	$5 \cdot 10^{-3}$	$7 \cdot 10^{-3}$
Events fully contained in the inner detector					
in FV	**56**	$80.6 \pm 0.3^{+7.3}_{-8.0}$	52	35	29
1-ring	32	48.4±6.7	28	18	17
μ-like	**30**	44.0±6.8	24	15	14
e-like	2	4.4±1.7	4	3	3
>1-ring	24	32.2±5.3	24	17	13

based on near detector total flux measurements and beam simulations.

4 Physics Interpretation of These Results

While a full quantative physics interpretation of these results is not yet possible, several important statements can be made. First, based solely on the observed number of fully contained fiducial volume events a quantitative statement about the probability of this observation for a given scenario involving oscillations can be made since the error on this expectation is relatively well understood. Secondly, while errors on the calculation of an expected neutrino spectrum are not completely understood, an expected spectrum can be constructed and from this a meaningful qualitative statement can be made about the appearance of the observed spectral shape.

There are several ways to calculate a meaningful probability to represent the power of a data set to make a statement about a hypothesis. It is probably fair to say that experts agree that the most important thing to do is to be as explicit as possible about how a given quantitative statement is derived. We have tried many methods and all give consistent results. The result of one is explained below.

A hypothesis that *a priori* predicts a central value expected to be higher than that of competing scenarios is disfavored at a confidence level given by:

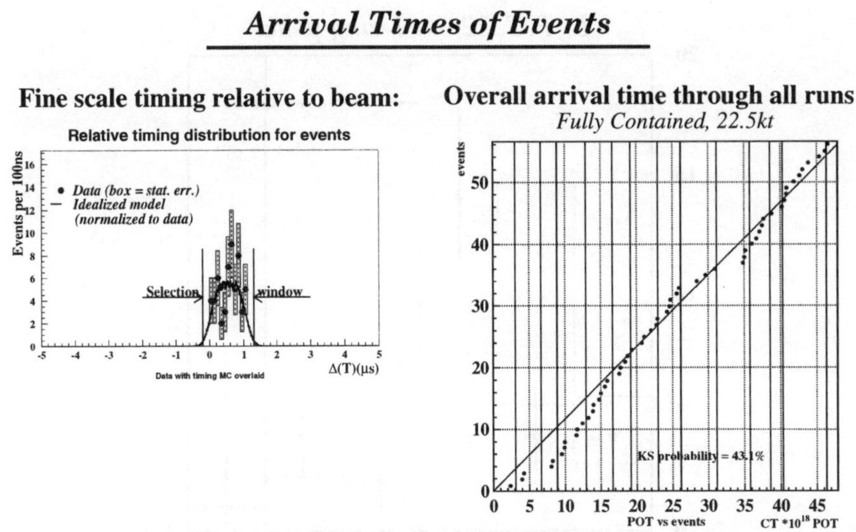

Figure 5. (Left)The fine time distribution relative to beam arrival for all 70 fully contained events (points). The histogram shows a simple simulation of the expected distribution using the best knowledge of the uncertainty on synchronization. (Right)The time of arrival of observed far detector contained events in units of cumulative protons on target. The solid vertical lines show the boundaries of different run periods, all but the first of which use the same beamline configuration.

$$\mathrm{C.L.} \equiv 1 - \int_0^\infty \left(\mathrm{e}^{-\lambda} \sum_{i=0}^{\mathrm{N}_\mathcal{O}} \frac{\lambda^i}{i!} \right) \times \mathcal{P}(\lambda) \mathrm{d}\lambda$$

where $\mathrm{N}_\mathcal{O}$ is the number observed, and $\mathcal{P}(\lambda) \approx \left(\sigma_\mathrm{L}\sqrt{2\pi}\right)^{-1} \mathrm{e}^{-\frac{1}{2}\left(\frac{(\lambda-\mu)}{\sigma_\mathrm{L}}\right)^2}$ is a gaussian approximation of the distribution of systematic errors for an expectation of μ with a lower side error of σ_L.

This prescription gives a probability of approximately 2% that the total number of fully contained fiducial volume events in the data sample is consistent with the hypothesis of no oscillations. This statement is independent of the information available in our measurement of the spectrum of the observed events. To make a quantitative statement using the spectrum, we need to understand the errors on the prediction of the spectrum (particularly the correlation of errors between bins of momentum). For now, only a qualitative statement can be made that the spectrum reconstructed from the K2K far

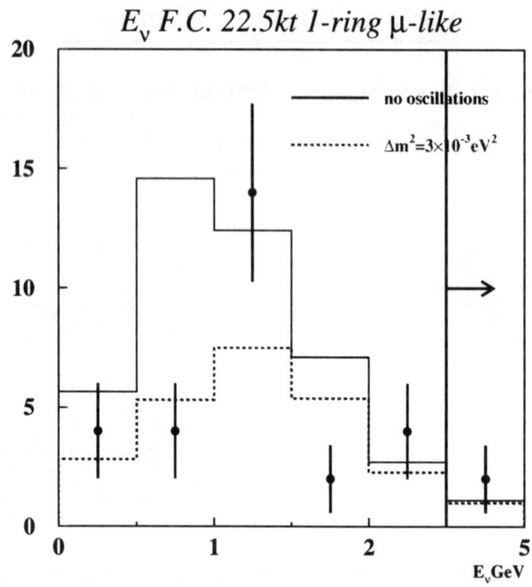

Figure 6. The neutrino spectrum reconstructed from data (points) and the prediction of simulations without (thick black solid line) and with (thin dot-dashed blue line) oscillations. The thick solid black line represents the expectations assuming no oscillation, and the thin dot-dashed blue line represents that assuming oscillations with the atmospheric neutrino best-fit parameters.

detector data set, while it lacks sufficient statistics, seems at least as consistent with the hypothesis of oscillations with the atmospheric neutrino best fit parameters as it does with the no oscillations hypothesis.

Only single non-showering ring events are most likely to be produced by quasi-elastic scatters, so only the 30 events in that subset of fully contained events in the fiducial volume are used to make a neutrino spectrum. The neutrino momentum is reconstructed from these events by assuming quasi-elastic kinematics using the well known direction of the beam. Figure 6 shows the spectrum derived from data events along with expectations with and without oscillations. The expectation from our simulations are normalized by the near detector measurement. (Later work will tune the simulation to more closely match the near detector measured spectral shape.) Since the systematic error on the expectation is still under study the only errors shown are the statistical errors on the data.

Finally, an analysis can be done on the full space of oscillation parameters using only the information of the total counts. Of course, the allowed or disallowed region derived with only one datum in a two parameter space will mapable to a lower dimension. Still an analysis can be done. Any such analysis must use a two-tailed probability distribution since the expectation in general may be above or below the observation. Furthermore, since the energy dependence of the errors is still under study, a more conservative treatment than was used in the null-hypothesis test above is in order. The result of such an analysis is shown in figure 7.

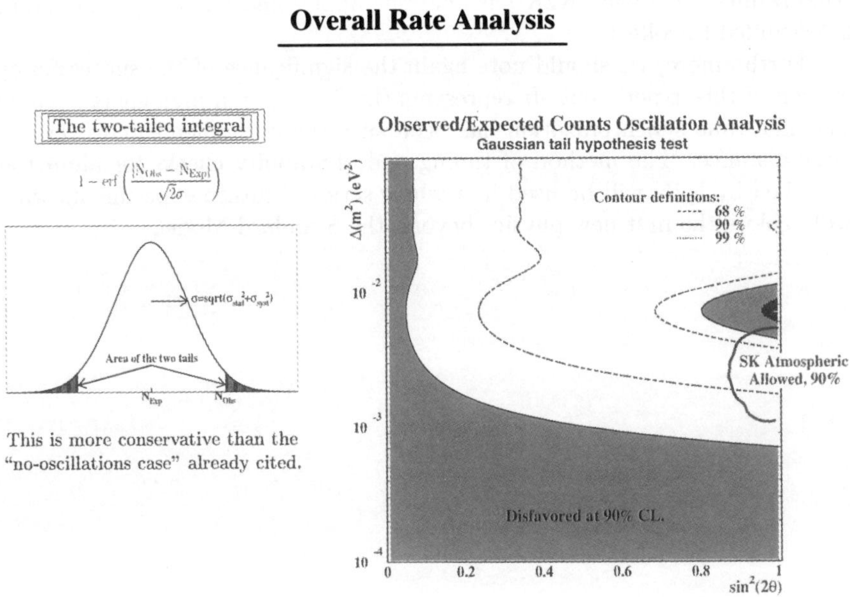

Figure 7. The disfavored region of oscillation parameter space by analysis of the total number of contained fiducial volume events only. (No spectral information is used.) In addition to the use of a 2-tailed probability as shown in the cartoon at right, the errors estimated for this analysis are also more conservative than those used for the null-hypothesis test. This is done to make the errors consistently conservative irrespective of spectral distortion by oscillation. The Super–Kamiokande atmospheric neutrino result is overlaid.

5 Conclusion

The hypothesis of no oscillations or other new physics is disfavored at approximately the 2σ level based on analysis of the total counts observed in K2K.

Spectral analysis gives more information, but a quantitative statement based on it is not yet available. Within a few years, K2K expects to gather more data and study systematic errors to make a firm statement about oscillations. With the full data set, K2K will cover a sensitive region of parameter space for oscillations which overlaps significantly with the atmospheric neutrino allowed region, giving it a good chance to confirm the nature of the effect. Figure 7 shows the first analysis of oscillation parameter space for two family mixing based on the total counts observed. A good agreement with atmospheric neutrino results is evident. K2K has analysed approximately 50% of the data it is scheduled to collect.

Furthermore, we should note again the significance of the successful operation of this experiment. It represents the first time in high energy physics that fine time synchronization has been successfully implemented over such large distance. The method of timing and of stability checks for aiming established by K2K will be used in a whole series of future experiments which will explore the first new physics beyond the Standard Model.

Acknowledgments

K2K is a collaboration of approximately 100 physicists from Japan, Korea, and the U.S. hosted by both the Japanese High Energy Physics Lab, KEK, and the Institute for Cosmic Ray Research. We gratefully acknowledge the cooperation of the Kamioka Mining and Smelting Company. This work has been supported by the Ministry of Education, Culture, Sports, Science and Technology, Government of Japan, the Japan Society for Promotion of Science, the U.S. Department of Energy, the Korea Research Foundation, and the Korea Science and Engineering Foundation. We thank the KEK and ICRR Directorates for their strong support and encouragement.

References

1. This first major physics paper from K2K is brief, but gives good detail about several of the subsystems and analysis features: "Detection of Accelerator Produced Neutrinos at a Distance of 250-km," S. H. Ahn *et al.* [K2K Collaboration], Phys. Lett. B **511**, 178 (2001) [arXiv:hep-ex/0103001].
   A few NIM papers about subsystems are cited below.
2. The full list of SK papers on atmospheric neutrinos is rather long. Here are some important ones:
   Y.Fukuda *et al.*, Phys. Lett. **B433**, 9 (1998); Y.Fukuda *et al.*, Phys. Lett. **B436**, 33 (1998);
   Y.Fukuda *et al.*, Phys. Rev. Lett. **81**, 1562 (1998).
3. F. Reines *et al.*, Phys. Rev. Lett. **15**, 429 (1965).
4. D. Casper *et al.*, Phys. Rev. Lett. **66**, 2561 (1991); R. Becker-Szendy *et al.*, Phys. Rev. **D46**, 3720 (1992).
5. K. S. Hirata *et al.*, Phys. Lett. **B205**, 416 (1988); K. S. Hirata *et al.*, Phys. Lett. **B280**, 146 (1992).
6. M. Aglietta, *et al.*, Europhys. Lett., '**8**(1989), 611.
7. C. Berger *et al.* [Frejus Collaboration], Phys. Lett. B **245**, 305 (1990).
8. W. W. Allison *et al.* [Soudan-2 Collaboration], Phys. Lett. B **449**, 137 (1999) [arXiv:hep-ex/9901024].
9. The best source of up-to-date information for this future experiment is its web page: `http://www-numi.fnal.gov:8875/`
10. Similarly, there are several web pages dedicate d to various parts of the proposals for a CERN to Gran Sasso experiment. A stable, general, and well maintained one is: `http://proj-cngs.web.cern.ch/proj-cngs/` which has details about the CERN site and links to pages about far

detectors proposed.

11. Two important experimental papers are:
S. Fukuda *et al.* [Super-Kamiokande Collaboration], ar neutrino data," Phys. Rev. Lett. **86**, 5656 (2001);
Q. R. Ahmad *et al.* [SNO Collaboration], trinos at the Sudbury Neutrino Observatory," Phys. Rev. Lett. **87**, 071301 (2001).
A lot of information on theory for solar neutrinos is available at: `http://www.sns.ias.edu/~jnb/`.
12. A. Piepke [KamLAND Collaboration], "KamLAND: A reactor neutrino experiment testing the solar neutrino anomaly," Nucl. Phys. Proc. Suppl. **91**, 99 (2001).
13. H. G. Berns and R. J. Wilkes, IEEE Nucl. Sci. **47**, 340 (2000)
14. H.Noumi *et al.*, Nucl. Instr. and Meth. **A398**, 399 (1997).
15. M.Ieiri *et al.*, Proc. 1st Asian Particle Accelerator Conference (1998).
16. H. Sato, Proc. Particle Accelerator Conference (1999);
K. Takayama, ICFA Beam Dynamics Newsletter No.20, (1999).
17. A. Suzuki *et al.* [K2K Collaboration], iment," Nucl. Instrum. Meth. A **453**, 165 (2000) [arXiv:hep-ex/0004024].
18. T. Ishii *et al.* [K2K MRD GROUP Collaboration], e," arXiv:hep-ex/0107041.
19. For a review of this topic, there is a section detailing the studies in the UNO "white paper" distributed at the Snowmass,'01 conference, " Physics Potential and Feasibility of UNO," preprint #:SBHEP01-3. The full report is available on the web at: `http://superk.physics.sunysb.edu/uno/`.

THE MINOS EXPERIMENT

M. D. MESSIER
High Energy Physics Laboratory,
Harvard University
42 Oxford Street
Cambridge, MA 02138, USA
E-mail: messier@huhepl.harvard.edu

The MINOS experiment is currently under construction. When complete, the experiment will use protons from the Fermilab Main Injector to create a pure beam of muon neutrinos. Two detectors will be placed in this beam, one at the Fermilab site and a second underground at a distance of 735 km. These detectors will make a high statistics measurement of the neutrino energy spectrum at both sites. Through a detailed comparison of the neutrino spectra at the near and far sites, MINOS expects to make roughly a 10% measurement of the oscillation parameter Δm^2, to limit non-oscillation explanations of the atmospheric muon neutrino deficits, and to constrain oscillations of muon neutrinos to other neutrinos besides tau neutrinos. The experiment will begin to take data early in 2005.

1. Introduction

1.1. *Neutrino Oscillations*

If neutrinos have mass, then the neutrino electroweak eigenstates are in general a superposition of the neutrino mass eigenstates:

$$\nu_\alpha = U_{\alpha i}\nu_i \tag{1}$$

where α =e,μ,τ, and $i = 1, 2, 3$. This mixing of the electroweak and mass eigenstates allows neutrinos to oscillation from one electroweak state to another as they propagate through space. The primary goal of the MINOS experiment is to make a detailed measurement of the oscillation pattern suggested by the atmospheric neutrino deficits[1], namely that ν_μ's to convert to ν_τ's with probability:

$$P_(\nu_\mu \to \nu_\tau) = \sin^2(2\theta)\sin^2(1.27\Delta m^2 L/E), \tag{2}$$

where L is the neutrino flight distance in km, E is the neutrino energy in GeV. From this measurement, MINOS expects to extract measurements of the neutrino mixing, $\sin^2(2\theta)$, and mass-splitting, Δm^2 (measured in eV^2).

This measurement will be accomplished by using protons from the Fermilab Main Injector to make an intense beam of neutrinos. Two detectors will be placed in this beam, one at the near site and one at the far site. The measurement of the oscillation probability will be made by comparing the neutrino spectrum at the near detector to the spectrum measured at the far detector. Any deficits of ν_μ relative to the expected event rate based on the near detector measurement, can be attributed to neutrino oscillations.

In addition to studies of $\nu_\mu \leftrightarrow \nu_\tau$, MINOS will also search for evidence of ν_e appearance at the atmospheric Δm^2 scale, and set limits on other possible "exotic" mechanisms of ν_μ disappearance such as oscillations to sterile neutrinos, neutrino decay, neutrino decoherence, or other mechanisms of neutrino disappearance.

2. The NuMI Neutrino Beam

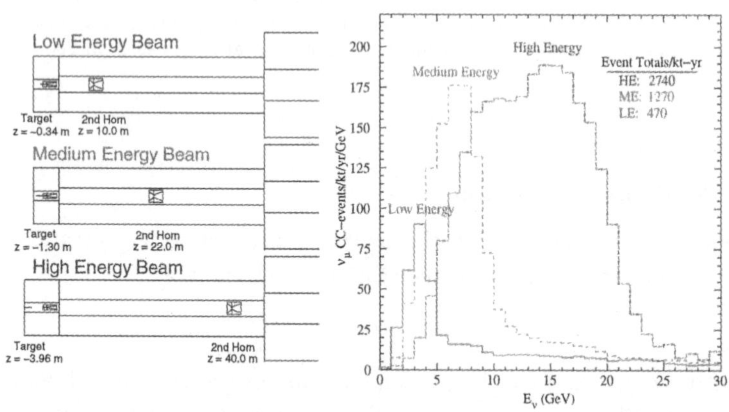

Figure 1. Left: The target and horn positions for the NuMI low, medium, and high-energy neutrino beams. Right: The neutrino event rate at the MINOS far detector for the three tunes of the NuMI beam.

The NuMI (Neutrinos at the Main Injector)[2] neutrino beam is produced via fast extraction (8 μs) of 120 GeV protons from the Fermilab Main Injector. The protons are directed to a 96 cm graphite target. The positive pions and kaons produce on this target are focused using two aluminum magnetic horns. The inner conductor of the horns is parabolic in shape to give mesons a p_T kick toward the beam axis which is proportional to the distance the meson is off axis, and hence act as a lens. To minimize pion absorption the inner conductor is kept thin 2 mm for the first horn, 3 mm for the second horn. The horn conductors carry a 200 kA and are pulsed in coincidence with each proton extraction, every 1.9 seconds. Prototype horns have been constructed at Fermilab an have taken roughly 3 million pulses, corresponding to roughly 1/3 of a year of NuMI operation.

Following the horns, mesons travel in a 675 m evacuated region where they decay to yield neutrinos. At the end of this region, the remaining mesons and interacted beam protons are absorbed in a composite aluminum/iron beam stop. Following the beam stop, muon monitoring counters are located in alcoves positioned on the beam axis. An additional monitor is placed upstream of the hadron absorber and is planned to be used during commissioning to center the proton beam on the axis to the far detector.

The focusing system is located in a 50 m target hall which allows for a great deal of flexibility. For example, the locations of the proton target and second horn can be changed relative to the location of the first horn position. This allows for beams of different energy neutrinos to be produced. Figure /reffig:numi-spect shows the locations of the target and second horn for the low-, medium-, and high-energy neutrinos beams, along with the expected event rates at the far detector for these configurations. As the low-energy beam has the most sensitivity to the oscillation parameters suggested by the atmospheric neutrino measurements it will be selected to start operations.

The excavation of the neutrino beam line is complete and work on the surface support buildings will begin shortly. The NuMI beam is expected to begin commissioning in early 2005 with data taking shortly thereafter.

3. The MINOS Detectors

The MINOS detectors[3] are optimized for muon tracking. The basic detector unit is an 2.54 cm thick iron plane with 1 cm thick solid plastic scintillator mounted on one face. The scintillator is mounted in 4 cm wide strips. The direction of the scintillator runs are clocked by 90° in alternate detector planes to provide 3-D track reconstruction. The far detector is built up of a total of 486 8 m×8 m octagonal planes which are divided into two "super-modules" of 243 planes each. When complete the total mass of the far-detector will be 5.4 kilo-tons.

In the scintillator strip, light is collected on a wave-length shifting fiber which runs down the center of the 4 cm strip. This light is guided to both strip ends and channeled to multi-anode Hamamastu M16 photomultipliers located in dark boxes mounted in crates at the edge of the detector. To save electronics, fibers are optically summed at the PMT with eight fibers mounted on a single PMT pixel. Summing the response at each end of the scintillator strip typically yields ten photoelectrons for a single muon crossing resulting in very efficient muon tracking.

The detectors are magnetized to an average field of 1.4 T using a current-carrying coil running through the center of iron planes. This field, enhances muon containment and allows measurements of muon sign and momentum. Using range and curvature muon momentum measurements can be made with 10% accuracy. Hadronic and electromagnetic energy resolutions are $65\%/\sqrt{E}$ and $23\%/\sqrt{E}$ respectively.

The near detector duplicates all the essential features of the far detector. However, due to the larger rates at the near site (roughly 10-20 neutrino interactions per spill compared to one neutrino interaction per 10,000 spills at the far detector) the electronics at the near site are required to be significantly faster than those for the far detector read out. The high neutrino rate has several advantages, however, requiring full instrumentation of only a relatively small region centered on the neutrino beam axis. The total near detector mass is 0.96 kton, roughly a factor of five less than far detector.

Figure 2. The partially assembled MINOS far detector.

Construction of the far detector is currently underway at the Soudan mine in Minnesota at a rate of roughly six to seven planes per week. Completion of the first of the two super-modules will be complete in July and magnetized at that time. Figure 2 shows a picture of the partially assembled detector. Commissioning of the far detector is underway using cosmic ray muons. Construction of the near detector is also underway on the surface to ready the detector for installation underground once the beam line installation is complete.

4. Physics Potential

4.1. $\nu_\mu \leftrightarrow \nu_\tau$

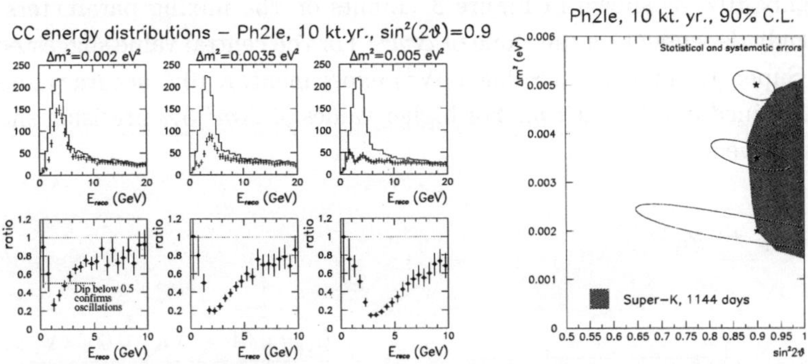

Figure 3. Left: The expected muon neutrino energy spectra at the far detector for a two-year MINOS run for different oscillation parameters. Right: The 90% C.L. confidence interval expected after a 2-year MINOS run for various values of Δm^2.

The main measurement to be made by the MINOS experiment is a precision measurement of the oscillation parameters which control the rate of $\nu_\mu \leftrightarrow \nu_\tau$ oscillations. These measurements rely on the detailed comparison of the MINOS near and far detector charged-current muon neutrino interaction rates as a function of energy. Due to the similarity of the MINOS near and far detectors many uncertainties due to detector efficiencies, resolutions, and neutrino cross-sections cancel in this comparison. The largest uncertainties in this measurement are likely to result from extrapolation of the muon neutrino flux from the near site to the far detector site. The largest uncertainties in this extrapolation result from the uncertainties in the production spectra of secondary mesons on the NuMI target. These uncertainties are expected to be greatly reduced by new measurements to be made within the next two years by the Fermilab E907[4] experiment. In addition to fundamental measurements of proton-nucleus interactions in the range 5-120 GeV using thin targets, E907 will also make a high statistics measurement of secondary production on the actually NuMI target using 120 GeV protons. This data will provide a direct input into simulations of the NuMI beam line reducing the absolute uncertainty in the neutrino production rate to roughly 5% and reduce the error in the near to far comparison to roughly 2%.

Figure 3 shows the expected neutrino spectra at the MINOS far detector after a 2-year run (7.6×10^{20} POT) and their comparison to the near detector spectra measurements for various values of the oscillation parameters. As shown in the figure, MINOS expects to resolve the first oscillation minimum which clearly establishes the presence of neutrino oscillations. Based on these spectra measurements, MINOS expects to measure Δm^2 to roughly 10% as shown in Figure 3. Limits on the mixing parameters are strongly dependent on the value of Δm^2. For the central value suggested by the Super–Kamiokande ($\simeq 0.003$ eV^2) experiment, a 10% accuracy should be obtained in a 2-year run. For higher values of Δm^2 5% precision should be achieved.

4.2. $\nu_\mu \leftrightarrow \nu_e$

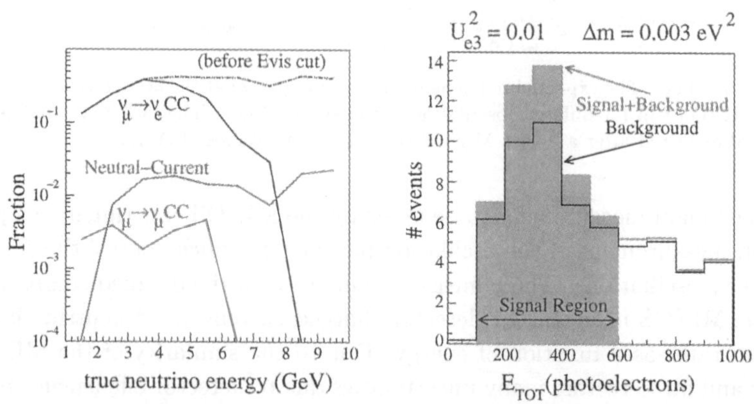

Figure 4. Left: The efficiency for accepting events of various categories into the ν_e sample. Right: The expected ν_e spectra for a two-year MINOS run. The expected appearance signal is superimposed on the the expected backgrounds.

4.3. $\nu_\mu \leftrightarrow \nu_e$

The MINOS experiment will also make a search for electron-neutrino appearance in the muon-neutrino beam. The far detector, which is optimized for tracking muons, and for large mass, is not fine grained enough to make an event-by-event distinction between electron-neutrino charged-

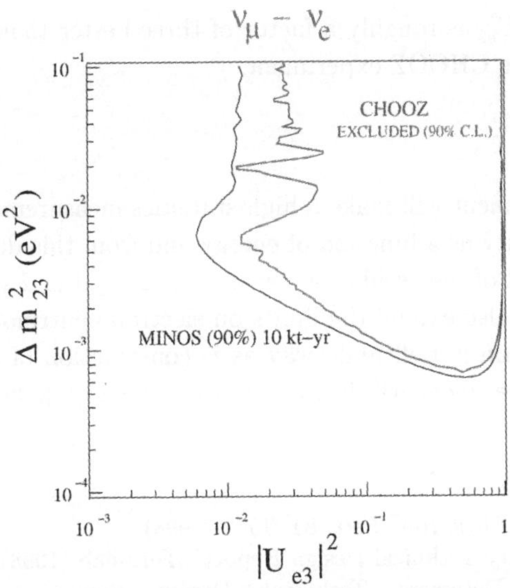

Figure 5. The 90% sensitivity of the MINOS experiment to subdominate oscillations of $\nu_\mu \leftrightarrow \nu_e$ oscillations.

current interactions and neutral-current interactions. Rather, electron neutrino events are selected based on a statistical analysis of the shower profile. The performance of the event identification is shown in Figure 4. The selection of electron-neutrino events can be made with roughly 30% efficiency while accepting only 2 % of the neutral-current interactions into the electron neutrino sample. Besides the neutral-current backgrounds, smaller contributions to the backgrounds also come from muon-neutrino charged-current events where most of the neutrino energy goes into the hadronic system making the muon hard to identify, and tau-neutrino interactions which produce both large hadron showers as well as electron showers from $\tau \to e$ decays. The expected electron-neutrino spectra for a 2-year MINOS run is shown in Figure 4. In the figure, an expected signal for U_{e3}^2 and $\Delta m^2 = 0.003$ eV^2 is superimposed on the background. For each set of oscillation parameters, an optimal signal region is selected in the visible energy spectrum. The 90% sensitivity of the MINOS experiment is calculated based on the number of events counted in this window over the expected backgrounds. The result of this calculation is shown in Figure 5. In the Δm^2 range suggested by the atmospheric neutrinos ($0.001 < \Delta m^2 < 0.005$ eV^2), the sensitivity of the MINOS experiment to non-zero values of U_{e3}^2 is roughly a factor of three better than the current best limit set by the CHOOZ experiment.

5. Conclusion

The MINOS experiment will make a high statistics measurement of the ν_μ oscillation probability as a function of energy and from this derive roughly a 10% measurement of the oscillation parameter Δm^2 after 2 years of operation. MINOS will also extend the limits on electron neutrino appearance. Detector construction is well underway as is construction of the neutrino beam line. The experiment will begin operation in early 2005.

References

1. Y. Fukuda et al., *Phys. Rev. Lett.* **81**, 1562 (1998).
2. "The NuMI Facility Technical Design Report", Fermilab (1998).
3. "The MINOS Detectors Technical Design Report," NuMI-L-337, http://www.hep.anl.gov/ndk/hypertext/minos_tdr.html.
4. "Proposal to Measure Particle Production in the Meson Area Using Main Injector Primary and Secondary Beams", Fermilab Proposal 907 (1998).

European accelerator-based neutrino projects

Mario Campanelli
DPNC University of Geneva, 24 Quai E.Ansermet, CH-1211 Geneva Switzerland
E-mail: Mario.Campanelli@cern.ch

Future neutrino projects in Europe will follow two distinct time lines. On the medium term, they will be dominated by the CERN-Gran Sasso long-baseline project, with two experiments OPERA and ICARUS, mainly concentrated on τ appearance. On the longer term, several projects are under discussion. A new proton driver at CERN that accelerates a 4 MW beam to 2.2 GeV of energy would open the possibility of a low-energy super-beam, possibly sent to the French laboratory under the Frejus. A new radioactive heavy ion facility could produce a pure ν_e beam, to be used independently or simultaneously with the super-beam. In the framework of R&D for Super-Beam and Neutrino Factory, the HARP experiment is studying hadron production at low energies on various targets.

1 Introduction

Traditionally, accelerator-based neutrino physics in Europe has been centred around CERN. Also for the future, this laboratory is expected to play a leading role in neutrino physics, despite the heavy commitment in the construction of LHC.

Due to the small atmospheric neutrino mass difference squared, the attention of the neutrino community is presently shifted towards long-baseline project. The CERN neutrino beam to Gran Sasso is higher in energy with respect to the Japanese [1] or American [2] projects, and will have the unique feature of being well above threshold for τ lepton production, in order to confirm the $\nu_\mu \to \nu_\tau$ nature of the atmospheric oscillations.

The OPERA detector is a dedicated experiment to search for τ appearance using topological informations, while ICARUS is a more versatile apparatus, that could reach similar τ identification capabilities, provided a sufficient number of modules is built.

On the longer term, the European projects are focused towards the possibility of building a super-beam based on a Superconducting Proton Linac (SPL), that also would be the first building block of a CERN-based neutrino factory. To estimate the flux of neutrinos produced in a super-beam, or the number of muons accelerated in a neutrino factory, a dedicated experiment (HARP) is presently under way. The Harp detector, built in a very short time, has taken data in summer 2001 and will have a second run in summer 2002 at the CERN PS. The goal is to measure total and differential cross sections for proton interactions with a large variety of targets, with energy in the range

3-15 GeV.

2 The Neutrino beam from CERN to Gran Sasso

As stated in the introduction, the CERN beam to Gran Sasso has the unique characteristic of being at a sufficiently high energy to produce a detectable number of τ leptons in charged current ν_τ interactions. This is a natural consequence of using a high-energy proton driver (400 GeV from the CERN SPS), having as a natural target the Gran Sasso lab near Rome (distant 732 km from CERN) and of the practical difficulties of building a near station, that makes hard to conceive disappearance measurements.

Civil engineering construction started on Oct 12, 2000; they will consist of 3 km of new tunnels and caverns: 700 m for the proton line to the target, 1 km of decay tube, plus various connections. Overall it will mean 45000 m^3 of rock removal, and 12000 m^3 of concrete added to reinforce the walls. The main properties of the CNGS beam are listed in table 2.

Baseline	732 km
Maximum depth	11.4 km
POT/extraction	2.4×10^{13}
POT/year	4.5×10^{19}
Flux at Gran Sasso	$3.5\times10^{12}/y/100m^2$
ν interactions	2500/kt/y
$< E_\nu >$	17 GeV
$\%\nu_\mu$	2.0
$\%(\nu_e + \bar{\nu}_e)$	0.8

Table 1: Main properties of the CNGS beam

The beam target will be 2 meters long, made of 10 cm long graphite (pure carbon) rods, interspaced to minimise absorption of pions with small transverse momentum with respect to the beam direction. Due to the strong thermical shocks, the whole system is cooled with helium. Secondaries produced in interactions with the target are focalised by a horn/reflector system. The horn will be 6.5 m long, with a diameter of 70 cm, and able to sustain a current of 150 kA for 1 ms, with a tolerance rate of 95% survival probability after 5×10^7 pulses. A horn prototype has already been built and tested, and survived 1.5 million pulses.

The beam transverse profile will be checked using two muon detectors at the end of the decay tunnel; the beam width at the far location is more than one kilometre, and approximately flat for a radius of about 500 m, much larger than the expected width of the detectors. Systematic uncertainties on the beam can arise from misalignment of the various components. As a reference, a 3% variation in the number of ν_μ in Gran Sasso could result from:

- the horn being off-axis by 6 mm
- the reflector being off-axis by 30 mm
- the beam being off-axis by 1 mm on the target
- an overall misalignment of 0.5 mrad

Beam alignment can be further checked in the Gran Sasso site, looking for the muons produced by neutrino interactions in the upstream rock. However, we have to bear in mind that flux systematics is not that relevant when doing an appearance experiment with few candidate events.

Possible upgrade schemes for the PS/SPS complex, in order to increase the neutrino flux, have been extensively studied [3]. It was found out that most of the components are at their limits. A gain of a factor 1.5 seems feasible; however, a factor 3 seems the absolute maximum, even if large investments are made.

The Gran Sasso laboratory from INFN (Laboratori Nazionali del Gran Sasso, LNGS) is operational since 1987, and offers an easy access from a motorway tunnel, and 1500 meters of rock overburden. The completion of the first generation of experiments (e.g. MACRO) leaves space in two large underground hall (B and C), oriented towards CERN, to perform long-baseline neutrino experiments. The two experiments foreseen for the long-baseline project are OPERA [4], already fully approved as CNGS1, a dedicated experiment to look for τ kink in a hybrid emulsion system, and ICARUS [5], of which the first module has already been built and successfully tested, a general-purpose detector based on a liquid Argon TPC.

3 OPERA

The requirement to have a large mass and a space resolution of the order of 1 μ m has lead to the choice of hybrid emulsions as basic detector component. A basic cell is composed of 1 mm lead planes, separated by an emulsion layer made of two 50 mm emulsion sheets separated by 100 μm plastic base. 56 such lead-emulsion cells constitute a brick, 3000 bricks, together with an extruded scintillator plane for quasi-online identification of interesting interactions are a module, 24 modules, completed with a muon spectrometer, form a supermodule. The experiment will be made of 3 such supermodules, for a total mass of about 2 ktons. The emulsion surface will be 176000 m^2; as a comparison, that of CHORUS was about 500 m^2. Given the very large amount of emulsion surface, finding few τ events, for an average τ decay length of half a millimetre, is an extremely challenging task. The first step will be the use of the scintillator tracker plane behind the bricks to identify where neutrino interactions take place. These bricks are removed from the wall, with a rate of about 40 bricks per day. They are exposed for a short period to cosmic rays in a shallower location, always in Gran Sasso, to provide a reference for the alignment, and then the emulsions are developed. Then, a search for a neutrino vertex and a decay candidate is performed quasi-online by various scanning stations. The goal is to understand if the brick extracted is sufficient for τ search, or additional nearby bricks have to be extracted. Finally, about 500 squared meters of emulsions (a factor 100 more than at CHORUS) are sent to the scanning labs.

The scanning speed is a fundamental factor to ensure the success of the experiment. A factor 20 improvement with respect to the present speed is expected from the use of the S-UTS technology presently under development in the Japanese laboratory of Nagoya University. This system is based on a fast CCD camera (3000 frames per second) connected to microscopes, with continuous movement on the x-y plane. The z position is determined by the focusing plane of the microscopes, controlled by pizo-attenuators. The scanning speed envisaged is 20 sqcm per hour, and the amount of data reduction from pixel to track of the order of 10^5.

The strategy for τ search is based first on finding a vertex, i.e. a common starting point for tracks with $\tan\theta < 0.4$, that are followed back until not any more found in two successive layers. Including geometrical losses, a vertex finding efficiency of 80% is expected. Then the search is split in two, depending on weather the τ decays in the same lead layer (short decays) or in the next. In the first case, the requirement is to have the momentum of the primary track above 1 GeV, and an impact parameter larger than a value, dependent on the z position, between 5 and 20 μm. In case of long decays, the requirement is to have a kink angle of 20 mrad with respect to the initial track (the kink angle resolution is about 3 mrad). The signal efficiency for several event categories is shown in table 3, and the number of expected backgrounds in table 3.

In absence of signal, OPERA can exclude at 90% C.L. $\nu_\mu \rightarrow \nu_\tau$ oscillations for $\Delta m^2 < 1.2 \times 10^{-3} eV^2$ at full mixing, and $\sin^2 2\theta < 5.7 \times 10^{-3}$ at large Δm^2. The number of expected events for signal and background for various values of Δm^2 and maximal mixing are shown in table 3

	DIS long	QE long	DIS short	Overall
$\tau \to e$	2.7	2.3	1.3	3.4
$\tau \to \mu$	2.4	2.5	0.7	2.8
$\tau \to h$	2.8	3.5	-	2.9
Total	8.0	8.3	1.3	9.1

Table 2: τ identification efficiency for OPERA in various decay channels, for Deep Inelastic Scattering (DIS) and Quasi-Elastic (QE) events, in the long and short configuration

	$\tau \to e$	$\tau \to \mu$	$\tau \to h$	Total
Long decays	0.15	0.29	0.24	0.67
Short decays	0.03	0.04	-	0.07

Table 3: Background for τ identification in number of events for long and short decays.

4 ICARUS

The ICARUS detector is composed of a large liquid Argon TPC, that combines big mass and a granularity of the order of 1 mm, being a veritable electronic bubble chamber. Its resolution and particle identification capabilities allow this kind of detector to contribute to a vast and versatile physics program, that includes study of atmospheric neutrinos, solar neutrinos, nucleon decay and of course neutrinos from the long baseline beam.

Drifting electrons for distances of the order of 1 meter, needed to achieve a sufficient detector mass, requires an extreme purity of the liquid Argon: the tolerance is a concentration of electronegative impurities at the level of 0.1 parts per billion, that was finally achieved after 10 years of R&D.

A 600 ton module has been built during the last years in Pavia, and successfully tested this summer with cosmic ray data. Many long (20 meters) muon events were observed, as well as other interesting events such as electromagnetic showers, stopping muons and strange particles.

This 600 ton detector will be transported to Gran Sasso at the beginning of 2003. The ICARUS technology is very well-suited for τ search, using a kine-

Δm^2	1.2×10^{-3}	2.4×10^{-3}	5.4×10^{-3}	BG
	2.7	10.8	53.5	0.75

Table 4: Number of expected signal events (with efficiency included) and of background, after 5 years of CNGS operation for the OPERA detector

matic approach, complementary to the topological one of OPERA. However, 600 tons are not sufficient to produce enough events when the tight cuts necessary for large background suppression are applied. The ICARUS collaboration has recently proposed [6] an upgrade to 5 modules, for a total fiducial mass of 2.35 kton. Golden $\tau \to e$ events can be identified cutting on fiducial volume, electron momentum, visible energy, electron P_t and Q_t and missing momentum. The best sensitivity is obtained combining several variables in a common likelihood, and efficiencies and backgrounds for a 3 kton ICARUS (table 4) is similar to that of OPERA. Due to its good electron identification, ICARUS

τ decay mode	1.6	2.5	3.0	4.0	BG
$\tau \to e$	3.7	9	13	23	0.7
$\tau \to \rho$ DIS	0.6	1.5	2.2	3.9	< 0.1
$\tau \to \rho$ QE	0.6	1.4	2.0	3.6	< 0.1
Total	4.9	11.9	17.2	30.5	0.7

Table 5: Signal events for different values of Δm^2 (in units of $10^{-3} eV^2$) and expected number of background events after 5 years of data taking for a 3 kton ICARUS detector.

will also be able to perform a search for $\nu_\mu \to \nu_e$ oscillations, improving by some factors the present limit on $\sin^2 2\theta_{13}$.

5 The CERN-Frejus Super-Beam

In the next 15-20 years the main commitment of CERN will be the construction and operation of LHC and its detectors. However, the European neutrino community is very active in trying to shape a next-generation neutrino program, in the context of a future CERN-based neutrino factory. The first brick of such a machine would be a low energy (2.2 GeV) high power (4 MW) proton linac[7].

Independently of the neutrino factory, this machine would be a considerable improvement to the CERN accelerator complex, since it will increase PS beam intensity, triple the brilliance of the LHC beam, supply the present ISOLDE facility with 5 times more current (or up to 100 times for Super-ISOLDE), and allow the construction of a low-energy superbeam.

Focusing the pions in the energy range of their maximal production, the resulting mean neutrino energy is about 250 MeV, an ideal energy for a water Cerenkhov detector. At this energy, the maximum of the oscillation occurs for a baseline of about 130 Km, the distance between CERN and the Frejus tunnel. Due to the civil engineering work foreseen to build a second motorway gallery, it is conceivable to propose building there a large Cerenkhov detector, to be used for proton decay, supernova and beam neutrinos, like that proposed by the UNO collaboration [8]. Considering 400 kton of fiducial mass, a π^0/e separation with 0.1 % confusion and efficiency around 70%, the 90% C.L. sensitivity to $\sin^2 2\theta_{13}$ would be in the range of 5×10^{-4} after 5 years for the current values of Δm^2. A ten times smaller exposure would already allow measuring $\sin^2 \theta_{23}$ at 1% level and Δm^2_{23} with a precision of 10^{-4}. For CP violation studies, even in the case of a favourable choice of parameters, 400 kton are needed, for 2 years of running with neutrinos and 10 years with antineutrinos.

6 The beta-beam

A recent interesting possibility[9] is producing a neutrino beam from the decay of radioactive ions collected in a storage ring. Antineutrinos would be for instance produced by the decay $^6He^{++} \rightarrow ^6 Li^{+++}\bar{\nu}_e e^-$, and neutrinos from similar decays of ^{18}Ne. The advantage of this approach with respect to the "traditional" Neutrino factory is that despite the lower intensities of the stored beam, the much larger quality factor Γ/E_0 for ions produces a very collimated neutrino beam, leading to fluxes in the far detector similar to those of a super-beam, for comparable energies. The produced beam will be extremely clean and well-known, and of only the electron neutrino flavour. Moreover, it is possible to think of an experiment combining the beta-beam and the super-beam, that would produce an almost pure ν_μ beam, separated from the ν_e beta-beam using timing information. Such a system would have the two-beam feature of the neutrino factory, but it will not need a magnetic detector to distinguish oscillated and beam events. Therefore, larger detector with lower threshold can be used (like a water Cerenkov), putting them at shorter distance to minimise matter effect. A high-intensity combination of super-beam and beta-beam could search for T-violation in an extremely convincing way.

7 Other experiments

The HARP detector, presently running at CERN, is not strictly speaking a neutrino experiment, but it will provide measurement of hadron production at small and large angles for a variety of target material of interest for neutrino beams. In particular large angle pion production is important in the design of the focusing system of a Neutrino Factory and a Super-Beam. It will also use cryogenic targets made of liquefied gases for hadronic production in the atmosphere in the region of interest of atmospheric neutrinos, as well as the measurement of replicas of existing neutrino targets (K2K, MiniBOONE). Several million of events have been collected in summer 2001, and a similar statistics will be collected this year. Since several sources of data-taking inefficiency have been removed, the number of useful events is expected to increase by a factor of approximately 5.

The possibility of using the existing AD ring as a muon accumulator for a "baby" neutrino factory to measure low-energy neutrino cross section was studied [10]. The interesting result was that these muons are already accumulated in the present antiproton running mode, and thus a neutrino beam is already existing there. Unfortunately, quantitative estimate showed that obtaining a sufficient number of stored muons in parasitic mode is incompatible with the present setup of the running AD experiments.

8 Conclusions

In the next years, CERN will focus on building and operating LHC. However, a large community is eager to maintain an active neutrino physics program. The CNGS long-baseline beam will start operation in spring 2005; the OPERA detector is already fully approved and under construction; a first 600 ton ICARUS module has already bee built and tested, and a proposal for a factor 5 mass increase is under approval. For the future, HARP is studying hadron production in the region of interest of present and future neutrino beams, neutrino factories and atmospheric neutrinos; there is an active group studying the possibility of a second-generation super-beam (possibly in conjunction with a beta-beam), and trying to make the first step towards a European neutrino factory for the post-LHC period.

References

1. K2K collaboration, S.H. Ahn et al, hep-ex/0103001, submitted to Physics Letters B.
2. NuMI-B-93 J. Morfin et al., Conceptual Design Report for the NuMI beam.
3. CERN/PS 2001-041 (AE) CERN/SL 2001-32
4. M. Guler et al.CERN/SPSC 2000-028, SPSC/P318, LNGS P25/2000
5. P.Cennini *et al.* LNGS 94/99-I and 94/99-II
6. ICARUS TM/2001-08 LNGS-EXP 13/89 add. 2/01
7. CERN 2000-012
8. "Physics Potential and Feasibility of UNO" (preprint: SBHEP01-3)
9. P.Zucchelli, CERN-EP/2001-056 hep-ex/0107006
10. M.Campanelli, S.Navas,A.rubbia, hep-ph/0107221

THE JHF-KAMIOKA NEUTRINO PROJECT

TAKAAKI KAJITA

Research Center for Cosmic Neutrinos, Institute for Cosmic Ray Research, University of Tokyo, Kashiwa-no-ha 5-1-5, Kashiwa, Chiba 939-2255, Japan

E-mail: kajita@icrr.u-tokyo.ac.jp

The JHF-Kamioka neutrino project is a second generation long base line neutrino oscillation experiment that probes physics beyond the Standard Model by high precision measurements of the neutrino masses and mixing. A high intensity neutrino beam is produced by a high intensity proton synchrotron at JHF. The neutrino energy is tuned to the oscillation maximum at < 1 GeV for the baseline length of 295 km. The far detector is the 50 kton water Cherenkov detector, Super-Kamiokande.

1 Introduction

The discovery of neutrino oscillations in atmospheric neutrinos [1] (see also [2] for earlier data) has opened a new window to study physics beyond the Standard Model through neutrino masses and mixing. The present data from Super-Kamiokande [3] show that the mixing between the 2nd and the 3rd generation neutrinos (θ_{23}) is consistent with the full mixing ($\sin^2 2\theta_{23} > 0.90$ at 90%C.L.). Also the difference of the neutrino mass squared is constrained well ($1.6 \times 10^{-3}\text{eV}^2 < \Delta m^2_{23} < 3.6 \times 10^{-3}\text{eV}^2$). In addition, the solar neutrino problem was proven to be due to neutrino oscillations by precise measurements by SNO [4] and Super-Kamiokande [5]. The mixing between the first and the second generation neutrinos (θ_{12}) is also found to be large. These neutrino mixing angles that are significantly different from those of quarks may be a key to a deeper understanding of the physics beyond the Standard Model. However, we know that our understanding of the neutrino mixing matrix is not complete, because we have little knowledge on the first and the third generation mixing (θ_{13}).

Results from the first generation long baseline neutrino oscillation experiment, K2K [6], are also consistent with the atmospheric neutrino data. In a few years, the MINOS experiment [7] will start taking data. This experiment will significantly improve our knowledge on $\sin^2 2\theta_{23}$ and Δm^2_{23}. The CNGS project [8] will also start experiments in 2005. The main goal of this project is the confirmation of $\nu_\mu \to \nu_\tau$ neutrino oscillations by direct detection of τ leptons.

The main goals of the JHF-Kamioka neutrino project are observation of non-zero $\sin^2 2\theta_{13}$ and precise measurement of $\sin^2 2\theta_{23}$ and Δm^2_{23}.

2 Overview of the experiment

The JHF-Kamioka neutrino project is a proposed long baseline neutrino oscillation experiment using the JHF 50 GeV proton synchrotron (PS). The JHF accelerator complex was officially approved in 2001 by the Japanese government. The construction of the 50 GeV PS is in progress in JEARI (Japan Atomic Energy Research Institute) at Tokai, and will be completed in 2006. In the earliest case, the neutrino project can start in 2007. 5 years of experimental period is assumed for the first phase of this neutrino project. The 50 GeV PS is designed to deliver 3.3×10^{14} protons every 3.4 seconds. The beam power is 0.77 MW. A future upgrade of the beam power to 4 MW is considered. The far detector is the Super-Kamiokande 50 kton water Cherenkov detector. The baseline length of the experiment is 295 km. A 1 Mton water Cherenkov detector, Hyper-Kamiokande, is seriously considered as a far detector in the second phase of this neutrino project.

A feature of this experiment is the use of a low-energy, narrow band, high-intensity neutrino beam. The neutrino energy will be tuned to the maximum oscillation energy. For $\Delta m^2_{23} = 3.0\times10^{-3}\text{eV}^2$, it is 715 MeV. In the early stage of the studies of this project, three different neutrino beams were considered. They are wide band beam (WBB), narrow band beam (NBB) by selecting mono-energetic pions, and off-axis beam (OAB). OAB is another option to produce a narrow neutrino energy spectrum [9]. The beam optics is almost same as the WBB, but the axis of the beam is displaced by a few degrees from the far detector direction. Due to the two body decay kinematics of pions, the energy of neutrinos that hit the far detector is almost independent of the pion energy spectrum. The neutrino energy can be adjusted by choosing the angle between the pion beam direction and the direction to the detector (off-axis angle).

Detailed Monte Carlo simulations have been carried out to estimate the expected neutrino spectrum and the number of events. Figure 1 shows a comparison of the expected spectra by the three beam configurations. It is clear that the OAB has the highest flux at the maximum oscillation energy. Table 1 shows the expected number of events for no oscillations. From these studies, we have decided to use the OAB for the JHF-Kamioka neutrino project.

If the off-axis angle is determined, the neutrino energy distribution is essentially determined. On the other hand, our knowledge on Δm^2_{23} is not precise enough to pre-determine the off-axis angle uniquely. Because of these conditions, the decay pipe is designed to accommodate off-axis angles between 2 and 3 degrees.

Beam	ν_μ CC	ν_e CC	NC	Total / year
Wide band beam	5200	59	1800	7100
Off-axis 2 degree	2200	45	900	3200
Off-axis 3 degree	800	22	300	1100
Narrow band beam (E_π = 2GeV)	620	5.0	250	880

Table 1. Summary of the expected number of events per year for various beam options. The oscillation effect is not included.

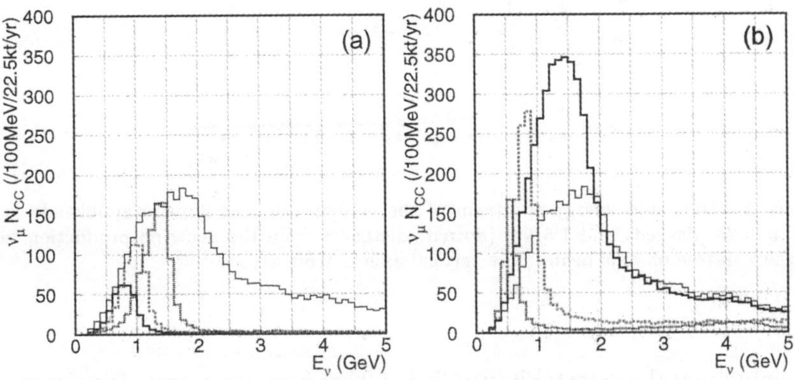

Figure 1. Neutrino energy spectra of charged current interactions. Thick solid, dashed and dotted histograms in (a) show NBB with 1.5, 2.0 and 3.0 GeV beam, and those in (b) show OAB with 1.0°, 2.0° and 3.0° off-axis angles, respectively. WBB is shown by a thin solid histogram in both (a) and (b).

3 The neutrino flux and the near detector

Another important element of the experiment is the front detector that monitors the neutrino flux without the oscillation effect. In order to achieve the designed goals of the experiment, it is very important to precisely understand the neutrino beam and the detection efficiency for various types of neutrino events. For this purpose, it is important that the near and the far detectors are designed as similar as possible so that various detector systematics cancel by taking the far-near ratio. The far detector, Super-Kamiokande, is a water Cherenkov detector. Therefore, we strongly think that the near detector should also be a water Cherenkov detector. By an event rate consideration, we estimate that the fiducial mass of the near detector should be about 100 tons,

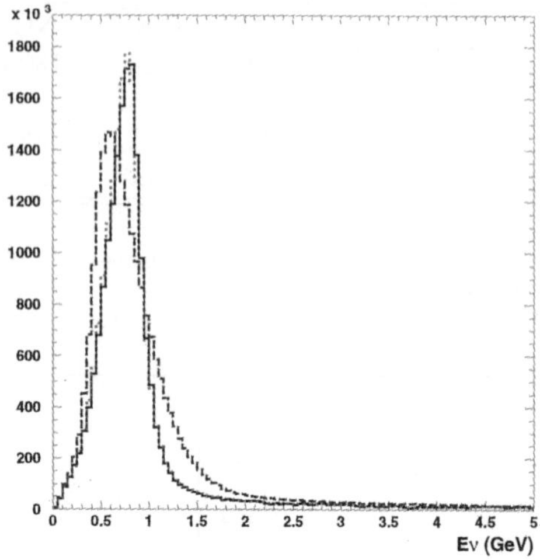

Figure 2. Calculated energy spectrum of the neutrino flux at Super-Kamiokande (solid) and at 0.28 (dashed) and 1.5 km (dotted) distances from the neutrino production target for the 2 degree off-axis beam. The vertical axis is arbitrary.

assuming that the detector is installed at 2 km from the target. We also require that the distance from the surface of the fiducial volume to the photomultiplier tubes (PMTs) is 2.0 m following the definition in Super-Kamiokande. In addition, the near detector has to measure the energy of muons up to about 1 GeV. For this reason, the distance from the surface of the fiducial volume to the PMTs should be about 5 m for the down-stream region. The fiducial volume can be 4 m in diameter and 8 m in length, and the total volume (including the volume for installing PMTs) can be about 9 m in diameter and 16 m in length. The total weight is about 1000 tons. In addition to the water Cherenkov detector, it might be important to install muon range counters at the down stream to measure the energies of high energy muons that shall be produced by the high energy tail of the flux. Furthermore, a fine-grained scintillator detector should be installed in front of the water Cherenkov detector in order to study the details of neutrino interaction kinematics.

It is known that the energy spectrum of the neutrinos at the Super-Kamiokande detector, which is 295 km away from the neutrino production point, shall not be identical to that at the front detector. For example, Figure 2 compares the calculated energy spectrum at Super-Kamiokande and at 0.28 and 1.5 km distances from the neutrino production target for the 2 degree off-axis beam. 0.28 km is the possible on-site neutrino detector position. The spectrum has lower energy peak at 0.28 km, while the 1.5 km and Super-Kamiokande spectra are almost identical. According to the beam Monte Carlo, the "far-near" ratio ($\equiv$ flux(far)/flux(near)$\times(L_{far}/L_{near})^2$) approaches to unity very quickly as L_{near} increases, where L_{far} (L_{near}) is the distance between the production target and the far (near) detector. For L_{near} >1.0 km, the deviation from unity is less than 10%.

Even if the uncertainty in the far-near ratio is large, it might be acceptable if it does not affect the physics results. From various considerations, 5% far-near systematic error could be a goal for various measurements. Let us discuss the physics sensitivities. In the following discussions, we assume that the true Δm^2_{23} is 3.0×10^{-3}eV^2. We also assume the 2 degree off-axis beam.

4 Physics

4.1 Measurement of $sin^2 2\theta_{23}$ and Δm^2

The neutrino energy can be reconstructed for quasi-elastic interactions assuming that the target nucleon is at rest;

$$E_\nu = \frac{m_N E_l - m_l^2/2}{m_N - E_l + P_l cos\theta_l} \tag{1}$$

Figure 3 shows the expected energy spectrum with and without neutrino oscillations assuming $\sin^2 2\theta_{23}$ =1.0 and Δm^2_{23} = 3.0×10^{-3}eV^2. Events with single μ-like Cherenkov ring are plotted. Since the peak flux is tuned to the maximum oscillation energy, most of the muon neutrinos must be oscillated to tau neutrinos. From this figure it is possible to estimate the sensitivities in $\sin^2 2\theta_{23}$ and Δm^2_{23}. See Figure 4 (upper). The expected accuracy is 1% for $\sin^2 2\theta_{23}$ and 1×10^{-4}eV^2 for Δm^2_{23}, if the true $\sin^2 2\theta_{23}$ value is 1.0. However, at the preparation stage we must also consider a case that the true $\sin^2 2\theta_{23}$ is less than 1.0. (The present lower limit from Super-Kamiokande is 0.90 at 90% C.L..) The maximum oscillation effect should occur at the 600 to 800 MeV energy bin for Δm^2_{23} = 3×10^{-3}eV^2. Since this bin is the most important one for the determination of $\sin^2 2\theta_{23}$, the statistical and systematic errors are estimated based on the number of events in this bin. Figure 4 (lower) shows the estimated statistical and systematic errors in the measurement of $\sin^2 2\theta_{23}$. If the far-near systematic error is assumed to be 10%, the statistical error is larger than the systematic error at $\sin^2 2\theta_{23}$ = 1.0, and the total error is about 1%. However, for smaller $\sin^2 2\theta_{23}$ values, systematic error

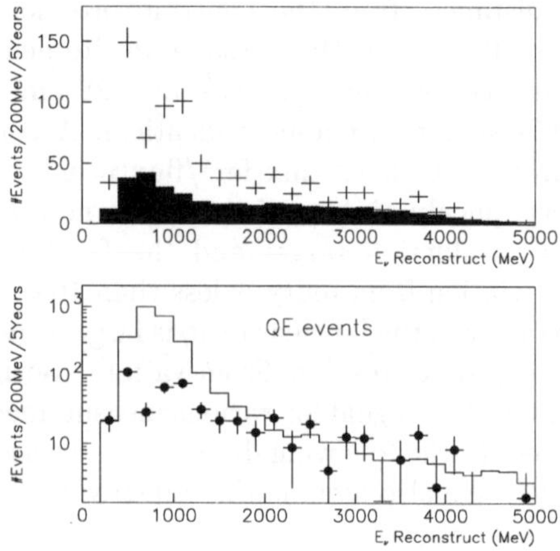

Figure 3. Upper: Reconstructed neutrino energy distribution for 5 year operation of the JHF-Kamioka experiment with the 2 degree off-axis beam. Events with single μ-like Cherenkov ring are plotted. The shaded histogram shows the contribution of non-quasi-elastic events. $\mathrm{Sin}^2 2\theta_{23}$ =1.0 and Δm^2_{23} =3.0×10^{-3}eV^2 are assumed. The maximum oscillation is expected to occur at the 0.6 to 0.8 GeV energy bin. Lower: Reconstructed E_ν distribution after subtracting the non-quasi-elastic events. The histogram and the dots show the non-oscillation and oscillation cases.

dominates the statistical error. Because the systematic error is proportional to the number of observed events in the 600 to 800 MeV energy bin, the systematic error rapidly gets bigger for smaller $\sin^2 2\theta_{23}$. In order not to limit the measurement by the far-near systematic error, it is essential to understand the far-near systematic effect to an accuracy of better than 5%.

4.2 Search for non-zero $sin^2 2\theta_{13}$

The JHF neutrino beam has a small ν_e contamination (0.2% at the energy of the peak flux). Furthermore, the ν_e appearance signal is maximized by tuning the neutrino energy at its oscillation maximum. Thus, this experiment has an excellent opportunity to discover non-zero θ_{13}.

The signal should be searched for in the single-ring e-like events. The possible background processes are single muon events with the muon being misidentified as an electron, contamination of ν_e in the beam, and NC (mostly π^0) events. Among them, the most serious one is the NC background. Special cuts have been developed to reject these events (see [10] for details of the cuts).

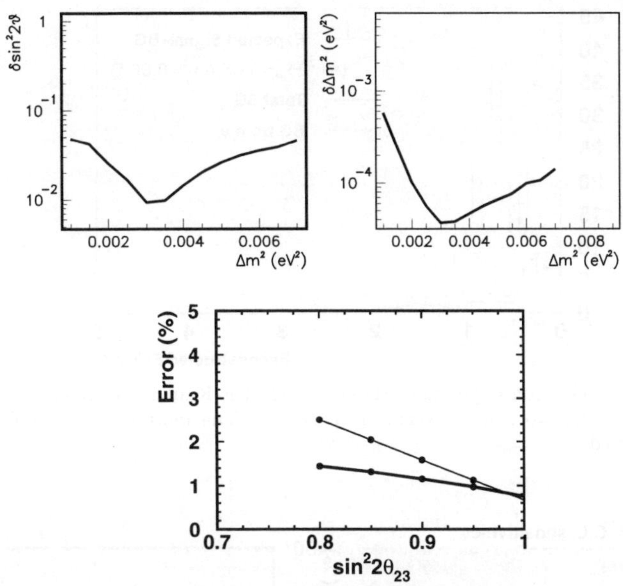

Figure 4. Upper: Sensitivity in $\sin^2 2\theta_{23}$ and Δm^2_{23} after 5 years of operation of the JHF-Kamioka experiment with the 2 degree off-axis beam. Lower: Expected statistical(thick line) and systematic(thin line) errors in the $\sin^2 2\theta_{23}$ measurement after 5 years of operation of the experiment for various values of $\sin^2 2\theta_{23}$. 10% far-near ratio uncertainty is assumed for the systematic error.

Figure 5 shows the expected energy distribution for the signal and background. Table 2 summarizes the expected number of signal and background events for two different values of $\sin^2 2\theta_{13}$. The sensitivity of this experiment to $\sin^2 2\theta_{13}$ is about 0.006. In this paper, we also consider a case that the true $\sin^2 2\theta_{13}$ is relatively large and that the $\sin^2 2\theta_{13}$ value is measurable. From Table 2, it is easy to roughly estimate the statistical and systematic errors. Again, only the

$\sin^2 2\theta_{13}$	ν_μ(CC+NC)	Beam ν_e	Osc'd ν_e	Total (signal + BG)
0.1	11.1	11.1	123.2	145.5
0.01	11.1	11.1	12.3	34.5

Table 2. Summary of the expected electron appearance signal and background for the measurement of $\sin^2 2\theta_{13}$. $\Delta m^2_{13} = 3.0 \times 10^{-3}\mathrm{eV}^2$ and $\sin^2 2\theta_{23} = 1.0$ are assumed. The matter and the CP violation effects are neglected.

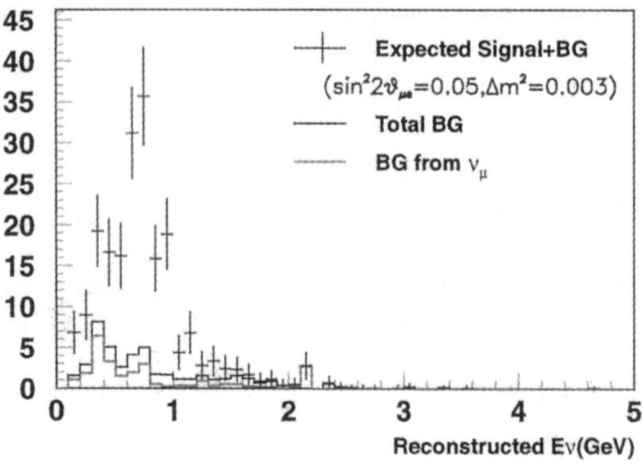

Figure 5. Expected electron appearance signal in the JHF-Kamioka neutrino project. It is assumed that $\sin^2 2\theta_{13}$ =0.1 and 5 years of operation. The matter and the CP violation effects are neglected.

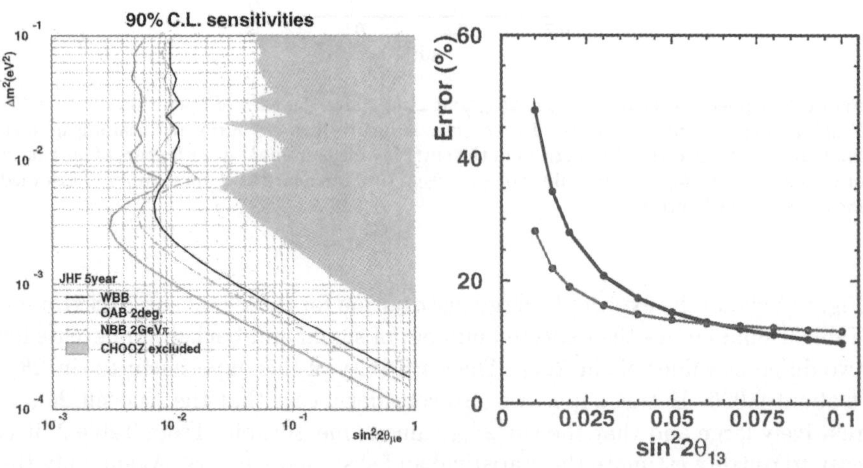

Figure 6. Left: 90% C.L. sensitivity in $\sin^2\theta_{\mu e} \equiv 0.5 \cdot \sin^2 2\theta_{13}$ after 5 years of operation. Right: Expected statistical(thick line) and systematic(thin line) errors in the $\sin^2 2\theta_{13}$ measurement after 5 years of the operation of the JHF-Kamioka experiment for various values of $\sin^2 2\theta_{13}$. 10% far-near ratio uncertainty is assumed for the systematic error. Intrinsic uncertainties due to CP violation and sign of Δm^2_{23} are neglected.

far-near ratio uncertainty is considered for the systematic error. Figure 6(left) shows the expected sensitivity in $\sin^2 2\theta_{\mu e}$ ($\equiv 0.5 \times \sin^2 2\theta_{13}$). Figure 6 (right) shows the estimated statistical and systematic errors for various $\sin^2 2\theta_{13}$ values, where the far-near systematic error of 10% is assumed. $\sin^2 2\theta$ =0.01 and 0.1 approximately correspond to the sensitivity of this experiment and the lower limit from the CHOOZ [11] and Palo Verde [12] experiments, respectively. In order not to limit the measurement by the systematics, the far/near systematic error must be controlled to an accuracy of better than 5%.

5 Phase 2 of the experiment

In the 2nd phase of the JHF-Kamioka neutrino project, the proton intensity delivered by the 50 GeV PS is planned to increase to 4 MW. Also a 1 Megaton water Cherenkov detector, Hyper-Kamiokande, is considered as the far detector. With these upgrades, it is possible to measure the CP phase in the lepton sector, provided that the solution of the solar neutrino problem is the Large Mixing Angle solution. However, I will not discuss in any details, because M. Aoki [13] discussed details of the CP violation physics with the JHF-Kamioka neutrino project.

6 Recent progress

In this article, I have discussed the physics sensitivity based on Ref. [10]. We feel that the event rate is not high enough for various measurements. Therefore, after writing Ref. [10], we have studied a possibility of a longer decay pipe for obtaining a higher flux. In Ref. [10], the length of the decay pipe was assumed to be 80 m. Based on the recent study, we decided to use 130 m decay pipe. The flux increased by about 40% at the peak energy of the flux. Therefore, the physics sensitivity described in this article could be achieved in about 3.5 years of operation (or a better sensitivity is expected for 5 years of operation).

7 Summary

The JHF-Kamioka neutrino project is a second generation neutrino oscillation experiment. The goals in the first phase are a precise measurement of $\sin^2 2\theta_{23}$ and Δm^2_{23}, and a discovery of non-zero θ_{13}. The expected accuracy of the measurement is $\delta(\sin^2 2\theta_{23}) \sim 0.01$ and $\delta(\Delta m^2_{23}) \sim 1 \times 10^{-4}\mathrm{eV}^2$. If $\sin^2 2\theta_{13}$ is larger than 0.01 it is possible to observe non-zero θ_{13}. In the second phase of this project, it is possible to measure the CP violation phase, provided that the solution to the solar neutrino problem is the Large Mixing Angle solution.

This project is a large project. It is essential that the experimental team is formed by physicists from all over the world who are interested in this project.

Acknowledgments

The author would like to thank the members of the JHF-Kamioka neutrino working group for many useful discussions and suggestions.

References

1. Y. Fukuda *et al.*, Phys. Lett. B **433** (1998) 9; Phys. Lett. B **436** (1998) 33; Phys. Rev. Lett. **81** (1998) 1562; Phys. Rev. Lett. **82** (1999) 2644; Phys. Lett. B **467** (1999) 185; S. Fukuda *et al.*, Phys. Rev. Lett. **85** (2000) 3999.
2. K.S. Hirata *et al.*, Phys. Lett. B **205** (1988) 416; Phys. Lett. B **280** (1992) 146; Y. Fukuda *et al.*, Phys. Lett. B **335** (1994) 237; S. Hatakeyama *et al.*, Phys. Rev. Lett. **81** (1998) 2016.
3. M. Messier, for the Super-Kamiokande collaboration, in these proceedings.
4. Q.R.Ahmad *et al.*, Phys. Rev. Lett. **87** (2001) 071301; S. Oser, for the SNO collaboration, in these proceedings.
5. S. Fukuda *et al.,* Phys. Rev. Lett. **86** (2001) 5651; Phys. Rev. Lett. **86** (2001) 5656; Y. Koshio, for the Super-Kamiokande collaboration, in these proceedings.
6. S.H. Ahn *et al.*, Phys. Lett. B **511** (2001) 178; J. Hill, for the K2K collaboration, in these proceedings.
7. M. Messier, for the MINOS collaboration, in these proceedings.
8. M. Campanelli, in these proceedings.
9. D. Beavis *et al.*, Proposal, BNL AGS E-889 (1995).
10. Y. Itow *et al.*, 2001, Letter of Intent "The JHF-Kamioka Neutrino Project", hep-ex/0106019.
11. M. Apollonio *et al.*, Phys. Lett. B **420** (1998) 397; Phys. Lett. B **466** (1999) 415.
12. F. Boehm *et al.*, Phys. Rev. Lett. **84** (2000) 3764; Phys. Rev. D **62** (2000) 072002.
13. M. Aoki, in these proceedings.

Measurement of CP-violating phase at a long baseline neutrino experiment with Hyper-Kamiokande

MAYUMI AOKI
Theory Group, KEK,
Tsukuba-shi, Ibaraki, Japan
E-mail: mayumi.aoki@kek.jp

As a joint project of KEK and Japan Atomic Energy Research Institute (JAERI), the High Intensity Proton Accelerator (HIPA) will be completed by the year 2007. The long baseline (LBL) neutrino oscillation experiment between HIPA and Super-Kamiokande with the baseline length of $L = 295$ km has been proposed and is expected to measure the mass-squared difference and the mixing angle precisely. In this report, as a sequel to that experiment, we study the potential of the LBL experiment between HIPA and Hyper-Kamiokande (HK), a proposed 1 Mton water-Čerenkov detector, with $L = 295$ km to measure the CP violating phase δ_{MNS} in the lepton-flavor-mixing matrix. By allocating 2 Mton·years for the ν_μ narrow-band beam (NBB) and 4 Mton·years for $\overline{\nu}_\mu$ NBB, we find that the HIPA-to-HK experiment can determine δ_{MNS} with the accuracy of about 30°. When $\delta_{\text{MNS}} = 90°$ or 270°, CP violation can be established at more than 3σ-level.

1 Introduction

In the next five years several long baseline (LBL) neutrino oscillation experiments [1,2,3,4] will start measuring the neutrino oscillation parameters. One of those experiments is proposed in Japan with High Intensity Proton Accelerator (HIPA) [5] and Super-Kamiokande (SK) [4]. The baseline length of this experiment is $L = 295$ km. The HIPA has a 50 GeV proton accelerator, which will be completed by the year 2007 in the site of JAERI (Japan Atomic Energy Research Institute) as the joint project of KEK and JAERI. The proton beam of HIPA can deliver neutrino beams of a few GeV range, whose intensity is two orders of magnitude higher than that of the KEK PS beam for the K2K experiment. High precision measurements of the mass-squared difference and the mixing angle will be achieved by HIPA-to-SK experiment [4].

Those LBL experiments of the near future, however, are unlikely to determine the CP violating phase δ_{MNS}, in the Maki-Nakagawa-Sakata (MNS) lepton-flavor-mixing matrix [6]. In this report, we study the capability of an LBL experiment between HIPA and Hyper-Kamiokande (HK), a next generation megaton-level water-Čerenkov detector proposed to be built in Kamioka [7]. We show that δ_{MNS} can be determined in such experiments by using ν_μ and $\overline{\nu}_\mu$ narrow-band-beams (NBB's) from the HIPA.

2 Neutrino oscillation parameters

The current results of the following neutrino oscillation experiments give the constraints on the MNS matrix elements and the mass-squared differences. An analysis of the atmospheric-neutrino data from the SK experiment [8] gives $\sin^2 2\theta_{_{\rm ATM}} = 0.88-1.0$ and $\delta m^2_{_{\rm ATM}} = (1.6-4.0)\times 10^{-3}$ (eV^2). The result of the CHOOZ reactor experiments [9] gives the constraint $\sin^2 2\theta_{_{\rm CHOOZ}} < 0.1$ including the corresponding atmospheric-neutrino region. The data of the solar-neutrino experiment favour the large mixing angle (LMA) MSW solution of the solar neutrino problem [10], $\sin^2 2\theta_{_{\rm SOL}} = 0.7 - 0.9$ and $\delta m^2_{_{\rm SOL}} = (3.0 - 15) \times 10^{-4}$ (eV^2).

The MNS matrix $U_{_{\rm MNS}}$ is defined by the three-neutrino-mixing through the charged current (CC) weak interaction as

$$\nu_\alpha = \sum_{i=1}^{3} (U_{_{\rm MNS}})_{\alpha i}\ \nu_i\,. \tag{1}$$

Here $\alpha = e, \mu, \tau$ are the lepton-flavor indices and ν_i $(i = 1,2,3)$ denotes the neutrino mass-eigenstate with the mass m_i. The MNS matrix has three mixing angles and one phase as an independent parameters, excepting for the Majorana phases to which the neutrino oscillation are not sensitive. Since the present neutrino oscillation experiments constrain directly on the elements, U_{e2}, U_{e3}, and $U_{\mu 3}$, it most convenient to adopt the parameterization where these three matrix elements are the independent parameters. Without losing generality, we can take U_{e2} and $U_{\mu 3}$ to be real and non-negative and U_{e3} to be complex number, $U_{e3} = |U_{e3}|e^{-i\delta_{\rm MNS}}$. All the other matrix elements of the $U_{_{\rm MNS}}$ are then determined by the unitary conditions.

We adopt the relationship between the mass-squared differences and those from the observations as

$$\delta m^2_{_{\rm SOL}} = \left|\delta m^2_{12}\right| \ll \left|\delta m^2_{13}\right| = \delta m^2_{_{\rm ATM}}\,. \tag{2}$$

Under this relationship, the independent mixing angles of the MNS matrix can be determined by the above neutrino oscillation experiments as

$$2|U_{e3}|^2 = 1 - \sqrt{1 - \sin^2 2\theta_{_{\rm CHOOZ}}}\,, \tag{3a}$$

$$2(U_{\mu 3})^2 = 1 - \sqrt{1 - \sin^2 2\theta_{_{\rm ATM}}}\,, \tag{3b}$$

$$2(U_{e2})^2 = 1 - |U_{e3}|^2 - \sqrt{\left(1 - |U_{e3}|^2\right)^2 - \sin^2 2\theta_{_{\rm SOL}}}\,, \tag{3c}$$

while the remaining independent parameter δ_{MNS} is unconstrained. The neutrino oscillation is also sensitive to the mass hierarchy. There exist two possible types of them. One has a spectrum $m_1 < m_2 \ll m_3$, which is called 'normal hierarchy', and another $m_3 \ll m_1 < m_2$, which is called 'inverted hierarchy'. There is an indication that the normal hierarchy is favored against the inverted one from the Super-Nova 1987A observation[11], so that we perform our analysis by assuming the normal hierarchy for simplicity.

The transition probability of ν_μ to ν_e ($\overline{\nu}_\mu$ to $\overline{\nu}_e$) in the vacuum is given by

$$P_{\nu_\mu \to \nu_e (\overline{\nu}_\mu \to \overline{\nu}_e)} \to 4|U_{e3}|^2|U_{\mu 3}|^2 \sin^2 \frac{\Delta_{13}}{2} - 2|U_{e2}|^2|U_{\mu 2}|^2 \Delta_{12} \sin \Delta_{13}$$
$$-2J_r \cos \delta_{\text{MNS}} \Delta_{12} \sin \Delta_{13} - (+)4J_r \sin \delta_{\text{MNS}} \Delta_{12} \sin^2 \frac{\Delta_{13}}{2} + O(\Delta_{12}^2), \quad (4)$$

where Δ_{ij} is

$$\Delta_{ij} \equiv \frac{m_j^2 - m_i^2}{2E_\nu} L \simeq 2.534 \frac{\delta m_{ij}^2(\text{eV}^2)}{E_\nu(\text{GeV})} L(\text{km}), \quad (5)$$

and J_r is reduced Jarlskog parameter

$$J_r = \frac{U_{e1}U_{e2}U_{\mu 3}U_{\tau 3}}{1 - |U_{e3}|^2} U_{e3}. \quad (6)$$

We have assumed that $\Delta_{12} \ll \Delta_{13} \sim 1$, which corresponds our HIPA-to-HK LBL experiment with $L = 295$ km and $\langle E_\nu \rangle = 1 \sim 1.5$ GeV. In the matter background, the neutrino-flavor oscillation is governed by the Schrödinger equation

$$i\frac{\partial}{\partial t}\begin{pmatrix}\nu_e\\ \nu_\mu\\ \nu_\tau\end{pmatrix} = \frac{1}{2E_\nu}\left[H_0 + \begin{pmatrix}a\,0\,0\\ 0\,0\,0\\ 0\,0\,0\end{pmatrix}\right]\begin{pmatrix}\nu_e\\ \nu_\mu\\ \nu_\tau\end{pmatrix}, \quad (7)$$

where H_0 is the Hamiltonian in the vacuum and a is the matter effect term

$$a = 2\sqrt{2}G_F n_e E_\nu = 7.56 \times 10^{-5}(\text{eV}^2)\left(\frac{\rho}{\text{g/cm}^3}\right)\left(\frac{E_\nu}{\text{GeV}}\right). \quad (8)$$

Here n_e is the electron density of the matter, G_F is the Fermi constant, and ρ is the matter density. For the oscillation of the anti-neutrino, eq.(7) should be changed $a \to -a$ and $U \to U^*$.

3 HIPA-to-HK experiment

We present the assumption about the detector and the beams of our study. The HK detector [7] in Kamioka-site is assumed a water-Čerenkov detector of fiducial volume 1 Mton, which is a capable of distinguishing between $e^{\pm}$ CC events and $\mu^{\pm}$ CC events. In our simple analysis, we do not make use of the capability of measuring the particle charges and hadron energies, then only use the total numbers of the produced $e^{\pm}$ and $\mu^{\pm}$ events.

We assume 0.77 MW of proton beam power from the HIPA and 10^{21} protons on target as a typical 1 year operation. As a beam option of the HIPA, three low-energy NBB's are used. Two are ν_μ enriched NBB's, NBB(2GeV) and NBB(3GeV), whose pion momentum $\langle p_\pi \rangle$ is respectively selected as $\langle p_\pi \rangle =$ 2GeV and 3 GeV. The other is $\overline{\nu}_\mu$ enriched NBB with $\langle p_\pi \rangle$ = 2GeV, NBB($\overline{\text{2GeV}}$).

Besides the primarily $\nu_\mu(\overline{\nu}_\mu)$ flux, the secondary $\nu_e, \overline{\nu}_\mu$ and $\overline{\nu}_e$ ($\overline{\nu}_e, \nu_\mu$ and ν_e) fluxes are contained in NBB(2GeV) and NBB(3GeV) (NBB($\overline{\text{2GeV}}$)). We show in Table 1 the expected numbers of total interactions of NBB's for 1Mton·year at L = 295 km in absence of oscillations. Although the compositions of the ν_μ and $\overline{\nu}_\mu$ fluxes in the NBB(2GeV) and that of $\overline{\nu}_\mu$ and ν_μ fluxes in the NBB($\overline{\text{2GeV}}$) are respectively almost the same, the ratio of $\nu_\mu N^{tot}/\overline{\nu}_\mu N^{tot} \sim 8.5\%$ for NBB($\overline{\text{2GeV}}$) is more than 10 times larger than the ratio of $\overline{\nu}_\mu N^{tot}/\nu_\mu N^{tot} \sim 0.70\%$ for NBB(2GeV). The reason is a factor of three smaller $\overline{\nu}_\mu$ cross section than the ν_μ one off water target at the relevant energies.

	$\nu_\mu N^{tot}$	$\nu_e N^{tot}$	$\overline{\nu}_\mu N^{tot}$	$\overline{\nu}_e N^{tot}$
ν_μ NBB(2GeV)	39000	300	270	18
ν_μ NBB(3GeV)	62000	420	290	22
$\overline{\nu}_\mu$ NBB($\overline{\text{2GeV}}$)	1100	64	13000	87

Table 1: Expected numbers of total interactions of NBB's for 1Mton·year at HK (L = 295 km) in absence of oscillations.

3.1 Signals

Although both numbers of the $e^{\pm}$ and $\mu^{\pm}$ events can be used as signals, it is the e-like events that is useful to give the constraint on the δ_{MNS}. The numbers of e-like events from the NBB(2GeV) and NBB(3GeV), $N(e, \text{2GeV})$

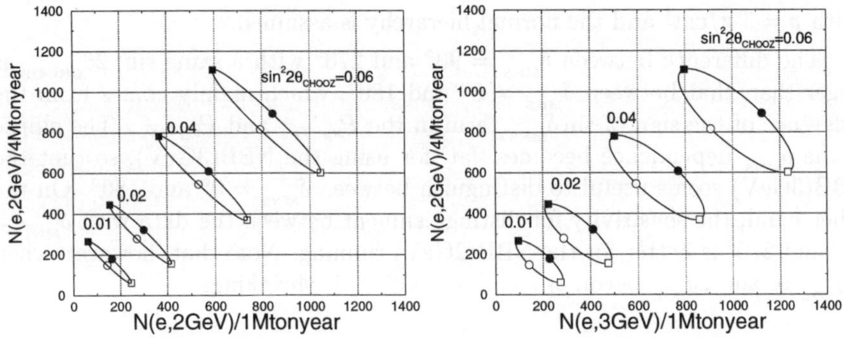

Figure 1: Numbers of the e-like signals. The horizontal axes in the left- and the right-hand figures are respectively $N(e, 2\text{ GeV})$ and $N(e, 3\text{ GeV})$ with 1 Mton·year running, while both vertical axes are $N(e, \overline{2\text{GeV}})$ with 4 Mton·years running. We take four values of $\sin^2 2\theta_{\text{CHOOZ}} = 0.06, 0.04, 0.02$, and 0.01, and vary δ_{MNS} from 0° to 360°. Solid-circle, solid-square, open-circle, and open-square show $\delta_{\text{MNS}} = 0°$, 90°, 180°, and 270°, respectively. The other parameters are given by eqs.(10).

and $N(e, 3\text{GeV})$, and the NBB($\overline{2\text{GeV}}$), $N(e, \overline{2\text{GeV}})$, are calculated as

$$N(e, 2(3)\text{GeV}) = MN_A \int dE_\nu \Phi_{2(3)\text{GeV}} \sigma_{e^-}^{\text{CC}} P_{\nu_\mu \to \nu_e} \,, \tag{9a}$$

$$N(e, \overline{2\text{GeV}}) = M' N_A \int dE_\nu \Phi_{\overline{2\text{GeV}}} \sigma_{e^+}^{\text{CC}} P_{\overline{\nu}_\mu \to \overline{\nu}_e} \,. \tag{9b}$$

Here M and M' are the statistical significance. $N_A = 6.017 \times 10^{23}$ is the Avogadro number and Φ is the flux of the primarily neutrino in the relevant NBB. The $\sigma_{e^-}^{\text{CC}}$ and $\sigma_{e^+}^{\text{CC}}$ are the cross sections of e^- and e^+ CC interactions, which are obtained by assuming a pure water target. Since the $\sigma_{e^+}^{\text{CC}}$ are smaller than the $\sigma_{e^-}^{\text{CC}}$, we assume that $M = 1$Mton·year and $M' = 4$Mton·years to obtain enough signals for our analysis.

We show the numbers of the e-like signals in Fig. 1. The horizontal axes in the left- and the right-hand figures are respectively $N(e, 2\text{ GeV})$ and $N(e, 3\text{ GeV})$, while both vertical axes are $N(e, \overline{2\text{GeV}})$. We take four values of $\sin^2 2\theta_{\text{CHOOZ}}$, $\sin^2 2\theta_{\text{CHOOZ}} = 0.06, 0.04, 0.02$, and 0.01, and vary δ_{MNS} from 0° to 360°. The solid-circle, solid-square, open-circle, and open-square show $\delta_{\text{MNS}} = 0°$, 90°, 180°, and 270°, respectively. The other parameters are

$$\sin^2 2\theta_{\text{ATM}} = 1.0 \,, \qquad \delta m^2_{\text{ATM}} = 3.5 \times 10^{-3}\ \text{eV}^2 \,, \tag{10a}$$

$$\sin^2 2\theta_{\text{SOL}} = 0.8 \,, \qquad \delta m^2_{\text{SOL}} = 10 \times 10^{-5}\ \text{eV}^2 \,, \tag{10b}$$

with $\rho = 3$ g/cm^3 and the normal hierarchy is assumed.

The difference between $\delta_{\text{MNS}} = 90°$ and $270°$ with a same $\sin^2 2\theta_{\text{CHOOZ}}$ is larger than that between $\delta_{\text{MNS}} = 0°$ and $180°$, which mainly comes from the difference of the sign of $\sin\delta_{\text{MNS}}$ term in the $P_{\nu_\mu\to\nu_e}$ and $P_{\overline{\nu}_\mu\to\overline{\nu}_e}$. The ellipse of the δ_{MNS} dependence becomes 'fat' for using the NBB(3GeV), so that the NBB(3GeV) seems useful to distinguish between $\delta_{\text{MNS}} = 0°$ and $180°$. On the other hand, the sensitivity for distinguishment between the data with $\delta_{\text{MNS}} = 90°$ and $270°$ is better for the NBB(2GeV) running. Note that, however, when δm^2_{SOL} or $\sin^2 2\theta_{\text{SOL}}$ becomes small, the ellipse is shrinking.

3.2 Backgrounds

The secondary fluxes of the NBB's give the background contributions to the e- and μ-like events. Also the e-like events receive contributions from the neutral current (NC) events where produced π^0's mimic electron shower in the water-Čerenkov detector. Using the estimations in the K2K experiments, we assume that the π^0's give a background of 0.2 % of the $\nu_\mu N^{tot}$ ($\overline{\nu}_\mu N^{tot}$) for $\nu_\mu(\overline{\nu}_\mu)$ enriched NBB in absence of oscillations. The e/π^0 misidentification probability, P_{e/π^0}, is given by

$$P_{e/\pi^0} = 0.2 \times (1 \pm 0.1)\%\,, \tag{11}$$

with the 10 % error, which is accounted for as a systematic error.

In Fig. 2, we show the expected e-like numbers of the signals and the backgrounds for (a) NBB(2GeV) and (b) NBB(3GeV) with 1 Mton·year running, and (c) NBB($\overline{\text{2GeV}}$) with 4 Mton·years running. The horizontal axes show the $\sin^2 2\theta_{\text{CHOOZ}}$. The other parameters are taken as the same as eqs.(10) with $\delta_{\text{MNS}} = n \times 10°$ ($n = 1 \sim 36$). The solid-circle, open-circle, open-triangle, and open-square show the e-like event numbers coming from the ν_μ, $\overline{\nu}_\mu$ ν_e, and $\overline{\nu}_e$ fluxes, respectively. The signals for NBB(2GeV) and NBB(3GeV) (NBB($\overline{\text{2GeV}}$)) are then given by the solid(open)-circles. The other tokens show the backgrounds coming from the secondary fluxes. The open-diamond give π^0 background from the NC events.

The backgrounds coming from the ν_e flux and the NC events are large for all NBB's runnings. In the NBB($\overline{\text{2GeV}}$) case, the $\overline{\nu}_e$ flux give also large background because of its non-negligible contamination for the NBB($\overline{\text{2GeV}}$). It is noted that for the small $\sin^2 2\theta_{\text{CHOOZ}}$, the background events can be larger than the signals.

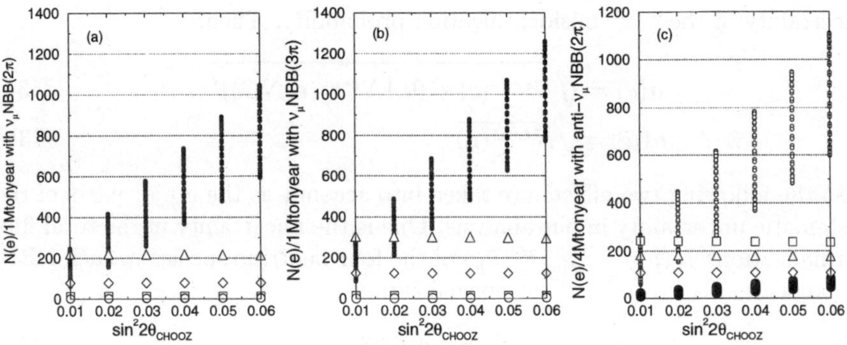

Figure 2: Expected e-like numbers of the signal and the backgrounds for (a) NBB(2GeV) and (b) NBB(3GeV) with 1 Mton·year running, and (c) NBB($\overline{\text{2GeV}}$) with 4 Mton·years running. Solid-circle, open-circle, open-triangle, and open-square show the e-like event numbers coming from the ν_μ, $\overline{\nu}_\mu$ ν_e, and $\overline{\nu}_e$ fluxes, respectively. Open-diamond give π^0 background from the NC events.

4 Results of χ^2 analysis

In order to quantify the sensitivity to the CP violating phase, we have performed a χ^2 analysis by assuming the normal hierarchy and the LMA MSW solution of the solar neutrino problem. For a given set of the three-neutrino parameters, we obtain the predictions of the e- and μ-like event numbers, $N(e)$ and $N(\mu)$, including the signals and the backgrounds for each NBB. The statistical significance is 1Mton·year each for NBB(2GeV) and NBB(3GeV), and 4Mton·years for NBB($\overline{\text{2GeV}}$), which corresponds to 6 years running with HK.

The χ^2 function is determined as

$$\chi^2 = \sum_{\text{NBB's}} \left\{ \left(\frac{N^{fit}(e) - N^{true}(e)}{\sigma(e)} \right)^2 + \left(\frac{N^{fit}(\mu) - N^{true}(\mu)}{\sigma(\mu)} \right)^2 + \sum_{\nu} \left(\frac{f_\nu - 1.0}{0.03} \right)^2 \right\} + \left(\frac{\rho - 3.0}{0.1} \right)^2, \tag{12}$$

where N^{true} is calculated by the three-neutrino oscillation parameters whose values are assumed to be realized in Nature. The first sum is over three NBB, NBB(2GeV), NBB(3GeV), and NBB($\overline{\text{2GeV}}$), runnings. We assume the detector efficiencies of all events are 100% for simplicity. The statistical errors of e- and μ-like predictions are then the square roots of the observed event numbers. For the e-like event, we account for the systematic error coming from the uncertainty of the e/π^0 misidentification probability. Then,

$$\sigma(e) = \sqrt{N^{true}(e) + (0.1N^{true}(e,NC))^2}\,. \tag{13a}$$

$$\sigma(\mu) = \sqrt{N^{true}(\mu)}\,. \tag{13b}$$

Also the following two effects are taken into account as the major parts of the systematic uncertainty in our analysis. One is the uncertainty in the total flux normalizations $f_\nu(\nu = \nu_e, \nu_\mu, \overline{\nu}_e, \overline{\nu}_\mu)$ of the four neutrino species in each NBB. We allocate 3 % error for those normalizations,

$$f_\nu = 1 \pm 0.03. \tag{14}$$

Another is the 3% uncertainty in the matter density along the base-line

$$\rho = 3.0 \pm 0.1 \text{ g/cm}^3\,. \tag{15}$$

We display in Fig. 3 the allowed regions in $\sin^2 2\theta_{\text{CHOOZ}}$ v.s. δ_{MNS} plane for our HIPA-to-HK experiment. The input data, N^{true}, are calculated for eqs.(10) with the normal hierarchy at $\sin^2 2\theta^{true}_{\text{CHOOZ}} = 0.06$, 0.01 and for four values of $\delta^{true}_{\text{MNS}}$, 0° (upper-right), 90° (upper-left), 180° (lower-left), and 270° (lower-right). In each figure, the input values of $\sin^2 2\theta^{true}_{\text{CHOOZ}}$ and $\delta^{true}_{\text{MNS}}$ are shown by a solid-circle for $\sin^2 2\theta^{true}_{\text{CHOOZ}} = 0.06$ and a solid-square for $\sin^2 2\theta^{true}_{\text{CHOOZ}} = 0.01$. The χ^2 fit has been performed for N^{fit} which are calculated by the following regions with the normal hierarchy:

$$0.7 \le \sin^2 2\theta_{\text{SOL}} \le 0.9\,, \qquad 3.0\times10^{-5} \le \delta m^2_{\text{SOL}}(\text{eV}^2) \le 15\times10^{-5}\,, \tag{16}$$

and the other four parameters, $\sin^2 2\theta_{\text{ATM}}, \delta m^2_{\text{ATM}}, \delta_{\text{MNS}}$ and $\sin^2 2\theta_{\text{CHOOZ}}$, are freely varied. The regions where χ^2_{min} <1, 4, and 9 are depicted by solid-, dashed-, and dotted-boundaries, respectively. The thick-lines stand for the $\sin^2 2\theta^{true}_{\text{CHOOZ}} = 0.06$ and the thin-lines for $\sin^2 2\theta^{true}_{\text{CHOOZ}} = 0.01$.

For $\delta^{true}_{\text{MNS}} = 90°$ and 270°, the δ_{MNS} can be constrained to ±30° (±60°) at the 1σ (3σ) level even if $\sin^2 2\theta^{true}_{\text{CHOOZ}} = 0.01$. CP violation in the lepton sector can hence be established at more than 3σ level if $\delta_{\text{MNS}} = 90°$ or 270° for the LMA parameters of eq.(16) if $\sin^2 2\theta^{true}_{\text{CHOOZ}} > 0.01$. More detained analysis shows that CP violation can be established at 3σ level even for the $\sin^2 2\theta_{\text{CHOOZ}}$ value as little as 0.004. For $\delta^{true}_{\text{MNS}} = 0°$ and 180°, δ_{MNS} can be measured to ±15° accuracy at 1σ level for $\sin^2 2\theta^{true}_{\text{CHOOZ}} = 0.06$, but it seems difficult to constrain on δ_{MNS}. It is because that the data for $\delta^{true}_{\text{MNS}} = 0°$ or 180° can be easily reproduced by changing the value of the $\sin^2 2\theta_{\text{CHOOZ}}$, which can be seen in Fig. 1.

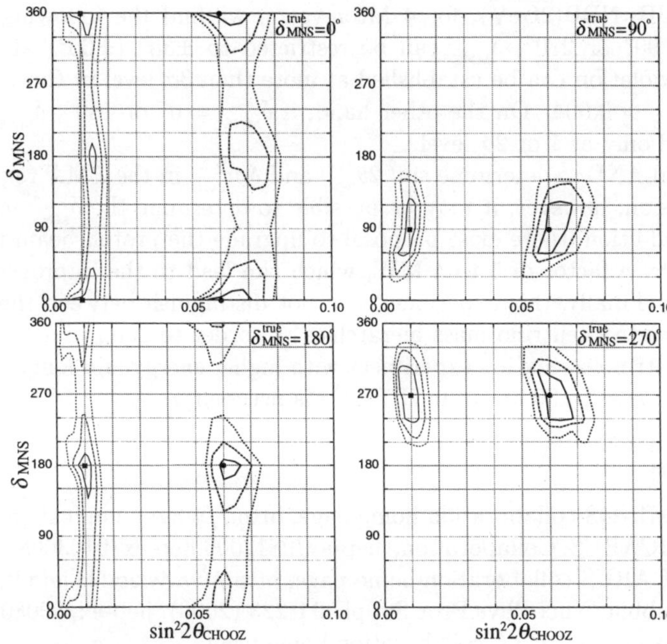

Figure 3: Allowed regions for our HIPA-to-HK experiment with the statistical significance of 1Mton·year each for NBB(2GeV) and NBB(3GeV), and 4Mton·years for NBB($\overline{2GeV}$). The input data, N^{true}, are calculated for eqs.(10) with the normal hierarchy at $\sin^2 2\theta_{\text{CHOOZ}}^{true} =$ 0.06 (solid-circles), 0.01 (solid-squares) and for four values of $\delta_{\text{MNS}}^{true}$, 0° (upper-right), 90° (upper-left), 180° (lower-left), and 270° (lower-right). The regions where χ^2_{min} <1, 4, and 9 are depicted by solid-, dashed-, and dotted-boundaries, respectively. Thick-lines stand for the $\sin^2 2\theta_{\text{CHOOZ}}^{true} = 0.06$ and thin-lines for $\sin^2 2\theta_{\text{CHOOZ}}^{true} = 0.01$.

5 Summary

In this report, we have studied the capability of measuring CP-phase angle δ_{MNS} by HIPA-to-HK LBL experiment with $L = 295$ km. We assume that the HK is a 1Mton water-Čerenkov detector which is capable of detecting the ν_e CC and ν_μ CC events almost perfectly, but cannot distinguish the lepton charges. We examine narrow-band beams (NBB) from HIPA with $\langle p_\pi \rangle$=2 GeV and 3 GeV, and calculate signals and backgrounds without requiring the HK capability of measuring the neutrino energy. We assume that the LMA MSW solution of the solar neutrino problem is realized and the neutrino-mass hierarchy is normal (the lower-mass two neutrinos participate in the solar neutrino oscillation). By using the ν_μ NBB's, NBB(2GeV) and NBB(3GeV), for 1 Mton·year each, and $\bar{\nu}_\mu$ NBB, NBB($\overline{2\text{GeV}}$), for 4 Mton·years, we find the following results. If $\delta_{\text{MNS}} = 90°$ or $270°$, δ_{MNS} can be restricted to $\pm30°$ ($\pm60°$) at 1σ (3σ) level. CP violation can be established at more than 3σ level in those cases, if $\sin^2 2\theta_{\text{CHOOZ}} > 0.004$. On the other hand, if $\delta_{\text{MNS}} = 0°$ or $180°$, δ_{MNS} can be constrained only at 1 or 2σ level.

If KamLAND [12] determine $\sin^2 2\theta_{\text{SOL}}$ and δm^2_{SOL} in the LMA region with good precision, however, it may be possible to determine the δ_{MNS} more precisely. In addition, there exist proposal to upgrade the proton beam power of the HIPA by a factor of 5 to 4 MW, which can lead to the improvements of our results. Finally, this experiment can not distinguish between the normal and the inverted neutrino mass hierarchies. In order to determine the hierarchy, the neutrino oscillation experiment with higher energy and longer baseline length (*e.g.* $L = 2,100$ km from HIPA [13]) is necessary.

References

1. The MINOS collaboration home page, http://www-numi.fnal.gov:8875/.
2. The ICARUS Collaboration, hep-ex/9812006,hep-ex/0103008; see also the ICARUS collaboration home page, http://www.aquila.infn.it/icarus/.
3. A. Rubbia, Nucl.Phys.Proc.Suppl. **91** 223 (2000), hep-ex/0008071; see also the OPERA collaboration home page, http://operaweb.web.cern.ch/operaweb/index.shtml.
4. The JHF Neutrino Working Group, hep-ex/0106019; see also the JHF Neutrino Working Group home page, http://neutrino.kek.jp/jhfnu.
5. See the HIPA home page, http://jkj.tokai.jaeri.go.jp/
6. M.Nakagawa Z.Maki and S.Sakata. Prog. Theor. Phys. **28**, 870 (1962).
7. M.Koshiba, Phys. Rep. **220** (229) 1992; K.Nakamura, *Neutrino Oscillations and Their Origin,* (Universal Academy Press, Tokyo, 2000), p. 359.
8. The Super-Kamiokande Collaboration Phys. Lett. **B433**, 9 (1998),Phys. Lett. **B436**, 33 (1998),Phys. Pev. Lett. **81**, 1562 (1998).
9. The CHOOZ Collaboration, Phys. Lett. **B420**, 397 (1998).
10. The Super-Kamiokande Collaboration Phys. Pev. Lett. **86**, 5656 (2001), Phys. Pev. Lett. **86**, 5651 (2001),hep-ex/0106064.
11. H. Minakata, H. Nunokawa, Phys. Lett. **B504**, 301 (2001).
12. See the KamLAND collaboration home page, http://www.awa.tohoku.ac.jp/html/KamLAND/index.html
13. M.Aoki, K.Hagiwara, Y.Hayato, T.Kobayashi, T.Nakaya, K.Nishikawa and N.Okamura, hep-ph/0112338.

Physics potential and present status of neutrino factories

OSAMU YASUDA
Department of Physics, Tokyo Metropolitan University
Minami-Osawa, Hachioji, Tokyo 192-0397, Japan
E-mail: yasuda@phys.metro-u.ac.jp

I briefly review the recent status of research on physics potential of neutrino factories with emphasis on measurements of the CP phase.

1 Introduction

The observation of atmospheric neutrinos (See, e.g., Ref.[1]) and solar neutrinos (See, e.g., Ref.[2]) gives the information on the mass squared differences and the mixings, which can be written in the three flavor framework of neutrino oscillations as ($|\Delta m^2_{32}|$, θ_{23}) and (Δm^2_{21}, θ_{12}), where I have adopted the standard parametrization [3] for the 3×3 MNSP [5,6] matrix. On the other hand, the CHOOZ result [4] tells us that $|\theta_{13}|$ has to be small ($\sin^2 2\theta_{13} \lesssim 0.1$). So the MNSP matrix looks like

$$U_{MNSP} \simeq \begin{pmatrix} c_\odot & s_\odot & \epsilon \\ -s_\odot/\sqrt{2} & c_\odot/\sqrt{2} & 1/\sqrt{2} \\ s_\odot/\sqrt{2} & -c_\odot/\sqrt{2} & 1/\sqrt{2} \end{pmatrix},$$

where I have used $\theta_{23} \simeq \pi/4$, $\sin^2 2\theta_{12} \equiv \sin^2 2\theta_\odot \simeq 0.8$ and $|\epsilon| \ll 1$.

The next thing to do is to determine θ_{13}, the sign of Δm^2_{32} and the CP phase δ. During the past few years a lot of research have been done on the possibilities of future long baseline experiments. One is a super-beam experiment and the other one is a neutrino factory. The former is super intense conventional neutrino beam which is obtained from pion decays while the latter is from muon decays in a storage ring. The background fraction f_B in the case of super-beams [7] is of order 10^{-2}, while in the case of a neutrino factory [8] it is or order 10^{-5}. The advantage of a neutrino factory is such low background fraction and neutrino factories are expected to enable us to determine θ_{13} and the sign of Δm^2_{32} (δ) for $\sin^2 2\theta_{13} \gtrsim 10^{-5}$ (10^{-3}), respectively.

In this talk I will mainly discuss measurements of the CP phase δ at neutrino factories and will try to clarify the reason why some group obtains different results for the optimized muon energy and baseline.

2 Measurements of the CP phase at neutrino factories

Measurement of the CP phase in neutrino oscillations is difficult not only because CP violating contribution in the oscillation probability is in general small but also because there is matter effect. Namely, the dependence of the probabilities for ν and $\bar{\nu}$ on δ and A is given by $P(\nu_\mu \to \nu_e) = f(E, L; \theta_{ij}, \Delta m^2_{ij}, \delta; A)$ and $P(\bar{\nu}_\mu \to \bar{\nu}_e) = f(E, L; \theta_{ij}, \Delta m^2_{ij}, -\delta; -A)$, where f is a certain function and $A \equiv \sqrt{2} G_F N_e$ stands for the matter effect. The quantity obtained from experiments on $\bar{\nu}$ is not $f(\cdots, -\delta; A)$, so that direct comparison between $f(\cdots, \delta; A)$ and $f(\cdots, -\delta; A)$ is impossible in a strict sense. The situation here is different from the the $K^0 - \bar{K}^0$ system where the quantity $N(K_L \to 2\pi)/N(K_L \to 3\pi)$ immediately gives us evidence for CP violation. Hence I have to compromise and adopt a kind of indirect measurement of CP violation, i.e., I deduce the values of δ etc. by comparing the energy spectra of the data and of the theoretical prediction with neutrino oscillations assuming the three flavor mixing. In determining δ, there are other oscillation parameters as well as the density of the Earth whose values are not exactly known, so that I have to take into account correlations of errors of these parameters.[a] Thus I introduce the following quantity to see the significance of the case with nonvanishing δ:

$$\Delta\chi^2 \equiv \min_{\overline{\theta_{k\ell}}, \overline{\Delta m^2_{k\ell}}, \bar{A}} \sum_j \left\{ \frac{\left[N_j(\nu_e \to \nu_\mu) - \bar{N}_j(\nu_e \to \nu_\mu)\right]^2}{\sigma_j^2} \right.$$
$$+ \frac{\left[N_j(\bar{\nu}_e \to \bar{\nu}_\mu) - \bar{N}_j(\bar{\nu}_e \to \bar{\nu}_\mu)\right]^2}{\sigma_j^2}$$
$$+ \frac{\left[N_j(\nu_\mu \to \nu_\mu) - \bar{N}_j(\nu_\mu \to \nu_\mu)\right]^2}{\sigma_j^2}$$
$$\left. + \frac{\left[N_j(\bar{\nu}_\mu \to \bar{\nu}_\mu) - \bar{N}_j(\bar{\nu}_\mu \to \bar{\nu}_\mu)\right]^2}{\sigma_j^2} \right\}, \qquad (1)$$

where $N_j(\nu_\alpha \to \nu_\beta) \equiv N_j(\nu_\alpha \to \nu_\beta; \theta_{k\ell}, \Delta m^2_{k\ell}, \delta, A)$, $\bar{N}_j(\nu_\alpha \to \nu_\beta) \equiv \bar{N}_j(\nu_\alpha \to \nu_\beta; \overline{\theta_{k\ell}}, \overline{\Delta m^2_{k\ell}}, \bar{\delta} = 0, \bar{A})$ stand for the numbers of events of the data and of the theoretical prediction with a vanishing CP phase, respectively, and σ_j^2 stands for the error which is given by the sum of the statistical and systematic errors. At neutrino factories appearance and disappearance channels for ν and $\bar{\nu}$ are observed, and I have included the numbers of events of all the channels in

[a] Correlations of errors at neutrino factories were studied in Ref. [9,10,11,12,13].

(1) to gain statistics. In the present analysis, $N_j(\nu_\alpha \to \nu_\beta)$ is substituted by theoretical prediction with a CP phase δ and $\Delta\chi^2$ obviously vanishes if $\delta = 0$.[b] The quantity $\Delta\chi^2$ reflects the strength of the correlation of the parameters, i.e., if $\Delta\chi^2$ turns out to be very small for a certain value of $\bar{A}$ then the correlation between δ and A would be very strong and in that case there would be no way to show $\delta \neq 0$. To reject a hypothesis "$\delta = 0$" at the 3σ confidence level I demand

$$\Delta\chi^2 \geq \Delta\chi^2(3\sigma CL) \tag{2}$$

where the right hand side stands for the value of χ^2 which gives the probability 99.7% in the χ^2 distribution with a certain degrees freedom, and $\Delta\chi^2(3\sigma CL)$=20.1 for 6 degrees freedom. From (2) I get the condition for the detector size to reject a hypothesis "$\delta = 0$" at 3σCL.

On the other hand, Koike et al.[11] claim that another quantity

$$\Delta\tilde{\chi}^2 \equiv \min_{\overline{\theta_{k\ell}}, \overline{\Delta m^2_{k\ell}}, \overline{A}} \sum_j \frac{1}{\sigma_j^2} \left[\frac{N_j(\nu_e \to \nu_\mu)}{N_j(\bar{\nu}_e \to \bar{\nu}_\mu)} - \frac{\bar{N}_j(\nu_e \to \nu_\mu)}{\bar{N}_j(\bar{\nu}_e \to \bar{\nu}_\mu)} \right]^2 \tag{3}$$

should be used since this particular combination improves the correlation between δ and A. In my opinion, however, both analyses with $\Delta\chi^2$ and $\Delta\tilde{\chi}^2$ are based on indirect measurements of CP violation anyway and there is no reason why one has to discard other information on δ. In fact it turns out that (3) is a combination in which the correlation between δ and A improves for low energy and worsens at high energy.

It should be pointed out here that indirect measurements of CP violation are also considered in the $B^0 - \bar{B}^0$ system.[c] To measure the phase ϕ_1, direct CP violating process $B(\bar{B}) \to J/\psi\, K_s$ is used [14,15], while those to measure ϕ_2 and ϕ_3 are $B \to 2\pi$ [16] and $B \to DK$ [17], which are not necessarily CP–odd processes. Their strategy is to start with the three flavor framework, to use the most effective process, which may or may not be CP–odd, to determine the CP phases and to check unitarity or consistency of the three flavor hypothesis.

The optimized muon energy E_μ and baseline L for the measurement of CP the phase at a neutrino factory have been investigated by several groups by taking into consideration the correlations of δ and all other parameters and the results are summarized in Table 1. The results in Ref.[13] are given in Fig. 2, which shows that the more uncertainty I have in the density, the shorter baseline I have to choose because the correlation between δ and A

---

[b] This is the reason why the quantity in (1) is denoted as $\Delta\chi^2$ instead of absolute χ^2. $\Delta\chi^2$ represents deviation from the best fit point rather than the goodness of fit.

[c] The term "indirect" in the B system is different from that in neutrino oscillations.

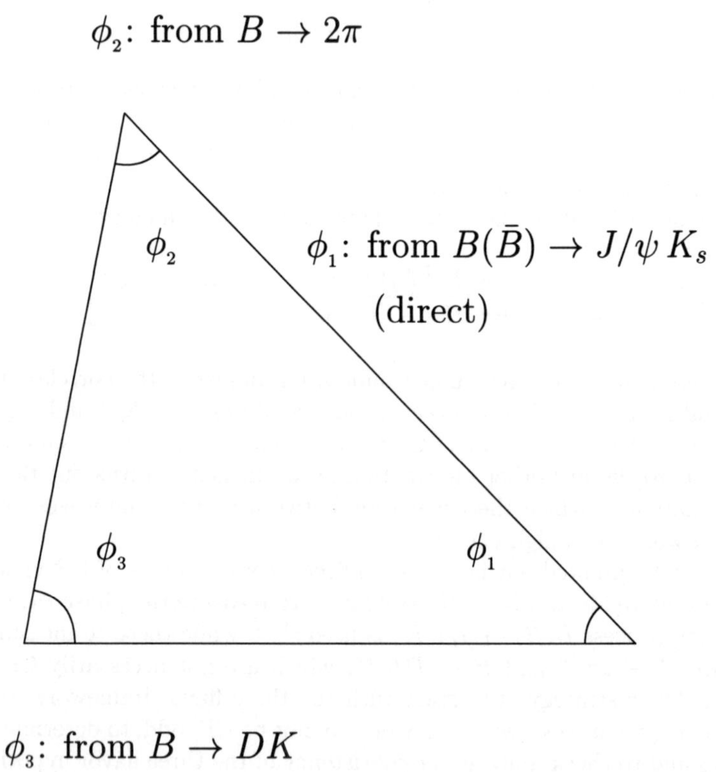

Figure 1: Unitarity triangle in the B meson system. The phase ϕ_1 is measured by a direct CP violating process $B(\bar{B}) \to J/\psi\, K_s$, whereas the most promising way to determine ϕ_2 and ϕ_3 is through $B \to 2\pi$ and $B \to DK$, which are not direct CP violating processes.

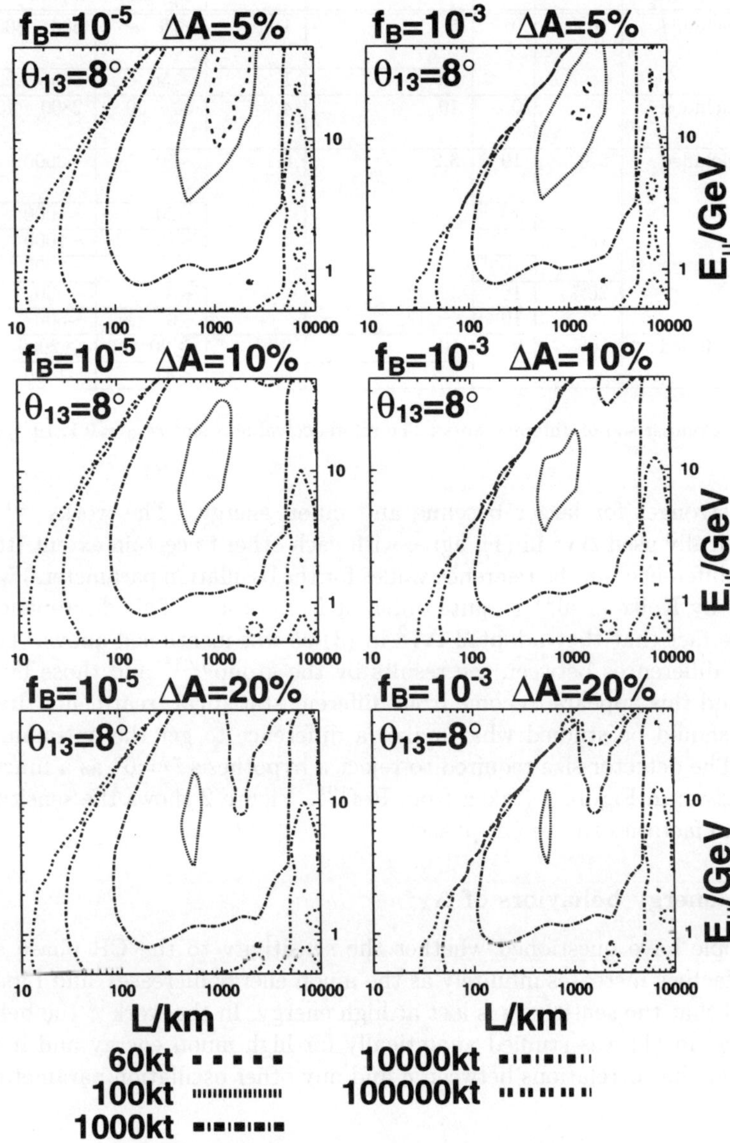

Figure 2: The contour plot of equi-number of data size required to reject a hypothesis $\bar{\delta} = 0$ at 3σ with the background fraction $f_B = 10^{-5}$ or 10^{-3} and the uncertainty of the matter effect ΔA=5%, 10% or 20%.

Ref.	correlations of θ_{ij},m^2_{ij}	$\lvert\Delta A/A\rvert$	f_B	$\Delta m^2_{21}/10^{-5}\text{eV}^2$	E_{th}/GeV	optimized E_μ/GeV	optimized L/km
KOS [11]	included	10%	0	5	1	$\lesssim 6$	600 – 800
				10	1	$\lesssim 50$	500 – 2000
FHL [12]	included	0	0	10	4	30 – 50	2800 – 4500
PY [13]	included	5%	10^{-5}	3.2	0.1	$\sim$50	$\sim$3000
			10^{-3}			$\sim$20	$\sim$1000
		10%	10^{-5}			$\sim$15	$\sim$1000
			10^{-3}			$\sim$10	$\sim$800
		20%	10^{-5}			$\sim$8	$\sim$500
			10^{-3}			$\sim$6	$\sim$500
H [18]	included	10%	0	10	0.1	$\gtrsim 20$	$\gtrsim 2000$
				3.5		$\sim$25	$\sim$1500

Table 1: Comparison of different works The reference value is $\sin^2 2\theta_{13} = 0.1$, $10^{21}\,\mu$s.

becomes stronger for larger baseline and muon energy. The works [12,13,18], which basically used $\Delta\chi^2$ in (1), agree with each other to certain extent (there are some differences on the reference values for the oscillation parameters) while the result by Koike et al.[11] is quite different from others. This discrepancy is due to the fact that they adopted $\Delta\tilde{\chi}^2$ in (3) as was mentioned above. There are slight differences between the results by the group [12,18] and those by the other [13] and this appears to come from different statistical treatments. In the future it should be studied what makes a difference to get the optimum set (E_μ, L). The detector size required to reject a hypothesis "δ=0" as a function of θ_{13} is given in Figure 2 (taken from Ref.[19]). Figure 2 shows the sensitivity of neutrino factories to the CP phase.

3 High energy behaviors of $\Delta\chi^2$

Some people have questioned whether the sensitivity to the CP phase at a neutrino factory increases infinitely as the muon energy increases, and Lipari[20] concluded that the sensitivity is lost at high energy. In the work [13] the behaviors of $\Delta\chi^2$ in (1) was studied analytically for high muon energy and it was shown after the correlations between δ and any other oscillation parameter or

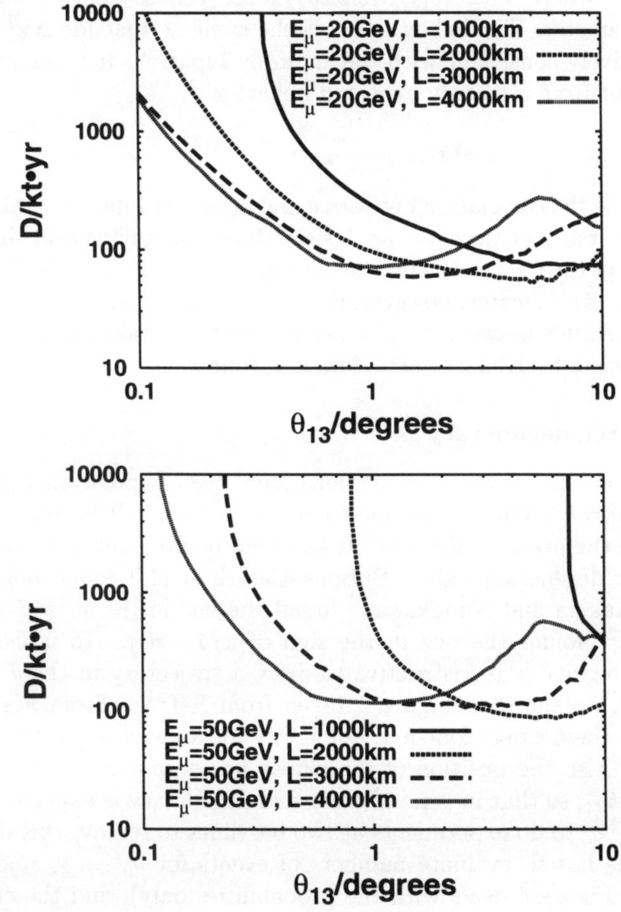

Figure 3: Data size (kt·yr) required to reject a hypothesis of $\delta = 0$ at 3σ when the true value is $\delta = \pi/2$, in the case of a neutrino factory with 10^{21} useful muon decays per year and a background fraction $f_B = 10^{-3}$.

A is taken into account that

$$\Delta\chi^2 \propto \left(\frac{J}{\sin\delta}\right)^2 \frac{1}{E_\mu}\left(\sin\delta + \text{const}\frac{\Delta m_{32}^2 L}{E_\mu}\cos\delta\right)^2 \tag{4}$$

for large E_μ, where $J \equiv (c_{13}/8)\sin 2\theta_{12}\sin 2\theta_{13}\sin 2\theta_{23}\sin\delta$ stands for the Jarlskog parameter. The behavior (4) is the same as that for $\Delta\tilde{\chi}^2$ in (3) and it is qualitatively consistent with the claim by Lipari [20]. It is remarkable that (4) is different from a naively expected behavior

$$\Delta\chi^2_{\text{naive}} \propto E_\mu(\cos\delta - 1)^2.$$

This is because the correlation between δ and other parameters is taken into account. Koike et al.[11] criticize the analysis with $\Delta\chi^2$ by saying that this quantity looks mainly at the CP conserving part at high muon energy ($10\text{GeV} \lesssim E_\mu \lesssim 50\text{GeV}$). The behavior (4) indicates, however, that CP violating part becomes dominant *after* the correlation between δ and other parameters is taken into consideration and the criticism by Koike et al.[11] does not apply.

4 Parameter degeneracy

The discussions in the previous sections have been focused on rejection of a hypothesis "$\delta = 0$". Once δ is found to be nonvanishing, it becomes important to determine the precise value of δ. It has been pointed out that various kinds of parameter degeneracy exist. Burguet-Castell et al.[21] found degeneracy in $(\delta,\ \theta_{13})$, Minakata and Nunokawa [22] found the one in the sign of Δm^2_{32}, and Barger et al.[23] found the one in the sign of $\pi/4 - \theta_{23}$. To understand the four-fold ambiguity it is instructive to draw a trajectory in the $P(\nu_\mu \to \nu_e)$ – $P(\bar{\nu}_\mu \to \bar{\nu}_e)$ plane (See Figure 4 taken from Ref.[24]). From Figure 4 it is obvious that there exists four-fold ambiguity for a given set of the oscillation parameters. Also the position of the ellipse has degeneracy in interchanging $\theta_{23} \leftrightarrow \pi/2 - \theta_{23}$, so that in general eight-fold degeneracy is expected. [23] It was proposed [21,22,23] to do experiments at two baselines to remove this degeneracy. In reality one has to evaluate numbers of events for $\nu_\mu \to \nu_e$ and $\bar{\nu}_\mu \to \bar{\nu}_e$ (Notice that Figure 4 deals with the probabilities only), and the correlations of errors have to be taken into consideration as well, so determination of the oscillation parameters is even more difficult than what Figure 4 indicates.

5 Summary

People in the field reached consensus that neutrino factories can measure the CP phase δ with the detector size larger than $10^{21}\mu\cdot100\text{kt}\cdot\text{yr}$ for $\sin^2 2\theta_{13} \gtrsim 10^{-3}$

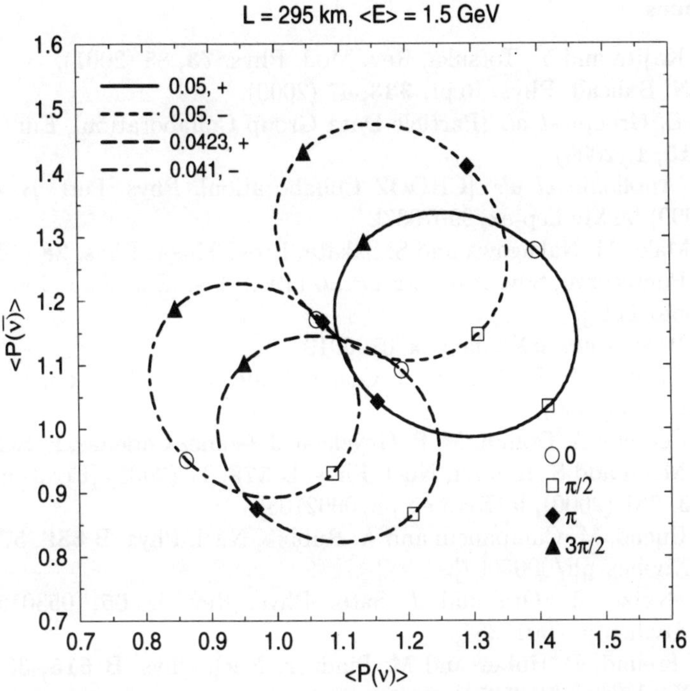

Figure 4: Illustration of the clover-leaf ambiguity in terms of CP trajectory diagram. Four solutions exist for given values of $P(\nu_\mu \to \nu_e)$ and $P(\bar{\nu}_\mu \to \bar{\nu}_e)$.

unless $|\delta|$ is small. The detailed study on the optimized muon energy and baseline still needs to be done.

Acknowledgments

I would like to thank Noriaki Kitazawa for useful discussions on CP violation in the B meson system and for providing references on the subject. This research was supported in part by a Grant-in-Aid for Scientific Research of the Ministry of Education, Science and Culture, #12047222, #13640295.

References

1. T. Kajita and Y. Totsuka, Rev. Mod. Phys. **73**, 85 (2001).
2. J. N. Bahcall, Phys. Rept. **333**, 47 (2000).
3. D. E. Groom *et al.* [Particle Data Group Collaboration], Eur. Phys. J. C **15**, 1 (2000).
4. M. Apollonio *et al.* [CHOOZ Collaboration], Phys. Lett. B **466**, 415 (1999) [arXiv:hep-ex/9907037].
5. Z. Maki, M. Nakagawa and S. Sakata, Prog. Theor. Phys. **28** (1962) 870.
6. B. Pontecorvo, Sov. Phys. JETP **26** (1968) 984 [Zh. Eksp. Teor. Fiz. **53** (1968) 1717].
7. Y. Itow *et al.*, arXiv:hep-ex/0106019.
8. A. Cervera, F. Dydak and J. Gomez Cadenas, Nucl. Instrum. Meth. A **451**, 123 (2000).
9. A. Cervera, A. Donini, M. B. Gavela, J. J. Gomez Cadenas, P. Hernandez, O. Mena and S. Rigolin, Nucl. Phys. B **579**, 17 (2000) [Erratum-ibid. B **593**, 731 (2000)] [arXiv:hep-ph/0002108].
10. A. Bueno, M. Campanelli and A. Rubbia, Nucl. Phys. B **589**, 577 (2000) [arXiv:hep-ph/0005007].
11. M. Koike, T. Ota and J. Sato, Phys. Rev. D **65**, 053015 (2002) [arXiv:hep-ph/0011387].
12. M. Freund, P. Huber and M. Lindner, Nucl. Phys. B **615**, 331 (2001) [arXiv:hep-ph/0105071].
13. J. Pinney and O. Yasuda, Phys. Rev. D **64**, 093008 (2001) [arXiv:hep-ph/0105087].
14. A. B. Carter and A. I. Sanda, Phys. Rev. D **23**, 1567 (1981).
15. I. I. Bigi and A. I. Sanda, Nucl. Phys. B **193**, 85 (1981).
16. M. Gronau and D. London, Phys. Rev. Lett. **65**, 3381 (1990).
17. M. Gronau and D. Wyler, Phys. Lett. B **265**, 172 (1991).
18. P. Huber, private communication.
19. J. Pinney, [arXiv:hep-ph/0106210].
20. P. Lipari, Phys. Rev. D **64**, 033002 (2001) [arXiv:hep-ph/0102046].
21. J. Burguet-Castell, M. B. Gavela, J. J. Gomez-Cadenas, P. Hernandez and O. Mena, Nucl. Phys. B **608**, 301 (2001) [arXiv:hep-ph/0103258].
22. H. Minakata and H. Nunokawa, JHEP **0110**, 001 (2001) [arXiv:hep-ph/0108085].
23. V. Barger, D. Marfatia and K. Whisnant, arXiv:hep-ph/0112119.
24. H. Minakata and H. Nunokawa, arXiv:hep-ph/0111131.

PROTON DECAY IN THE SEMI-SIMPLE UNIFICATION MODEL

MASAAKI FUJII AND TAIZAN WATARI

Department of Physics, University of Tokyo, Hongo 7-3-1, Bunkyou-ku, 113-0033, Tokyo, Japan

We derive a prediction of proton life time based on models of SU(5) unified theories called "semi-simple unification" in this article/talk. Life Time is typically $3\times10^{34}-10^{35}$yrs. in the "semi-simple unification models". This article/talk is based on a paper Phys.Lett.**B527**106(2002), along with extension of the result obtained there.

1 Introduction

Supersymmetric (SUSY) Grand Unified Theories (GUT's) have a number of theoretical motivations and is suggested by the precise measurements of the standard-model gauge coupling constants. However, it would be the proton decay signal that makes us believe the GUT's. Various models of the SUSY GUT's provide different predictions on the decay rate of the proton.

Realistic models of unified theories should naturally provide a mass of order of the GUT scale ($\sim 10^{16}$GeV) to the colored SU(5)-partner of the doublet Higgses, keeping the two Higgs doublets almost massless. Moreover, experimental bounds on the partial decay rate of the proton such as $\tau(p \to K^+\bar{\nu}) \gtrsim 6.7 \times 10^{32}$yr.(90 % C.L.)[1] require that the proton decay through the dimension 5 operators[2] should be suppressed in the realistic models.

The semi-simple unification [3,4] is one of models that solve the above two problems in a natural way. A discrete R symmetry in this model guarantees the massless property of the Higgs doublets and the absence of the dimension 5 proton decay operators. Colored Higgs particles obtain their mass terms through the missing partner mechanism [5]. In this article/talk, we calculate the proton-decay rate through the gauge-boson exchange based on this model.

We conclude that the life time of proton is typically 3×10^{34}-10^{35}yrs., which shows that the proton decay can be observed in the next-generation water Čerenkov detectors [6].

2 Brief Review of the Semi-Simple Unification Model

We briefly review the semi-simple unification model that uses $SU(5)_{GUT} \times U(3)_H$ gauge group. This gauge group is spontaneously broken down to the $SU(3)_C \times SU(2)_L \times U(1)_Y$ of the MSSM. Three families of quarks and leptons

$(\mathbf{5}^* + \mathbf{10})$ and Higgs $(H^i(\mathbf{5}) + \bar{H}_i(\mathbf{5}^*))$ supermultiplets are singlets of the $\mathrm{U}(3)_{\mathrm{H}}$ gauge group and transform under the $\mathrm{SU}(5)_{\mathrm{GUT}}$ as in the standard $\mathrm{SU}(5)_{\mathrm{GUT}}$ unification models. Some more fields are required for the spontaneous GUT symmetry breaking: $X^\alpha{}_\beta(\alpha, \beta = 1, 2, 3)$ transforming as $(\mathbf{1}, \mathbf{adj.}{=}\mathbf{8}+\mathbf{1})$ under the $\mathrm{SU}(5)_{\mathrm{GUT}} \times \mathrm{U}(3)_{\mathrm{H}}$ gauge group, and $Q^\alpha_i(i = 1, \cdots, 5) + Q^\alpha_6$ and $\bar{Q}^i{}_\alpha(i = 1, \cdots, 5) + \bar{Q}^6{}_\alpha$ transforming as $(\mathbf{5}^* + \mathbf{1}, \mathbf{3})$ and $(\mathbf{5} + \mathbf{1}, \mathbf{3}^*)$. Indices $\alpha, \beta = 1, 2, 3$ are for the $\mathrm{U}(3)_{\mathrm{H}}$ and $i = 1, \cdot, 5$ for the $\mathrm{SU}(5)_{\mathrm{GUT}}$. $X^\alpha{}_\beta$ is also expressed as $X^c(t_c)^\alpha{}_\beta(c = 0, 1-8)$, where $t^a(a = 1-8)$ are Gell'mann matrices[a] and $t_0 \equiv \mathbf{1}_{3\times 3}/\sqrt{6}$. Superpotential is given by[3]

$$
\begin{aligned}
W = {} & \sqrt{2}\lambda_{3\mathrm{H}} \bar{Q}^i{}_\alpha X^a (t_a)^\alpha_\beta Q^\beta_i + \sqrt{2}\lambda'_{3\mathrm{H}} \bar{Q}^6{}_\alpha X^a (t_a)^\alpha_\beta Q^\beta_6 \\
& + \sqrt{2}\lambda_{1\mathrm{H}} \bar{Q}^i{}_\alpha X^0 (t_0)^\alpha_\beta Q^\beta_i + \sqrt{2}\lambda'_{1\mathrm{H}} \bar{Q}^6{}_\alpha X^0 (t_0)^\alpha_\beta Q^\beta_6 \\
& - \sqrt{2}\lambda_{1\mathrm{H}} v^2 X^\alpha_\alpha \\
& + h' \bar{H}_i \bar{Q}^i{}_\alpha Q^\alpha_6 + h \bar{Q}^6{}_\alpha Q^\alpha_i H^i \\
& + y_{\mathbf{10}} \mathbf{10} \cdot \mathbf{10} \cdot H + y_{\mathbf{5}^*} \mathbf{5}^* \cdot \mathbf{10} \cdot \bar{H} + \cdots,
\end{aligned}
\tag{1}
$$

where the parameter v is of order of the GUT scale, $y_{\mathbf{10}}, y_{\mathbf{5}^*}$ are Yukawa coupling constants of the quarks and leptons, and $\lambda_{3\mathrm{H}}, \lambda'_{3\mathrm{H}}, \lambda_{1\mathrm{H}}, \lambda'_{1\mathrm{H}}, h', h$ are dimensionless coupling constants. One can see that the above superpotential has $\mathbf{Z}_4$ R symmetry under a charge assignment given in Table 1, and this symmetry forbids enormous mass term $W = \bar{H}H$ for the Higgs doublets[b]. The bifundamental representation Q^α_i and $\bar{Q}^i{}_\alpha$ acquire vacuum expectation value, $\langle Q^\alpha{}_i \rangle = v\delta^\alpha_i$ and $\langle \bar{Q}^i{}_\alpha \rangle = v\delta^i_\alpha$, because of the second and the third line in (1). Then, the mass terms of the colored Higgs multiplets arise from the fourth line in (1) in the GUT-breaking vacuum. The $\mathbf{Z}_4$ R symmetry is not broken even after the GUT symmetry is broken. One can also see that this $\mathbf{Z}_4$ R symmetry forbids the dangerous dimension 5 proton decay operators $W = \mathbf{10} \cdot \mathbf{10} \cdot \mathbf{10} \cdot \mathbf{5}^*$.

One of the characteristic features of this model is that the gauge coupling constants of the $\mathrm{U}(3)_{\mathrm{H}}$ should be strong compared with that of the $\mathrm{SU}(5)_{\mathrm{GUT}}$ gauge group. Otherwise, the approximate SU(5)-unification at the GUT scale would be violated through the contributions from the $\mathrm{U}(3)_{\mathrm{H}}$ gauge group in the tree-level matching relations:

$$
\left(\frac{1}{\alpha_3} \equiv \frac{1}{\alpha_C} \right) = \frac{1}{\alpha_{\mathrm{GUT}}} + \frac{1}{\alpha_{3\mathrm{H}}}, \tag{2}
$$

---

[a] A normalization condition $\mathrm{tr}(t_a t_b) = \delta_{ab}/2$ is understood. Note that the normalization of the t_0 is determined so that it also satisfies $\mathrm{tr}(t_0 t_0) = 1/2$.

[b] μ-term can be obtained through the Giudice-Masiero mechanism[7].

$$\left(\frac{1}{\alpha_2} \equiv \frac{1}{\alpha_L}\right) = \frac{1}{\alpha_{\rm GUT}} \tag{3}$$

and

$$\left(\frac{1}{\alpha_1} \equiv \frac{3/5}{\alpha_Y}\right) = \frac{1}{\alpha_{\rm GUT}} + \frac{2/5}{\alpha_{\rm 1H}}, \tag{4}$$

where $\alpha_C, \alpha_L, \alpha_Y, \alpha_{\rm GUT}, \alpha_{\rm 3H}$ and $\alpha_{\rm 1H}$ are fine structure constants of the three MSSM gauge groups, SU(5)$_{\rm GUT}$, SU(3)$_{\rm H}$ and U(1)$_{\rm H}$.

Another feature is that the cut-off scale of this model lies below the Planck scale ($M_{pl} = 2.4 \times 10^{18}$GeV). The gauge coupling constant of the U(1)$_{\rm H}$ is already large at the GUT scale ($\sim 10^{16}$GeV), and it becomes larger until it gets infinity through the asymptotic-non-free running. Naive 1-loop analysis shows that the cut-off scale Λ can be higher by one order of magnitude than the GUT scale (10^{17}GeV) but it cannot be as high as 10^{18}GeV. We consider that the cut-off scale Λ is in between this region.

The last remark is that the field content of the newly introduced sector has a multiplet structure of the $\mathcal{N} = 2$ SUSY [8], and the 1st-3rd lines of the superpotential (1) exhibit the form of interactions in the $\mathcal{N} = 2$ SUSY gauge theories[9]. We come back to this issue later.

One can also consider a model using the SU(5)$_{\rm GUT}\times$ U(2)$_{\rm H}$ gauge group instead of the SU(5)$_{\rm GUT}\times$ U(3)$_{\rm H}$. The matter content of the SU(5)-breaking sector consists of $X^{\alpha}{}_{\beta}$ (**1**,**adj.**= **3**+**1**), $Q^{\alpha}_{i,6}$(**5***+**1**,**2**) and $\bar{Q}^{i,6}_{\alpha}$(**5**+**1**,**2***) ($\alpha = 1, 2$). In this model, ordinary Higgs multiplets $H(\mathbf{5})^i$ and $\bar{H}(\mathbf{5}^*)_i$ are not necessary and the 4th line of the superpotential (1) is absent; composite fields $\bar{Q}^i_{\alpha} Q^{\alpha}{}_6$ and $\bar{Q}^6{}_{\alpha} Q^{\alpha}{}_i$ play the role of $H^i(\mathbf{5})$ and $\bar{H}_i(\mathbf{5}^*)$. These composites provide the two Higgs doublets while they do not contain Higgs triplets as their 1-particle degrees of freedom, and that is how the doublet-triplet splitting problem is solved in this model.

All three remarks on the SU(5)$_{\rm GUT}\times$U(3)$_{\rm H}$ model also hold in this SU(5)$_{\rm GUT}\times$U(2)$_{\rm H}$ model. The tree level matching relations of the gauge couplings Eqs.(2-4) are replaced by

$$\left(\frac{1}{\alpha_3} \equiv \frac{1}{\alpha_C}\right) = \frac{1}{\alpha_{\rm GUT}}, \tag{5}$$

$$\left(\frac{1}{\alpha_2} \equiv \frac{1}{\alpha_L}\right) = \frac{1}{\alpha_{\rm GUT}} + \frac{1}{\alpha_{\rm 2H}}, \tag{6}$$

and

$$\left(\frac{1}{\alpha_1} \equiv \frac{3/5}{\alpha_Y}\right) = \frac{1}{\alpha_{\rm GUT}} + \frac{3/5}{\alpha_{\rm 1H}}, \tag{7}$$

where $\alpha_{2\mathrm{H}}$ is the fine structure constant of the $\mathrm{SU}(2)_{\mathrm{H}}$ gauge group.

3 1-Loop Corrections at the GUT Scale

Since the $\mathrm{U}(1)_{\mathrm{H}}$ gauge coupling constant is infra-red free, $1/\alpha_{1\mathrm{H}}$ contribution is larger than the $1/\alpha_{3\mathrm{H}}$ or $1/\alpha_{2\mathrm{H}}$ in the tree-level matching relations Eqs.(2-4) or Eqs.(5-7). Therefore, the resulting difference between the three gauge couplings of the MSSM implies that the tree-level matching scale (*i.e.*, the actual "GUT scale") is below the α_1-α_2 unification scale, $\sim 2\times 10^{16}$GeV (see Fig.1). Thus, the proton decay will be faster compared with the rate using the 2×10^{16}GeV. We derive the precise prediction of the proton decay in what follows.

In the analysis at the next-to-leading order [10], 1-loop renormalization group (RG) running and 1-loop threshold corrections of the GUT model are taken into account. The three MSSM gauge coupling constants just below the GUT scale are expressed in terms of the gauge coupling constants and various masses in the spectrum of the GUT model, including the mass M_v of the GUT gauge boson which induces the proton decay.

Gauge coupling constants of the MSSM is given by

$$
\begin{aligned}
\frac{1}{\alpha_3}(\mu) &= \frac{1}{\alpha_{\mathrm{GUT}}}(\Lambda) + \frac{1}{\alpha_{3\mathrm{H}}}(\Lambda) + \frac{1}{2\pi}\ln\left(\left(\frac{\Lambda}{\mu}\right)^{-3}\left(\frac{\Lambda}{M_v}\right)^{-4}\left(\frac{\Lambda^2}{M_{\bar{c}}M_c}\right)\left(\frac{M_{8v}}{M_{8c}}\right)^{6}\right), \\
\frac{1}{\alpha_2}(\mu) &= \frac{1}{\alpha_{\mathrm{GUT}}}(\Lambda) \qquad\qquad\qquad\quad + \frac{1}{2\pi}\ln\left(\left(\frac{\Lambda}{\mu}\right)\left(\frac{\Lambda}{M_v}\right)^{-6}\right), \\
\frac{1}{\alpha_1}(\mu) &= \frac{1}{\alpha_{\mathrm{GUT}}}(\Lambda) + \frac{2/5}{\alpha_{1\mathrm{H}}}(\Lambda) + \frac{1}{2\pi}\ln\left(\left(\frac{\Lambda}{\mu}\right)^{\frac{33}{5}}\left(\frac{\Lambda}{M_v}\right)^{-10}\left(\frac{\Lambda^2}{M_{\bar{c}}M_c}\right)^{\frac{2}{5}}\right)
\end{aligned}
\tag{8}
$$

in the $\mathrm{SU}(5)_{\mathrm{GUT}}\times\mathrm{U}(3)_{\mathrm{H}}$ model. "μ" is a renormalization point, which is taken to be just below the GUT scale, M_v, M_c, $M_{\bar{c}}$, M_{8v}, M_{8c} are masses of particles around the GUT scale (see Table 2) and $\alpha_{\mathrm{GUT,3H,1H}}(\Lambda)$ are fine structure constants of the gauge groups $\mathrm{SU}(5)_{\mathrm{GUT}}$, $\mathrm{SU}(3)_{\mathrm{H}}$ and $\mathrm{U}(1)_{\mathrm{H}}$, respectively, at the cut-off scale Λ. Similar expressions are obtained in the $\mathrm{SU}(5)_{\mathrm{GUT}}\times\mathrm{U}(2)_{\mathrm{H}}$ model, using the mass spectrum of this model (Table 3).

The 1-loop analysis is valid only when the higher-loop effects are negligible, and this condition is satisfied when the $\mathcal{N}=2$ SUSY relations

$$
\frac{(\lambda_{3\mathrm{H}})^2}{4\pi} \simeq \frac{(\lambda'_{3\mathrm{H}})^2}{4\pi} \simeq \alpha_{3\mathrm{H}},
$$

$$\frac{(\lambda_{1\mathrm{H}})^2}{4\pi} \simeq \frac{(\lambda'_{1\mathrm{H}})^2}{4\pi} \simeq \alpha_{1\mathrm{H}}, \qquad (9)$$
$$\frac{(\lambda_{2\mathrm{H}})^2}{4\pi} \simeq \frac{(\lambda'_{2\mathrm{H}})^2}{4\pi} \simeq \alpha_{2\mathrm{H}},$$

are satisfied [11]. These relations are not fine-tunings of parameters, but rather a symmetric limit; an $\mathcal{N}$=2 SUSY enhances in the GUT breaking sector in this limit. Therefore, we assume these relations in our analysis. Then, in this case, the contributions from $\mathrm{SU(3)_C}$ octets ($\mathrm{SU(2)_L}$ triplets) decouple from the threshold corrections in the $\mathrm{SU(5)_{GUT}\times U(3)_H}$ model (in the $\mathrm{SU(5)_{GUT}\times U(2)_H}$ model, respectively) because of a degeneracy between the octets (between the triplets, respectively).

By now, due to the simple spectrum of the models, it is quite easy to extract the mass of the GUT gauge boson:

$$M_v = \sqrt{\frac{\mu^3}{\Lambda}} \exp\left(-\frac{2\pi}{24}\left(\frac{2}{\alpha_3}+\frac{3}{\alpha_2}-\frac{5}{\alpha_1}\right)(\mu)\right) \times\sqrt{\frac{M_{8v}}{M_{8c}}} \exp\left(-\frac{2\pi}{12}\left(\frac{1}{\alpha_{1\mathrm{H}}}-\frac{1}{\alpha_{3\mathrm{H}}}\right)(\Lambda)\right). \qquad (10)$$

in the $\mathrm{SU(5)_{GUT}\times U(3)_H}$ model and

$$M_v = \sqrt{\frac{\mu^3}{\Lambda}} \exp\left(-\frac{2\pi}{24}\left(\frac{2}{\alpha_3}+\frac{3}{\alpha_2}-\frac{5}{\alpha_1}\right)(\mu)\right) \times\left(\frac{M_{3v}}{M_{3c}}\right)^{\frac{1}{3}} \exp\left(-\frac{2\pi}{8}\left(\frac{1}{\alpha_{1\mathrm{H}}}-\frac{1}{\alpha_{2\mathrm{H}}}\right)(\Lambda)\right). \qquad (11)$$

in the $\mathrm{SU(5)_{GUT}\times U(2)_H}$ model. The cut-off scale in the denominator is a direct consequence of the infra-red free running of the $\mathrm{U(1)_H}$ (and also of the $\mathrm{SU(2)_H}$) gauge coupling constant(s). This negative power dependence on the cut-off scale leads to light GUT gauge boson, as was expected from a naive argument at the beginning of this section. The last 2 factors in the above expressions are negligible due to the the $\mathcal{N}$=2 SUSY relation Eq.(9) ($M_{8v}/M_{8c} \simeq 1$, $M_{3v}/M_{3c} \simeq 1$) and to the smallness of the $1/\alpha_{3\mathrm{H},1\mathrm{H},2\mathrm{H}}(\Lambda)$.

We calculate the SUSY threshold corrections to obtain the MSSM gauge coupling constants (at weak scale) according to the method in Ref.[12]. 2-loop RG is used to obtain the MSSM couplings just below the GUT scale. The life time of proton is calculated using the formula in Ref.[10], with an up-dated value of a hadron form factor in Ref.[13]. Our result is summarized in Fig.2, which shows contour plot of proton life time as a function of SUSY breaking parameters (we assumed minimal SUGRA SUSY breaking in our analysis).

4 Conclusions

Now, we can estimate the proton life time for various SUSY particle spectra. We neglect possible two uncertainties expressed by the last 2 factors in Eqs.(10,11). Effects of such uncertainties and other sources of uncertainties are discussed in Ref.[11]. Here, we also set the cut-off scale to be 10^{17}GeV; In most part of SUSY breaking parameter space, the three gauge coupling constants unify approximately at around 10^{16}GeV and hence the cut-off scale Λ is expected to be no less than 10^{17}GeV. Therefore, we obtain a conservative upper bound of the proton life time, using the lowest cut-off scale (see Eqs.(10,11)).

As we can see from these contours, the proton life time is in the range $3\times10^{34}-10^{35}$ yr. in most part of the parameter space regardless of choices of $\tan\beta$, A_0. We find the minimum of the proton life time is no less than 3×10^{34} yr. in whole parameter space, which is well above the current experimental limit by the Super-Kamiokande, 5.0×10^{33} yr. (90% C.L.)[14]. The thick contour lines corresponding to the life time of proton 7×10^{34} yr. represent the 3σ discovery limit of the 1Mt (fiducial volume) detector after ten years running [6].

Therefore, in the semi-simple unification model, we have an intriguing possibility to confirm the existence of the GUT in nature by observing the proton decay in the next-generation Mt water Čerenkov detectors, such as Hyper-Kamiokande and TITAND. In the optimistic cases where some enhancement factors arises in the decay rate of proton (see Ref.[11]), we have a chance to detect the proton decay also in UNO [15] ($\sim$ 500kt fiducial volume) experiment.

For more details, readers are recommended to see Ref.[11].

References

1. [SuperKamiokande Collaboration], *Phys. Rev. Lett.* **83**, 1529 (1999).
2. N. Sakai and T. Yanagida, *Nucl. Phys.* B **197**, 533 (1982);
   S. Weinberg, *Phys. Rev.* D **26**, 287 (1982).
3. K.-I. Izawa and T. Yanagida, Prg.Theor.Phys. **97**, 913 (1997).
4. Ref.[3] is based on a idea developed in preceding papers such as :
   T. Yanagida, *Phys. Lett.* B **344**, 211 (1995);
   J. Hisano and T. Yanagida, Mod.Phys.Lett.A **10**, 3097 (1995);
   T. Hotta, K.-I. Izawa and T. Yanagida, *Phys. Rev.* D **53**, 3913 (1996),*Phys. Rev.* D **54**, 6970 (1996).
5. A. Masiero, D. Nanopoulos, K.Tamvakis and T. Yanagida, *Phys. Lett.* B

115, 380 (1982).
6. M. Koshiba, Phys.Rept. **220**, 229 (1992); Y. Suzuki *et al.* [TITAND Working Group Collaboration], arXiv:hep-ex/0110005.
7. G. Giusice and A. Masiero, *Phys. Lett.* B **206**, 480 (1988).
8. J. Hisano and T. Yanagida of Ref.[4].
9. Y. Imamura, T. Watari and T. Yanagida, *Phys. Rev.* D **64**, 065023 (2001).
10. J. Hisano, H. Murayama and T. Yanagida, *Nucl. Phys.* B **402**, 46 (1993).
11. M. Fujii and T. Watari, *Phys. Lett.* B **527**, 106 (2002).
12. D. M. Pierce, J. A. Bagger, K. T. Matchev and R. j. Zhang, *Nucl. Phys.* B **491**, 3 (1997).
13. S. Aoki et al.,*Phys. Rev.* D **53**, 014506 (2000).
14. [Super-Kamiokande Collaboration],*Phys. Rev. Lett.* **81**, 3319 (1998).
15. C. K. Jung, arXiv:hep-ex/0005046.

Table 1. Charge assignment of the $\mathbf{Z}_4$ R-symmetry is given. **1** denotes a right handed neutrino. H and $\bar{H}$ are not in the matter content of the $\mathrm{SU}(5)_{\mathrm{GUT}}\times\mathrm{U}(2)_{\mathrm{H}}$ model.

Fields	**5***,**10**,**1**	$(H\ ,\ H)$	X^{α}_{β}	Q_i,Q^i	Q_6	Q^6
$\mathbf{Z}_4$ R charge	1	0	2	0	2	-2

Table 2. Particle spectrum around the GUT scale in the $\mathrm{SU}(5)_{\mathrm{GUT}}\times\ \mathrm{U}(3)_{\mathrm{H}}$ model. The first line denotes the representation under the MSSM gauge group. In the second line, m.vect. denotes $\mathcal{N}=1$ massive vector multiplet and $\chi+\chi^{\dagger}$ a pair of $\mathcal{N}=1$ chiral and anti-chiral multiplet. Mass of each multiplet is given in terms of gauge couplings and parameters in the superpotential (1) in the fourth line, and given in the third line is the expression of the mass used in the text. Multiplets with masses M_{1v} and M_{1c}, M_{8v} and M_{8c} can be regarded as $\mathcal{N}=$ 2-SUSY partner of each other in the $\mathcal{N}=$ 2-SUSY limit.

$(\mathbf{3},\mathbf{2})^{-\frac{5}{6}}$	$(\mathbf{3},\mathbf{1})^{-\frac{1}{3}}$	$(\mathbf{3},\mathbf{1})^{-\frac{1}{3}}$
m.vect.	$\chi+\chi^{\dagger}$	$\chi+\chi^{\dagger}$
$M_v=$	$M_c=$	$M_{\bar{c}}=$
$\sqrt{2}g_{\mathrm{GUT}}v$	hv	$h'v$

$(\mathbf{1},\mathbf{1})^0$	$(\mathbf{1},\mathbf{1})^0$	$(\mathbf{adj.},\mathbf{1})^0$	$(\mathbf{adj.},\mathbf{1})^0$
m.vect.	$\chi+\chi^{\dagger}$	m.vect.	$\chi+\chi^{\dagger}$
$M_{1v}=$	$M_{1c}=$	$M_{8v}=$	$M_{8c}=$
$\sqrt{2(g_{1\mathrm{H}}^2+2g_{\mathrm{GUT}}^2/5)}v$	$\sqrt{2}\lambda_{1\mathrm{H}}v$	$\sqrt{2(g_{3\mathrm{H}}^2+g_{\mathrm{GUT}}^2)}v$	$\sqrt{2}\lambda_{3H}v$

Table 3. Particle spectrum of the $SU(5)_{GUT}\times U(2)_H$ model. Multiplets with masses M_{1v} and M_{1c}, M_{3v} and M_{3c} can be regarded as $\mathcal{N}=2$-SUSY partner of each other in the $\mathcal{N}=2$-SUSY limit.

$(\mathbf{3},\mathbf{2})^{-\frac{5}{6}}$	$(\mathbf{1},\mathbf{1})^0$	$(\mathbf{1},\mathbf{1})^0$	$(\mathbf{1},\mathbf{adj.},)^0$	$(\mathbf{1},\mathbf{adj.},)^0$
m.vect.	m.vect.	$\chi+\chi^\dagger$	m.vect.	$\chi+\chi^\dagger$
$M_v=$ $\sqrt{2}g_{GUT}v$	$M_{1v}=$ $\sqrt{2(g_{1H}^2+3g_{GUT}^2/5)}v$	$M_{1c}=$ $\sqrt{2}\lambda_{1H}v$	$M_{3v}=$ $\sqrt{2(g_{2H}^2+g_{GUT}^2)}v$	$M_{3c}=$ $\sqrt{2}\lambda_{2H}v$

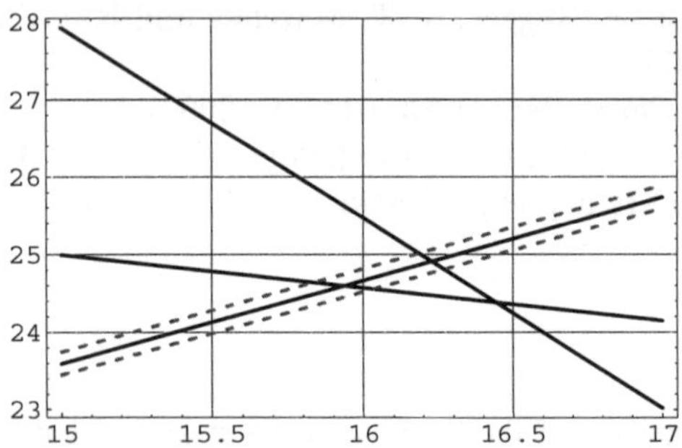

Figure 1. Approximate $SU(5)_{GUT}$ relation between the three MSSM gauge coupling constants and deviation from it. $\log_1 0(\mu/1\text{GeV})$ in horizontal axis and $1/\alpha_{1,2,3}$ in vertical axis. 1σ error bar of the QCD coupling are also described. SUSY threshold corrections are calculated using the spectrum of mSUGRA model with $m_0=250$GeV, $M_{1/2}=500$GeV, $A_0=0$ and $\tan\beta=10$. The sign of μ-term is taken to be negative.

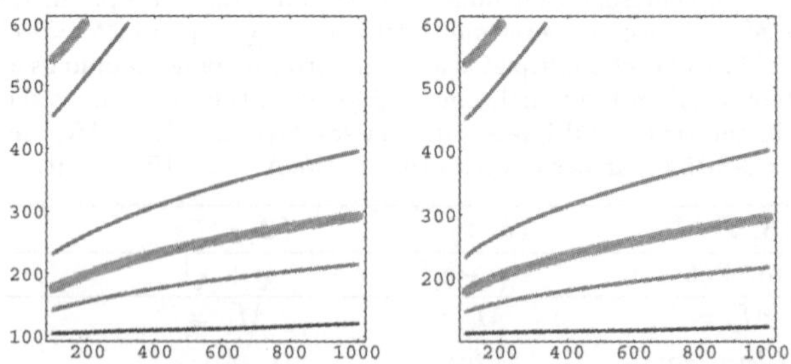

Figure 2. Contour plots of the proton life time in m_0-$M_{1/2}$ space for $\mu>0$. $(\tan\beta, A_0)$ =(10,0) in the figure at the left-hand side, and (30,300 GeV) in that at the right-hand side. 4 contour lines corresponds to the proton life time 5×10^{34}yr., 7×10^{34}yr., 10^{35}yr. and 2×10^{35}yr. from inside to outside.

LEPTOGENESIS VIA LH_U FLAT DIRECTION

MASAAKI FUJII

Department of Physics, University of Tokyo, Tokyo 113-0033, Japan
E-mail: fujii@hep-th.phys.s.u-tokyo.ac.jp

The leptogenesis via the LH_u flat direction is the minimal scenario to generate the observed baryon asymmetry in the supersymmetric framework. This scenario is ready to work in the minimal supersymmetric standard model (MSSM) with tiny but nonzero masses of the left-handed neutrinos in the superpotential. The most promising feature of this leptogenesis is a direct connection between the resultant baryon asymmetry and the neutrino physics. The observed baryon asymmetry requires the mass of the lightest neutrino to be much smaller than the mass scale indicated from the atmospheric and solar neutrino oscillations. Such a small mass of the lightest neutrino leads to a high predictability on the rate of the neutrinoless double beta ($0\nu\beta\beta$) decay.

1. Introduction

The origin of the observed baryon asymmetry in the present universe is one of the most fundamental problems in particle physics as well as in cosmology. Unfortunately, there is no way to generate the observed baryon asymmetry within the Standard Model (SM). The electroweak baryogenesis scenario in the SM requires a very small mass of the Higgs boson to generate the required asymmetry, which is already experimentally excluded. Therefore, only the existence of the baryon asymmetry in the present universe is sufficient to conclude that we need a physics beyond the SM.

In fact, we already know an important ingredient beyond the SM. It is the existence of tiny but nonzero masses of the neutrinos, which is now confirmed by various experiments. Another highly motivated ingredient beyond the SM is the supersymmetry (SUSY). Especially, the minimal supersymmetric standard model (MSSM) naturally solves the hierarchy problem and predicts a beautiful unification of the three gauge coupling constants of the SM gauge groups. Furthermore, it also provides a promising candidate of dark matter in the universe. The lightest supersymmetric particle (LSP) is stable under the conservation of the R-parity, and its thermal relic abundance results in a cosmologically interesting mass density of dark matter in a naturally way.

Therefore, it is very interesting and important to find a minimal and a natural scenario to generate the observed baryon asymmetry in the MSSM with the neutrino mass terms. As we will see, this is the leptogenesis via the LH_u flat direction based on the Affleck-Dine mechanism [1]. Interestingly enough, in this scenario, the resultant baryon asymmetry is determined almost only by the mass of the lightest neutrino, and it is independent of the reheating temperatures of inflation among others [2,3]. The direct connection between the observed baryon asymmetry and the neutrino physics, which is low energy physics, is the most promising feature of this scenario.

In the leptogenesis via the LH_u flat direction, as we will see, the observed baryon asymmetry predicts the mass of the lightest neutrino is much smaller than the mass scale implied by the solar neutrino oscillation experiments. This leads to a high predictability on the rate of the $0\nu\beta\beta$ decay [3,4]. It is quite interesting that the predicted value of the mass parameter $m_{\nu_e\nu_e}$ is in the accessible range of future experiments [5,6,7,8,9,10] of the $0\nu\beta\beta$ decay. In this letter, we briefly review the leptogenesis via the LH_u flat direction and summarise its predictions on the $0\nu\beta\beta$ decay.

2. Leptogenesis via the LH_u flat direction

In this section we briefly review the leptogenesis via the LH_u flat direction [2,3]. Let us start by writing down the effective dimension-five operator in the superpotential,

$$W = \frac{1}{2M_i}\left(L_i H_u\right)\left(L_i H_u\right) , \tag{1}$$

which induces small neutrino masses m_{ν_i} [11] after the neutral component of the Higgs field H_u obtains its vacuum expectation value $\langle H_u \rangle = 174\ \mathrm{GeV} \times \sin\beta$,[a]

$$m_{\nu_i} = \frac{\langle H_u \rangle^2}{M_i}, \tag{2}$$

where we have taken a basis in which the neutrino mass matrix is diagonal. We adopt the following supersymmetric D-flat direction [12]:

$$\widetilde{L} = \frac{1}{\sqrt{2}}\begin{pmatrix} \phi \\ 0 \end{pmatrix}, \quad H_u = \frac{1}{\sqrt{2}}\begin{pmatrix} 0 \\ \phi \end{pmatrix} . \tag{3}$$

[a] $\tan\beta$ is defined as $\tan\beta \equiv \langle H_u \rangle / \langle H_d \rangle$, where H_u and H_d are the Higgs fields which provide masses for the up- and down-type quarks, respectively.

Here and hereafter, we suppress the family index i. As we will see below, the flat direction which generates the lepton asymmetry most effectively corresponds to the *lightest* neutrino.

By virtue of the flatness of the potential, the ϕ field can obtain a large expectation value along the flat direction given in Eq. (3) during the inflation. All the evolutions of the ϕ field which determines the resultant baryon asymmetry can be obtained from the following scalar potential of the ϕ field:

$$\begin{aligned} V(\phi) = {} & m_\phi^2|\phi|^2 + \frac{m_{3/2}}{8M}\left(a_m\phi^4 + H.c.\right) \\ & - c_H H^2|\phi|^2 + \frac{H}{8M}\left(a_H\phi^4 + H.c.\right) \\ & + \sum_{f_k|\phi|<T} c_k f_k^2 T^2|\phi|^2 + a_g\alpha_S(T)^2 T^4 \ln\left(\frac{|\phi|^2}{T^2}\right) \\ & + \frac{1}{4M^2}|\phi|^6 . \end{aligned} \tag{4}$$

Here, the potential terms in the first line comes from the SUSY breaking at the zero temperature $T = 0$, which are mediated by supergravity, and we take $m_\phi \simeq m_{3/2}|a_m| \simeq 1$ TeV, hereafter.[b] The terms in the second line, depending on the Hubble parameter H, reflects the additional SUSY breaking effects caused by the finite energy density of the inflaton [13]. Hereafter, we take $c_H \simeq 1 (> 0)$ and $|a_H| \simeq 1$. The terms in the third line represent the effects of the finite temperature [14,2,15,3]. Here, f_k represent Yukawa and gauge coupling constants of the field ϕ, α_S is the gauge coupling constant of the $SU(3)_C$, and c_k and a_g are coefficients of order unity.[c] Finally, the last term directly comes from the superpotential in Eq. (1).

During the inflation, the negative Hubble mass term $-c_H H^2|\phi|^2$ causes an instability of the flat direction field ϕ around the origin ($\phi \simeq 0$), and ϕ field runs to one of the following four minima of the potential:

$$|\phi| \simeq \sqrt{MH}\,, \tag{5}$$

$$\arg(\phi) \simeq \frac{-\arg(a_H) + (2n+1)\pi}{4}\,, \qquad n = 0\cdots 3\,, \tag{6}$$

which are the balance points between the Hubble-induced terms and the $|\phi|^6$ term.

After the inflation, the amplitude of the ϕ field starts to be reduced following the gradual decrease of the Hubble parameter H [see Eq. (5)], and

[b] We assume gravity mediated SUSY breaking models.

[c] The list of c_k and f_k is given in Ref. [2]. In the case of LH_u flat direction, a_g is given by $a_g = 9/8$.

eventually, either thermal terms $T^2|\phi|^2$ or $T^4\ln(|\phi|^2)$, or the soft mass term $m_\phi^2|\phi|^2$ dominate the scalar potential, which causes a coherent oscillation of the ϕ around the origin. The Hubble parameter at this time $H = H_{\rm osc}$ is given by the following equation:

$$c_H H^2 = m_\phi^2 + \sum_{f_k|\phi|<T} c_k f_k^2 T^2 + a_g \alpha_S(T)^2 \frac{T^4}{|\phi|^2} \,. \tag{7}$$

At this stage, the difference between the phases of the complex couplings a_m and a_H leads to a phase rotational motion of ϕ, which results in generation of the lepton asymmetry, $n_L = (1/2)i(\dot{\phi}^*\phi - \phi^*\dot{\phi})$.

The estimation of the resultant baryon asymmetry is rather straightforward. The equation of motion for the lepton number density is written as

$$\dot{n}_L + 3Hn_L \simeq \frac{m_{3/2}}{2M}{\rm Im}(a_m\phi^4), \tag{8}$$

where we have neglected a contribution coming from the Hubble induced A-term potential which is the second term in the second line in Eq. (4). This is because that the AD field traces the volley of this potential and hence the torque to rotate it is mainly supplied by the ordinary A-term proportional to the gravitino mass.

By integrating the Eq. (8), we obtain the produced lepton asymmetry:[d]

$$\frac{n_L}{s} = \frac{MT_R}{12M_*^2}\left(\frac{|a_m|m_{3/2}}{H_{\rm osc}}\right)\delta_{\rm eff} \,, \tag{9}$$

where $H_{\rm osc}$ is given as the solution of Eq. (7), $M_* = 2.4 \times 10^{18}$ GeV is the reduced Planck scale, $\delta_{\rm eff} \simeq \sin(4{\rm arg}\phi + {\rm arg}a_m)$ represents an effective CP-violating phase, and T_R is the reheating temperature of inflation. Decay of ϕ produces this lepton asymmetry in the thermal bath and a part of the produced lepton asymmetry is then converted into the baryon asymmetry [16] via the sphaleron effects [17], which is about $n_B \simeq -n_L/3$.

We show the result of detailed calculation [3] in Fig. 1, which presents the contour plot of the baryon asymmetry in the neutrino mass (m_ν) – (T_R) plane. As can be seen in Fig. 1, the baryon asymmetry is almost independent of the reheating temperature T_R, and is determined only by the mass of the lightest neutrino m_ν for $T_R \gtrsim 10^5$ GeV. This is caused by the T_R dependence of $H_{\rm osc}$ in Eq. (9). Consequently, the baryon asymmetry

---

[d]From this relation, you can understand that the flat direction corresponding to the lightest neutrino is the relevant one to generate the observed baryon asymmetry.

in the present universe $n_B/s \simeq (0.4\text{–}1) \times 10^{-10}$ indicates the mass of the lightest neutrino to be

$$m_\nu \simeq (0.1 - 3) \times 10^{-9} \text{ eV} \tag{10}$$

in a wide range of the reheating temperature 10^5 GeV $\lesssim T_R \lesssim 10^{12}$ GeV.[e] Note that the scale of M is related to the mass of the lightest neutrino by Eq. (2), $m_\nu = \langle H_u \rangle^2 / M$.

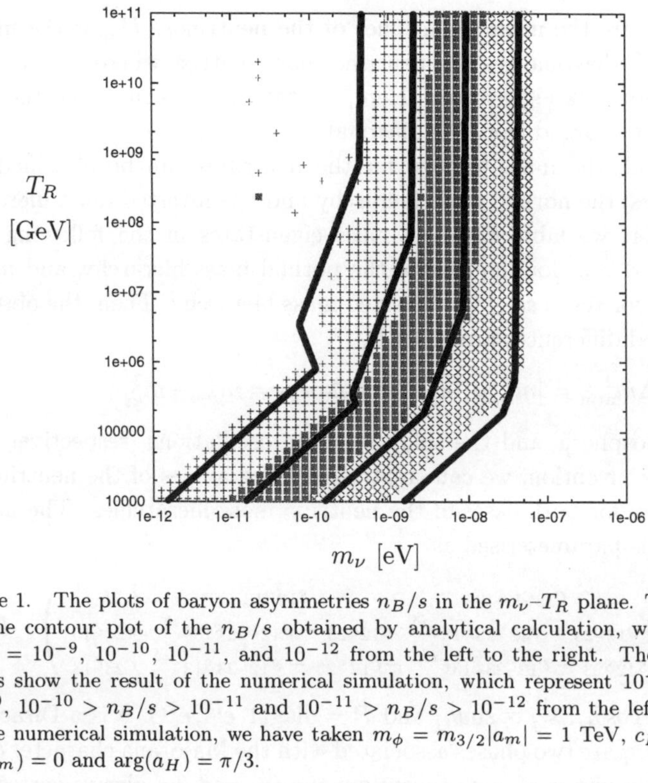

Figure 1. The plots of baryon asymmetries n_B/s in the m_ν–T_R plane. The solid lines are the contour plot of the n_B/s obtained by analytical calculation, which represent $n_B/s = 10^{-9}$, 10^{-10}, 10^{-11}, and 10^{-12} from the left to the right. The regions with points show the result of the numerical simulation, which represent $10^{-9} > n_B/s > 10^{-10}$, $10^{-10} > n_B/s > 10^{-11}$ and $10^{-11} > n_B/s > 10^{-12}$ from the left to the right. In the numerical simulation, we have taken $m_\phi = m_{3/2}|a_m| = 1$ TeV, $c_H = |a_H| = 1$, $\arg(a_m) = 0$ and $\arg(a_H) = \pi/3$.

[e] If the $U(1)_{B-L}$ symmetry is gauged in high energy scale, the resultant baryon asymmetry is enhanced in the "D-term stopping case" [18]. However, even in this case, the required mass of the lightest neutrino must satisfy $m_\nu \lesssim 10^{-5}$ eV to avoid the cosmological gravitino problem, and the prediction on the $0\nu\beta\beta$ decay is almost the same as that presented in this letter.

3. Prediction on the $0\nu\beta\beta$ decay

The neutrinoless double beta ($0\nu\beta\beta$) decay, if observed, is the strongest evidence for lepton number violation. In other words, it suggests the Majorana character of the neutrinos and thus the existence of the nonrenormalizable operator given in Eq.(1), which is a crucial ingredient for our leptogenesis to work.

The most important ingredient to determine the rate of the $0\nu\beta\beta$ decay is the effective mass of the electron-type neutrino, which is given by

$$|m_{\nu_e \nu_e}| = |U_{e1}^2 m_{\nu_1} + U_{e2}^2 m_{\nu_2} + U_{e3}^2 m_{\nu_3}|, \tag{11}$$

where m_{ν_i}'s are the mass eigenvalues of the neutrinos. $U_{\alpha i}$ is the mixing matrix which diagonalizes the neutrino mass matrix, where $\alpha = e,\ \mu,\ \tau$ represent the weak eigenstates. Here, we take a basis in which the mass matrix of the charged lepton is diagonal.

In general, the mass pattern for the neutrinos can be classified into the two cases; the normal mass hierarchy and the inverted mass hierarchy. In this letter, we label the each mass eigenstates in the following way; $m_{\nu_1} < m_{\nu_2} < m_{\nu_3}$ for the case of the normal mass hierarchy, and $m_{\nu_3} < m_{\nu_1} < m_{\nu_2}$ for the case of the inverted mass hierarchy. Then, the observed mass squared differences are given by

$$\Delta m^2_{\text{atm}} = |m^2_{\nu_3} - m^2_{\nu_2}|, \qquad \Delta m^2_{\text{sol}} = m^2_{\nu_2} - m^2_{\nu_1}\ , \tag{12}$$

for the atmospheric and the solar neutrino oscillations, respectively. By taking this convention, we can treat the mixing angles of the neutrinos in a similar way for both cases of the neutrino mass hierarchies. The mixing matrix $U_{\alpha i}$ is parameterised as

$$U_{\alpha i} = \begin{pmatrix} c_{12}c_{13} & s_{12}c_{13} & s_{13}e^{-i\delta} \\ -s_{12}c_{23} - c_{12}s_{23}s_{13}e^{i\delta} & c_{12}c_{23} - s_{12}s_{23}s_{13}e^{i\delta} & s_{23}c_{13} \\ s_{12}s_{23} - c_{12}c_{23}s_{13}e^{i\delta} & -c_{12}s_{23} - s_{12}c_{23}s_{13}e^{i\delta} & c_{23}c_{13} \end{pmatrix} \cdot P \tag{13}$$

where $c_{ij} \equiv \cos\theta_{ij}$, $s_{ij} \equiv \sin\theta_{ij}$ and $P = \text{diag}(1, e^{i\beta}, e^{i\gamma})$. δ is a Dirac-type phase and β, γ are two phases associated with the Majorana character of the neutrinos. By virtue of our convention, the θ_{23} and θ_{12} always correspond to the mixing angles for the atmospheric and the solar neutrino oscillations, respectively, regardless of the type of mass hierarchy for the neutrinos. In addition, the element of the mixing matrix U constrained by the CHOOZ experiment is always correspond to $(e, 3)$ element as $|U_{e3}| \lesssim 0.15$ [19].

The requirement for our leptogenesis to explain the observed baryon asymmetry is that the mass of the lightest neutrino is to be $m_\nu \sim 10^{-9}$ eV, which is negligibly small compared with the other two mass eigenvalues of the neutrinos. This immediately leads to the following upper and lower bounds on the electron-type neutrino mass:

$$\left||U_{e2}|^2 m_{\nu_2} - |U_{e3}|^2 m_{\nu_3}\right| \leq |m_{\nu_e\nu_e}| \leq |U_{e2}|^2 m_{\nu_2} + |U_{e3}|^2 m_{\nu_3} , \quad (14)$$

$$\left||U_{e1}|^2 m_{\nu_1} - |U_{e2}|^2 m_{\nu_2}\right| \leq |m_{\nu_e\nu_e}| \leq |U_{e1}|^2 m_{\nu_1} + |U_{e2}|^2 m_{\nu_2} \quad (15)$$

for the normal and the inverted mass hierarchies, respectively.

The predictions on the $0\nu\beta\beta$ decay in the case of the normal mass hierarchy have been investigated in detail in Ref. [3]. The upper and lower bounds on the electron-type neutrino mass in Eq. (14) can be written as

$$|m_{\nu_e\nu_e}|^{\max}_{\min} \simeq s_{12}^2\sqrt{\Delta m^2_{\rm sol}} \pm |U_{e3}|^2\sqrt{\Delta m^2_{\rm atm}} . \quad (16)$$

In Fig. 2, we present the numerical result for the case of the large mixing angle (LMA) solution.

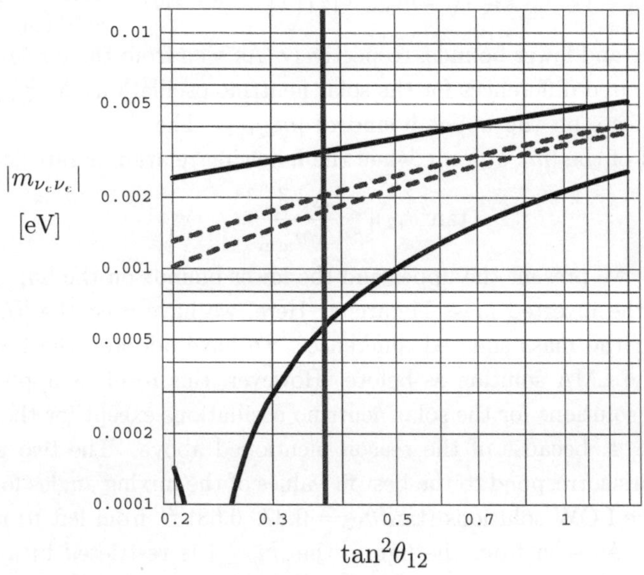

Figure 2. The upper and lower bands on the effective mass of the electron-type neutrino $|m_{\nu_e\nu_e}|$ in the case of the normal mass hierarchy.

The red (solid) and blue (dashed) lines correspond to the cases where $|U_{e3}| = 0.15$ and $|U_{e3}| = 0.05$, respectively. As for the mass squared differences, we have adopted the following best fit values [20,21]:

$$\Delta m^2_{\rm atm} = 3.2 \times 10^{-3}\ {\rm eV}^2 , \qquad \Delta m^2_{\rm sol} = 4.9 \times 10^{-5}\ {\rm eV}^2 . \quad (17)$$

The green (vertical) line denotes the best fit values of the mixing angle $\tan^2\theta_{12} = 0.37$. The behaviour of the bounds when we vary the mass squared differences $\Delta m^2_{\rm atm}$ and $\Delta m^2_{\rm sol}$ is easily seen from the Eq. (16). We see that the overall scale of the $|m_{\nu_e\nu_e}|$ is almost proportional to the $\sqrt{\Delta m^2_{\rm sol}}$. For example, when we vary the $\Delta m^2_{\rm sol}$ within the 95% C.L. allowed region of the LMA solution, $|m_{\nu_e\nu_e}|$ changes within about $\times/\div 1.7$ of the value presented in Fig. 2. As clearly shown in this figure, the $|m_{\nu_e\nu_e}|$ is predicted in a narrow range when the $|U_{e3}|$ is much smaller than the present bound.

Now, let us consider the case of the inverted mass hierarchy for the neutrinos. The Eq. (15) can be written in terms of the observables in neutrino oscillation experiments as

$$|m_{\nu_e\nu_e}|_{\rm max} \simeq (1-|U_{e3}|^2)\sqrt{\Delta m^2_{\rm atm}}\left(1-\frac{\Delta m^2_{\rm sol}}{2\Delta m^2_{\rm atm}}c^2_{12}\right), \tag{18}$$

$$|m_{\nu_e\nu_e}|_{\rm min} \simeq (1-|U_{e3}|^2)\sqrt{\Delta m^2_{\rm atm}}\; c^2_{12}\left|\left((1-\tan^2\theta_{12})-\frac{\Delta m^2_{\rm sol}}{2\Delta m^2_{\rm atm}}\right)\right| \tag{19}$$

for the upper and lower bounds, respectively. As seen from these relations, the mass squared differences for the solar neutrino oscillations $\Delta m^2_{\rm sol}$ gives negligible effects on the upper bound of $|m_{\nu_e\nu_e}|$. The $\Delta m^2_{\rm sol}$ affects the lower bound of the $|m_{\nu_e\nu_e}|$ only when the following relation is satisfied:

$$|(1-\tan^2\theta_{12})| \lesssim \frac{\Delta m^2_{\rm sol}}{\Delta m^2_{\rm atm}}. \tag{20}$$

In Fig. 3, we present the upper and the lower bounds on the $|m_{\nu_e\nu_e}|$ in the case of the inverted mass hierarchy. Here, we have used the $|U_{e3}| = 0.15$. As for the mass squared differences, we have adopted the best fit values for the LMA solution as before. However, this result is applicable to the other solutions for the solar neutrino oscillations except for the case of $\tan^2\theta_{12} \simeq 1$, because of the reason mentioned above. The two green (vertical) lines correspond to the best fit values of the mixing angles for the LMA and the LOW solutions ($\tan^2\theta_{12} = 0.37,\ 0.68$) [21] from left to right, respectively. As seen from the figure, the $|m_{\nu_e\nu_e}|$ is restricted in a very small range such as $0.01\ {\rm eV} \lesssim |m_{\nu_e\nu_e}| \lesssim 0.06\ {\rm eV}$ when $\tan^2\theta_{12} \lesssim 0.7$, which is clearly in the reach of the future $0\nu\beta\beta$ decay experiments. Even if we change the $\Delta m^2_{\rm atm}$ within the 99% C.L. allowed region, the overall scale of the $|m_{\nu_e\nu_e}|$ varies within only $\times/\div 1.5$ of the value presented in Fig. 3.

4. Conclusions

Leptogenesis via the LH_u flat direction is one of the most interesting scenarios to explain the observed baryon asymmetry, in which the observed

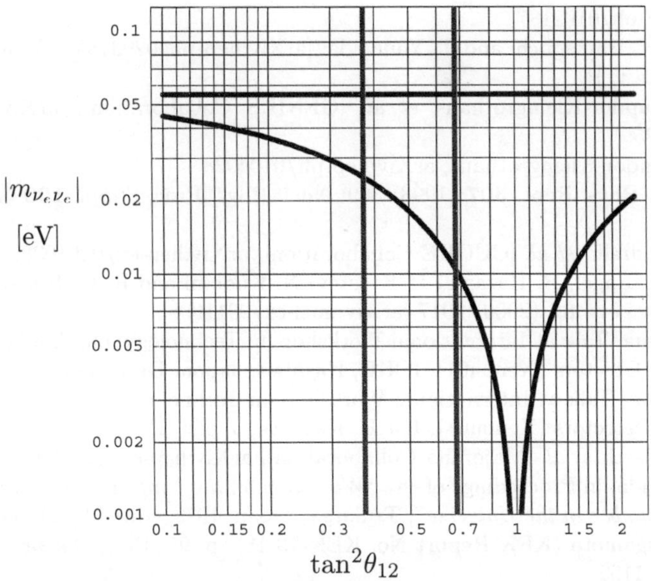

Figure 3. The upper and lower bands on the effective mass of the electron-type neutrino $|m_{\nu_e\nu_e}|$ in the case of the inverted mass hierarchy.

baryon asymmetry has a direct connection to the mass of the lightest neutrino regardless of the details of high energy physics. In this letter, we have derived predictions on the rate of the $0\nu\beta\beta$ decay, which is regarded as a low energy consequence of the present leptogenesis; the mass of the lightest neutrino must be $m_\nu \sim 10^{-9}$ eV.

Both in the cases of the normal and the inverted mass hierarchies for the neutrinos, the region of $|m_{\nu_e\nu_e}|$ is strongly restricted and it is accessible in the future $0\nu\beta\beta$ decay experiments [5,6,7,8,9,10]. Furthermore, predictions on the $|m_{\nu_e\nu_e}|$ become much more definite when the $|U_{e3}|$ is severely constrained (for the case of the normal hierarchy) or the mixing angle for the solar neutrino oscillation is restricted as $\tan^2\theta_{12} < 1$ (for the case of the inverted hierarchy) in the future experiments. The predictions on the $0\nu\beta\beta$ decay will play a crucial role on testing the leptogenesis via the LH_u flat direction.

References

1. I. Affleck and M. Dine, Nucl. Phys. B**249** (1985) 361.
2. T. Asaka, M. Fujii, K. Hamaguchi and T. Yanagida, Phys. Rev. D**62** (2000) 123514 [arXiv:hep-ph/0008041].
3. M. Fujii, K. Hamaguchi and T. Yanagida, Phys. Rev. D**63** (2001) 123513

[arXiv:hep-ph/0102187].
4. M. Fujii, K. Hamaguchi and T. Yanagida, [arXiv:hep-ph/0203189]. To appear in Phys. Lett. B.
5. H. V. Klapdor-Kleingrothaus *et al.* [GENIUS Collaboration], arXiv:hep-ph/9910205;
H. V. Klapdor-Kleingrothaus, arXiv:hep-ph/0104028.
6. E. Fiorini, Phys. Rept. **307** (1998) 309; Nucl. Phys. Proc. Suppl. **100** (2001) 332;
A. Alessandrello *et al.* [CUORE Collaboration], arXiv:hep-ex/0201038.
7. H. Ejiri, J. Engel, R. Hazama, P. Krastev, N. Kudomi and R. G. Robertson, Phys. Rev. Lett. **85** (2000) 2917 [arXiv:nucl-ex/9911008].
8. S. Moriyama, Talk at International Workshop on Technology and Application of Xenon Detectors (Xenon01), ICRR, Kashiwa, Japan, December 3-4, 2001.
9. S. Waldman, Talk at International Workshop on Technology and Application of Xenon Detectors (Xenon01), ICRR, Kashiwa, Japan, December 3-4, 2001.
10. C. E. Aalseth *et al.* [Majorana Collaboration], arXiv:hep-ex/0201021.
11. T. Yanagida, in Proceedings of the *"Workshop on the Unified Theory and the Baryon Number in the Universe"*, Tsukuba, Japan, 1979, edited by O. Sawada and A. Sugamoto (KEK Report No. KEK-79-18), p. 95; Prog. Theor. Phys. **64** (1980) 1103;
M. Gell-Mann, P. Ramond and R. Slansky, in *"Supergravity"*, edited by D.Z. Freedman and P. van Nieuwenhuizen (North-Holland, Amsterdam, 1979).
12. H. Murayama and T. Yanagida, Phys. Lett. B**322** (1994) 349 [arXiv:hep-ph/9310297].
13. M. Dine, L. Randall and S. Thomas, Phys. Rev. Lett. **75** (1995) 398 [arXiv:hep-ph/9503303]. Nucl. Phys. B**458** (1996) 291 [arXiv:hep-ph/9507453].
14. R. Allahverdi, B. A. Campbell and J. Ellis, Nucl. Phys. B**579** (2000) 355 [arXiv:hep-ph/0001122].
15. A. Anisimov and M. Dine, arXiv:hep-ph/0008058;
See also, A. Anisimov, arXiv:hep-ph/0111233.
16. M. Fukugita and T. Yanagida, Phys. Lett. B**174** (1986) 45.
17. V. A. Kuzmin, V. A. Rubakov and M. E. Shaposhnikov, Phys. Lett. B**155** (1985) 36.
18. M. Fujii, K. Hamaguchi and T. Yanagida, Phys. Rev. D**65** (2002) 043511 [arXiv:hep-ph/0109154].
19. M. Apollonio *et al.*, Phys. Lett. B**466** (1999) 415 [arXiv:hep-ex/9907037].
20. Y. Fukuda *et al.* [Super-Kamiokande Collaboration],
Phys. Lett. B**433** (1998) 9 [arXiv:hep-ex/9803006];
Phys. Lett. B**436** (1998) 33 [arXiv:hep-ex/9805006];
Phys. Rev. Lett. **81** (1998) 1562 [arXiv:hep-ex/9807003].
See also a recent data,
C. Yanagisawa, Nucl. Phys. Proc. Suppl. **95** (2001) 93.
21. G. L. Fogli, E. Lisi, D. Montanino and A. Palazzo, Phys. Rev. D**64** (2001) 093007 [arXiv:hep-ph/0106247];
G. L. Fogli, E. Lisi, A. Marrone, D. Montanino and A. Palazzo, arXiv:hep-ph/0201290, and references therein.

Anomalous $\mathcal{U}(1)$: Solving Various Puzzles Of MSSM And $SU(5)$ GUT

Qaisar Shafi
Bartol Research Institute, University of Delaware, Newark, DE 19716, USA
E-mail address: shafi@bartol.udel.edu

Zurab Tavartkiladze
Institute for Theoretical Physics, Heidelberg University, Philosophenweg 16, D-69120 Heidelberg, Germany
E-mail address: Z.Tavartkiladze@ThPhys.Uni-Heidelberg.DE
Institute of Physics, Georgian Academy of Sciences, Tbilisi 380077, Georgia

Abstract

We discuss how an anomalous $\mathcal{U}(1)$ symmetry when appended to MSSM and SUSY GUTs [e.g. $SU(5)$] can help overcome a variety of puzzles related to charged fermion masses and mixings, flavor changing processes, proton decay and neutrino oscillations. Proton lifetime for $SU(5)$ GUT, in particular, is predicted in a range accessible to the ongoing or planned searches.

1 Introduction: Some Puzzles of MSSM and Beyond

The standard model provides an excellent description of almost all present experimental data. Supersymmetry (SUSY) is highly motivated because it offers possibility of resolving the gauge hierarchy problem. Furthermore, superstring theories are believed to be good candidates for a unified description of gauge theories and quantum gravity. Therefore, for realistic model building including SUSY is a good way to proceed and SUSY GUTs provide an excellent example. However, supersymmetry introduces several problems and puzzles which require explanations. The puzzles can be divided in two categories: **1** puzzles which are common only to SUSY theories, i.e. those which appear due to SUSY extension (SUSY puzzles); and **2** The puzzles which exist also in non-SUSY theories and within SUSY extensions have the same status (non-SUSY puzzles). From puzzles **1**, **2** we list here only those which we will attempt to resolve.

1 SUSY puzzles

a) Problem of FCNC arises because with SUSY there is a new source for flavor violation. Apriori, without any specific arrangement there is no universality and an alignment (which would guarantee flavor conservation) in the soft SUSY breaking terms. This can create non-diagonal gaugino-fermion-sfermion interactions, leading to the FCNC [1].

b) $d = 5$ Baryon number violation;

With SUSY there is a new source for baryon number violation. Namely $d = 5$ operators [2]

$$\frac{\lambda}{M} qqql \ , \qquad \frac{\lambda'}{M} u^c u^c d^c e^c \ , \tag{1}$$

where M is some cutoff mass scale and λ, λ' are dimensionless couplings depending on the model. In GUTs, M usually stands for colored triplet masses. In minimal $SU(5)$ the nucleon decay mainly proceeds through the channel $p \to K\nu$ and its lifetime is estimated to be $10^{29\pm2}$ yr [3, 4], which is embarrassingly small in comparison with the latest SuperKamiokande (SK) limit $\tau_p^{\exp} \gtrsim 10^{33}$ yr [5]. Therefore, some mechanism (with reasonable extension) [4, 6, 7] must be applied to suppress these contributions. Once colored triplet induced $d = 5$ operators are properly suppressed, one has to take care of Planck scale $d = 5$ baryon number violating operators. In (1) with $M = M_{Pl} = 2.4 \cdot 10^{18}$ GeV, in order to satisfy the experimental bounds we need $\lambda, \lambda' \lesssim 10^{-8}$. This kind of suppression requires additional explanation [4, 8-13].

2 Non-SUSY puzzles

a) Problem of flavor - hierarchies between charged fermion masses and mixings;

In the charged fermion sector there are noticeable hierarchies within the fermion Yukawa couplings and the CKM matrix elements. Since the mass of the top quark is close to the electroweak symmetry breaking scale (~ 100 GeV), its Yukawa coupling is of order unity ($\lambda_t \sim 1$). As far as the Yukawa couplings of the b quark and τ lepton are concerned, their values could vary in a range $\lambda_b \sim \lambda_\tau \sim 10^{-2} - 1$, depending on the value of the MSSM parameter $\tan\beta$ ($\sim 1 - 60$). Introducing the dimensionless parameter $\epsilon \simeq 0.2$, one can express the observed hierarchies between the charged fermion Yukawa couplings as follows:

$$\lambda_t \sim 1 \ , \quad \lambda_u : \lambda_c : \lambda_t \sim \epsilon^6 : \epsilon^3 : 1 \ , \tag{2}$$

$$\lambda_b \sim \lambda_\tau \sim \lambda_t \frac{m_b}{m_t} \tan\beta \ , \quad \lambda_d : \lambda_s : \lambda_b \sim \epsilon^5 : \epsilon^2 : 1 \ , \tag{3}$$

$$\lambda_e : \lambda_\mu : \lambda_\tau \sim \epsilon^5 : \epsilon^2 : 1 \ , \tag{4}$$

while for the CKM matrix elements:

$$V_{us} \sim \epsilon \ , \quad V_{cb} \sim \epsilon^2 \ , \quad V_{ub} \sim \epsilon^3 \ . \tag{5}$$

In constructing models, one should arrange for a natural understanding of the hierarchies in (2)-(5).

b) Atmospheric and Solar Neutrino puzzles;

The latest atmospheric and solar neutrino data (see [14] and [15] respectively) seem to provide convincing evidence for the phenomena of neutrino oscillations. Ignoring the LSND data, the atmospheric and solar neutrino anomalies can be explained within the three states of active neutrinos. In this paper we will study oscillation scenarios without the sterile neutrinos, which are disfavored by the data [15,?].

The atmospheric neutrino data suggest oscillations of ν_μ into ν_τ, with the following oscillation parameters:

$$\mathcal{A}(\nu_\mu \to \nu_\tau) \equiv \sin^2 2\theta_{\mu\tau} \simeq 1 \ ,$$

$$\Delta m^2_{\rm atm} \simeq 3 \cdot 10^{-3}\ {\rm eV}^2 \ . \qquad (6)$$

The solar neutrino anomaly seems consistent with oscillation scenarios, amongst which the most likely seems to be the large angle MSW (LMA) oscillation of ν_e into $\nu_{\mu,\tau}$ [15], with the oscillation parameters:

$$\mathcal{A}(\nu_e \to \nu_{\mu,\tau}) \equiv \sin^2 2\theta_{e\mu,\tau} \approx 0.8 \ ,$$

$$\Delta m^2_{\rm sol} \sim 10^{-4}\ {\rm eV}^2 \ . \qquad (7)$$

The scenario of low MSW (LOW) oscillations of solar neutrinos require:

$$\sin^2 2\theta_{e\mu,\tau} \simeq 1.0 \ ,$$

$$\Delta m^2_{\rm sol} \simeq 8 \cdot 10^{-8}\ {\rm eV}^2 \ . \qquad (8)$$

Let us note that the small angle MSW and large angle vacuum oscillation solutions seem to be disfavored by data.

It is worth noting that within MSSM, the neutrinos acquire masses only through non-renormalizable $d = 5$ Planck scale operators $l_i l_j h_u^2/M_P$ which, for $\langle h_u^0 \rangle \sim 100$ GeV and $M_P = 2.4 \cdot 10^{18}$ GeV (reduced Planck mass) give $m_{\nu_i} \sim 10^{-5}$ eV. Therefore, for (6) or (7) we need physics beyond the MSSM. In order to generate the appropriate neutrino masses, we will introduce heavy right handed neutrino states $\mathcal{N}_i$. The 'light' left-handed neutrinos will acquire masses through the see-saw mechanism [16].

In building neutrino oscillation scenarios, the main challenge is to generate desirable magnitudes for neutrino masses and their mixings, and to understand why in some cases the mixing angles are large (and even maximal), while the quark CKM matrix elements (5) are suppressed. Below we will present a mechanism which successfully resolves all of these problems.

c) Wrong asymptotic relations originating from GUTs.

Within GUTs the quark-lepton families are embedded in unified multiplets and because of this, wrong asymptotic relations for masses and mixings are possible unless some care is exercised. In minimal $SU(5)$ chiral matter consists of anomaly free $10+\bar{5}$ supermultiplets per generation, where $10 = (q, u^c, e^c)$, $\bar{5} = (l, d^c)$. The couplings generating the up, down quark and charged lepton masses are respectively $10 \cdot 10 \cdot H + 10 \cdot \bar{5} \cdot \bar{H}$, ($H$, $\bar{H}$ are 'higgs' superfields in 5 and $\bar{5}$ representations). The second coupling gives $\hat{M}_f^0 = \hat{M}_d^0$ at GUT scale, which for the third generation yields the reasonable asymptotic relation $m_b^0 = m_\tau^0$, but for light generations it gives $\left(\frac{m_d}{m_s}\right)^0 = \left(\frac{m_e}{m_\mu}\right)^0$ which is unacceptable. For improving this picture, either a scalar 45 plet [17], or vector-like matter [18], [7], or some non-renormalizable operators [19] can be employed.

As we see, for solving the problems listed above, an extension of the minimal scheme is needed. Otherwise, in some cases, we should simply assume the presence of appropriately suppressed couplings and mass scales [for instance for **1**.b), **2**.a), b)]. The latter puzzle is tied to the so-called naturalness issue [20], namely why are some couplings and scales small, when apriori there is no reason to expect it? Below we discuss some extensions which provide natural mechanisms for resolving the above-listed problems.

2 Anomalous $\mathcal{U}(1)$ as a Flavor Symmetry and Mediator of SUSY Breaking

We introduce a $\mathcal{U}(1)$ gauge symmetry which acts as a flavor symmetry and provides for a natural understanding of the hierarchies between charged fermion masses and mixings. $\mathcal{U}(1)$ also turns out to be crucial for building textures of neutrino mass matrices that provide scenarios for simultaneous explanations of atmospheric and solar neutrino puzzles. It will turn out that $\mathcal{U}(1)$ is anomalous, which allows us to use it as a mediator of SUSY breaking [21], [22]. Thanks to this, the squarks and sleptons which correspond to the light generations can have masses of few $\cdot$ 10 TeV (this value can be acceptable also for $\tilde{b}^c$, $\tilde{\nu}_\tau$ and $\tilde{\tau}$ in the low $\tan\beta$ regime). Because of this, the FCNC can be adequately suppressed due to the decoupling [23]-[26], [22]. It turns out that together with this, some nucleon decay modes are strongly suppressed [26], [22]. Namely, diagrams with heavy squarks and/or sleptons inside the loops decouple. As far as the Planck scale $d = 5$ baryon number violating operators are concerned, they can also be adequately suppressed by the $\mathcal{U}(1)$ symmetry. Therefore, the advantages of $\mathcal{U}(1)$ symmetry are manyfold, and we will present an $SU(5)$ GUT to show how things work out.

The anomalous $\mathcal{U}(1)$ factors can arise from string theories [a]. Cancellation of the anomaly occurs through the Green-Schwarz mechanism

[a] Recently, ref. [13] presented an example where anomalous $\mathcal{U}(1)$ arises in 4D level through 5D orbifold compactification. The cancellation of anomalies occur through bulk Chern-Simons term.

[27]. Due to the anomaly the Fayet-Illiopoulos D-term $\xi\int d^4\theta V_A$ is always generated, where in string theory ξ is given by [28]

$$\xi = \frac{g_A^2 M_P^2}{192\pi^2}\mathrm{Tr}Q \ . \tag{9}$$

The D_A-term will have the form:

$$\frac{g_A^2}{8}D_A^2 = \frac{g_A^2}{8}\left(\Sigma Q_a|\varphi_a|^2 + \xi\right)^2 \ , \tag{10}$$

where Q_a is the 'anomalous' charge of φ_a superfield. For $\mathcal{U}(1)$ breaking we introduce the singlet superfield X with $\mathcal{U}(1)$ charge Q_X. Assuming $\xi > 0$ [$\mathrm{Tr}Q > 0$ in (9)], and taking

$$Q_X = -1 \ , \tag{11}$$

the cancellation of D_A in (10) and nonzero $\langle X\rangle$ are ensured ($\langle X\rangle = \sqrt{\xi}$). Further, we will take

$$\frac{\langle X\rangle}{M_P} \equiv \epsilon \simeq 0.2 \ , \tag{12}$$

where ϵ turns out to be an important expansion parameter. Let us note that an anomalous $\mathcal{U}(1)$ for understanding the hierarchies of fermion masses and mixings and a variety of neutrino oscillation scenarios has been discussed in several papers of [29-35, 7, 22, 18, 36, 11]

2.1 *Neutrino oscillations and quark-lepton masses*

Let us begin with the neutrino sector and we first discuss two ways of obtaining large/maximal neutrino mixings with the help of $\mathcal{U}(1)$ flavor symmetry. With two flavors of lepton doublets l_1 and l_2, one way of having large mixing is the so-called *democratic approach.* Here the $\mathcal{U}(1)$ symmetry does not distinguish the two flavors [29], i.e. they have the same $\mathcal{U}(1)$ charges $Q_{l_1} = Q_{l_2} = n$(positive integer number). In this case, the expected neutrino mass matrix will be:

$$\hat{m}_\nu = \begin{pmatrix} 1 & 1 \\ 1 & 1 \end{pmatrix}\bar{m}\epsilon^{2n} \ , \quad \bar{m} = \frac{h_u^2}{\bar{M}} \ , \tag{13}$$

with entries of order unity ($\bar{M}$ is some mass scale and we have assumed $Q_{h_u} = 0$). Therefore, naturally large $\nu_1 - \nu_2$ mixing is expected, $\sin^2 2\theta_{12} \sim 1$. Also, one can expect $m_{\nu_1} \sim m_{\nu_2}$, and if this mechanism is used for atmospheric neutrinos, somehow one has to keep one state light, in order to accomodate also the solar neutrino puzzle. This can be done [30,?] by introducing a single right handed neutrino $\mathcal{N}$. After integrating it out, due to degeneracy only one state acquires a non-zero mass. The remaining states can be used for the solar neutrino puzzle. An appropriate mass scale for the latter can be generated by introducing a relatively heavy right handed state $\mathcal{N}'$ with suppressed coupling to $\mathcal{N}$.

A different approach is the so-called *maximal mixing mechanism* [31], [36], [11]. It is realized by assigning different $\mathcal{U}(1)$ charges for the flavors l_1, l_2. Introducing two right handed states $\mathcal{N}_1, \mathcal{N}_2$ and the following $\mathcal{U}(1)$ charge prescriptions

$$Q_{l_1} = k + n \ , \quad Q_{l_2} = k \ , \quad Q_{h_u} = 0 \ ,$$

$$Q_{\mathcal{N}_1} = -Q_{\mathcal{N}_2} = k + k' \ , \tag{14}$$

with $k, n, k' > 0$, $n \geq k'$, the 'Dirac' and 'Majorana' coupling are given by:

$$\begin{array}{c} \\ l_1 \\ l_2 \end{array} \begin{array}{c} \begin{array}{cc} \mathcal{N}_1 & \mathcal{N}_2 \end{array} \\ \left(\begin{array}{cc} \epsilon^{2k+n+k'} & \epsilon^{n-k'} \\ \epsilon^{2k+k'} & 0 \end{array} \right) \end{array} h_u \ , \quad \begin{array}{c} \\ \mathcal{N}_1 \\ \mathcal{N}_2 \end{array} \begin{array}{c} \begin{array}{cc} \mathcal{N}_1 & \mathcal{N}_2 \end{array} \\ \left(\begin{array}{cc} \epsilon^{2(k+k')} & 1 \\ 1 & 0 \end{array} \right) \end{array} M_{\mathcal{N}} \ . \tag{15}$$

After integrating out the heavy $\mathcal{N}_1, \mathcal{N}_2$ states, the neutrino mass matrix is given by

$$\hat{m}_\nu = \left(\begin{array}{cc} \epsilon^n & 1 \\ 1 & 0 \end{array} \right) \bar{m} \ , \quad \bar{m} = \frac{h_u^2 \epsilon^{2k+n}}{M_{\mathcal{N}}} \ , \tag{16}$$

a quasi off-diagonal form, leading to a mixing angle

$$\sin^2 2\theta_{12} = 1 - \mathcal{O}(\epsilon^{2n}) \ , \tag{17}$$

which is close to maximal mixing. The form (16) is guaranteed by the appropriate zero entries in (15), which are ensured by $\mathcal{U}(1)$ symmetry. This mechanism turns out to be very convenient for achieving nearly maximal mixings between neutrino flavors within various realistic models, such as $SU(5)$ [18], $SO(10)$ [35], $SU(4)_c \times SU(2)_L \times SU(2)_R$ [11], etc.

Returning to our scheme, we attempt to obtain the bi-maximal texture through $\mathcal{U}(1)$ flavor symmetry. For this, we will combine the two mechanisms discussed above. Namely, the second and third lepton doublet states will have the same $\mathcal{U}(1)$ charges, which will lead to their large mixing. The state l_1 will have a suitable charge, one that ensures maximal $\nu_1 - \nu_2$ mixing.

Introducing two right handed $\mathcal{N}_{1,2}$ neutrino states and choosing $\mathcal{U}(1)$ charges as

$$Q_X = -1 \ , \ Q_{l_2} = Q_{l_3} = k \ , \ Q_{l_1} = k + n \ , \ Q_{h_u} = Q_{h_d} = 0 \ ,$$

$$Q_{\mathcal{N}_1} = -Q_{\mathcal{N}_2} = k + k' \ , \tag{18}$$

with

$$k, n, k' > 0 \ , \quad n \gtrsim k' \ , \tag{19}$$

the 'Dirac' and 'Majorana' couplings will have forms:

$$\begin{array}{c} \\ l_1 \\ l_2 \\ l_3 \end{array} \begin{array}{c} \begin{array}{cc} \mathcal{N}_1 & \mathcal{N}_2 \end{array} \\ \left(\begin{array}{cc} \epsilon^{2k+n+k'} & \epsilon^{n-k'} \\ \epsilon^{2k+k'} & 0 \\ \epsilon^{2k+k'} & 0 \end{array} \right) \end{array} h_u \ , \qquad \begin{array}{c} \\ \mathcal{N}_1 \\ \mathcal{N}_2 \end{array} \begin{array}{c} \begin{array}{cc} \mathcal{N}_1 & \mathcal{N}_2 \end{array} \\ \left(\begin{array}{cc} \epsilon^{2(k+k')} & 1 \\ 1 & 0 \end{array} \right) \end{array} M_{\mathcal{N}} \ . \tag{20}$$

After integrating out $\mathcal{N}_{1,2}$, we obtain the texture

$$\hat{M}_\nu \propto \left(\begin{array}{ccc} \epsilon^n & 1 & 1 \\ 1 & 0 & 0 \\ 1 & 0 & 0 \end{array} \right) m \ , \qquad m = \frac{\epsilon^{2k+n} h_u^2}{M_{\mathcal{N}}} \ , \tag{21}$$

In (21) coefficients of order unity are assumed. Without (1, 1) entry the (21) has $L_e - L_\mu - L_\tau$ symmetry, which also can be used [37] for obtaining bi-maximal texture. In our case deviation from (1, 1) zero entry is controlled by $\mathcal{U}(1)$ flavor symmetry [34]. The nonzero (1, 1) entry in (21) guarantees that $\Delta m_{12}^2 \neq 0$. Using (21) the oscillation parameters are:

$$\begin{gathered} \Delta m_{32}^2 \equiv m_{\rm atm}^2 = m^2 \sim 10^{-3} \ {\rm eV}^2 \ , \\ \mathcal{A}(\nu_\mu \to \nu_\tau) \sim 1 \ , \end{gathered} \tag{22}$$

$$\begin{gathered} \Delta m_{21}^2 \simeq 2 m_{\rm atm}^2 \epsilon^n \ , \\ \mathcal{A}(\nu_e \to \nu_{\mu,\tau}) = 1 - \mathcal{O}(\epsilon^{2n}) \ . \end{gathered} \tag{23}$$

Note that the model does not constrain n for the time being. So, LMA and LOW solutions for solar neutrinos can be realized. With prescription (18), the expected contribution from the charged lepton sector to the angles θ_{23}^l and θ_{12}^l will be ~ 1 and $\sim \epsilon^n$ respectively. These do not alter the expressions in (22), (23).

The $\mathcal{U}(1)$ charge selection in (18) nicely blends with the charged fermion sector. Indeed, considering the following prescription:

$$\begin{gathered} Q_{q_3} = 0 \ , \ \ Q_{q_2} = 2 \ , \ \ Q_{q_1} = 3 \ , \ \ Q_{d_3^c} = Q_{d_2^c} = p + k \ , \\ Q_{d_1^c} = p + k + 2 \ , \ \ Q_{u_3^c} = 0 \ , \ \ Q_{u_2^c} = 1 \ , \ \ Q_{u_1^c} = 3 \ , \\ Q_{e_3^c} = p \ , \ \ Q_{e_2^c} = p + 2 \ , \ \ Q_{e_1^c} = p + 5 - n \ , \end{gathered} \tag{24}$$

the structures of Yukawa matrices for up-down quarks and charged leptons are respectively:

$$\begin{array}{c} \\ q_1 \\ q_2 \\ q_3 \end{array} \begin{array}{c} \begin{array}{ccc} u_1^c & u_2^c & u_3^c \end{array} \\ \left(\begin{array}{ccc} \epsilon^6 & \epsilon^4 & \epsilon^3 \\ \epsilon^5 & \epsilon^3 & \epsilon^2 \\ \epsilon^3 & \epsilon & 1 \end{array} \right) \end{array} h_u \ , \tag{25}$$

$$\begin{array}{c} \\ q_1 \\ q_2 \\ q_3 \end{array} \begin{array}{c} \begin{array}{ccc} d_1^c & d_2^c & d_3^c \end{array} \\ \left(\begin{array}{ccc} \epsilon^5 & \epsilon^3 & \epsilon^3 \\ \epsilon^4 & \epsilon^2 & \epsilon^2 \\ \epsilon^2 & 1 & 1 \end{array} \right) \end{array} \epsilon^{p+k} h_d \ , \tag{26}$$

$$\begin{array}{c} \\ l_1 \\ l_2 \\ l_3 \end{array} \begin{array}{c} \begin{array}{ccc} e_1^c & e_2^c & e_3^c \end{array} \\ \left(\begin{array}{ccc} \epsilon^5 & \epsilon^{n+2} & \epsilon^n \\ \epsilon^{5-n} & \epsilon^2 & 1 \\ \epsilon^{5-n} & \epsilon^2 & 1 \end{array} \right) \end{array} \epsilon^{p+k} h_d \ . \tag{27}$$

Upon diagonalization of (25)-(27) it is easy to verify that the desired relations (2)-(5) for the Yukawa couplings and CKM matrix elements are realized. From (26), (27) we have

$$\tan\beta \sim \epsilon^{p+k} \frac{m_t}{m_b} \ . \tag{28}$$

As we previously mentioned, MSSM does not fix the values of n, k, p in (18), (24). Because of this, the solar neutrino oscillation scenario is not specified and both LMA and LOW are possible solutions.

2.2 D_A-term SUSY Breaking Suppression of FCNC and Nucleon Decay

The cancellation of D_A-term (10) was ensured by the VEV of X (at this stage we do not consider any superpotential couplings for X). Including a mass term for X in the superpotential,

$$W_m = \frac{m}{2} X^2 \ , \tag{29}$$

the cancellation of D_A will be partial, and SUSY will be broken due to non-zero F and D terms. Taking into account (10) and (29), we have

$$X^2 = \xi - \frac{4m^2}{g_A^2} \ , \quad \langle D_A \rangle = \frac{4m^2}{g_A^2} \ , \quad \langle F_X \rangle \simeq m\sqrt{\xi} \ . \tag{30}$$

From (10), taking into account (30), for the soft scalar masses squared (mass^2) we have

$$m_{\tilde{\varphi}_a}^2 = Q_a m^2 \ . \tag{31}$$

Thus, the scalar components of superfields which have non-zero $\mathcal{U}(1)$ charges gain masses through $\langle D_A \rangle$. We will assume that the scale m is in the range ~ 10 TeV. Those states which have zero $\mathcal{U}(1)$ charges will gain soft masses of the order of gravitino mass $m_{3/2}$ from the Kähler potential

$$m_{3/2} = \frac{F_X}{\sqrt{3}M_P} = m\frac{\epsilon}{\sqrt{3}} \ , \tag{32}$$

which, for $m = 10$ TeV, is relatively suppressed (~ 1 TeV). The gaugino masses also will have the same magnitudes

$$M_{\tilde{G}_i} \sim m_{3/2} \sim 1 \text{ TeV} \ . \tag{33}$$

The mass term (29) violates the $\mathcal{U}(1)$ symmetry and is taken to be in the 10 TeV range. Its origin may lie in a strong dynamics where m is replaced by the VEV of some superfield(s). We do not present here any examples of this and refer the reader to [21], [22] for detailed discussions. The important point is that the soft masses (31) of sparticles are controlled by the $\mathcal{U}(1)$ symmetry.

Turning now to the question of FCNC, we require that the 'light' quark-lepton superfields carry non-zero $\mathcal{U}(1)$ charges. This means that the soft masses of their scalar components are in the 10 TeV range, which automatically suppresses flavor changing processes such as $K^0 - \bar{K}^0$, $\mu \to e\gamma$ etc., thereby satisfying the present experimental bounds [1]. To prevent upsetting the gauge hierarchy, the third generation up squarks must have masses no larger than a TeV or so [24] (hence they have zero $\mathcal{U}(1)$ charge). The same applies to sbottom and stau for large $\tan\beta$ since, for $\lambda_b \sim \lambda_\tau \sim 1$, large masses ($\gtrsim 10$ TeV) of $\tilde{b}$ and $\tilde{\tau}$ would spoil the gauge hierarchy.

Although the tree level mass of the stop can be arranged to be in the 1 TeV range by the $\mathcal{U}(1)$ symmetry, the 2-loop contributions from heavy sparticles of the first two generations can drive the stop mass^2 negative [24]. This is clearly unacceptable, and one proposal for avoiding it [25] requires the existence of new states in the multi-TeV range. This type of extension makes the decoupled solution viable and realistic models can be built [22].

Let us now turn to some implications for proton decay. We assume that $d = 5$ baryon number violating operators arise from the couplings

$$qAqT + qBl\bar{T} \ , \tag{34}$$

after integration of color triplets $T, \bar{T}$ with mass $M_T \sim 2 \cdot 10^{16}$ GeV (we first consider triplet couplings with left-handed matter). After wino dressing of appropriate $d = 5$ operators, the resulting $d = 6$ operators causing proton to decay into a neutrino and charged lepton channels have the respective forms [3], [4]:

$$\frac{g_2^2}{M_T}\alpha(u_a d_b^i)(d_c^j \nu^k)\varepsilon^{abc} \ , \tag{35}$$

$$\frac{g_2^2}{M_T}\alpha'(u_a d_b^i)(u_c e^j)\varepsilon^{abc} , \tag{36}$$

where

$$\alpha = -[(L_d^+\hat{B}L_e)_{jk}(L_u^+\hat{A}L_d^*)_{mn}+(L_d^+\hat{A}L_u^*)_{jm}(L_d^+\hat{B}L_e)_{nk}]V_{mi}(V^+)_{n1}I(\tilde{u}^m,\tilde{d}^n)+$$

$$[(L_u^+\hat{A}L_d^*)_{1i}(L_u^+\hat{B}L_e)_{mk}-(L_d^+\hat{A}L_u^*)_{im}(L_u^+\hat{B}L_e)_{ik}]V_{mj}I(\tilde{u^m},\tilde{e^k}) , \tag{37}$$

$$\alpha' = [-(L_u^+\hat{A}L_d^*)_{1i}(L_d^+\hat{B}L_e)_{mj}+(L_u^+\hat{A}L_d^*)_{1m}(L_d^+\hat{B}L_e)_{ij}](V^+)_{m1}I(\tilde{d^m},\tilde{\nu^j})+$$
$$[(L_u^+\hat{B}L_e)_{1j}(L_u^+\hat{A}L_d^*)_{mn}+(L_u^+\hat{A}L_d^*)_{1m}(L_e^T\hat{B}^TL_u^*)_{jn}](V^+)_{m1}V_{ni}I(\tilde{d^m},\tilde{u}^n) . \tag{38}$$

L's are unitary matrices which rotate the left handed fermion states to diagonalize the mass matrices, and I's are functions obtained after loop integration and depend on the SUSY particle masses circulating inside the loop. For example [3],

$$I(\tilde{u},\tilde{d}) = \frac{1}{16\pi^2}\frac{m_{\tilde{W}}}{m_{\tilde{u}}^2-m_{\tilde{d}}^2}\left(\frac{m_{\tilde{u}}^2}{m_{\tilde{u}}^2-m_{\tilde{W}}^2}\ln\frac{m_{\tilde{u}}^2}{m_{\tilde{W}}^2}-\frac{m_{\tilde{d}}^2}{m_{\tilde{d}}^2-m_{\tilde{W}}^2}\ln\frac{m_{\tilde{d}}^2}{m_{\tilde{W}}^2}\right) , \tag{39}$$

with similar expressions for $I(\tilde{d},\tilde{\nu})$ and $I(\tilde{u},\tilde{e})$.

Consider those diagrams in which sparticles of the first two families participate. Since their masses are large ($\gtrsim$ 10 TeV) compared to the case with minimal $N=1$ SUGRA, we expect considerable suppression of proton decay. For minimal $N=1$ SUGRA, $m_{\tilde{u}}\sim m_{\tilde{d}}\sim m_{\tilde{W}}\sim m_{3/2}\sim$ 1 TeV, and (39) can be approximated by

$$I_0 \approx \frac{1}{16\pi^2}\frac{1}{m_{3/2}} . \tag{40}$$

In the $\mathcal{U}(1)$ mediated SUSY breaking scenario, expression (39) takes the form

$$I' \approx \frac{1}{16\pi^2}\frac{m_{\tilde{W}}}{m_{\tilde{q}}^2} \equiv \eta I_0 \tag{41}$$

The nucleon lifetime in this case will be enhanced by the factor $\frac{1}{\eta^2}\sim 10^4$.

Of course, there exist diagrams in which one sparticle from the third and one from the 'light' families participate. In this case, (39) takes the form

$$I'' \approx \frac{1}{16\pi^2}\frac{2m_{\tilde{W}}}{m_{\tilde{q}}^2}\ln\frac{m_{\tilde{q}}}{m_{\tilde{W}}} \equiv \eta' I_0 \tag{42}$$

and the corresponding proton lifetime will be $\sim\frac{1}{\eta'^2}\sim 500$ times larger. This suppression is enough to bring the proton lifetime near the experimental limit, providing an opportunity for testing this type of scenario in the near future.

As pointed out in [3], [4] (within minimal $N=1$ SUGRA), the contribution from diagrams in which sparticles from the third generation participate could be comparable with those arising from the light generation sparticle exchange. Namely, second term of (37) with $\tilde{t}$, $\tilde{\tau}$ inside the loop give a contribution comparable to diagrams with first two generation sparticles circulating inside the loop. Since minimal SUSY $SU(5)$ gives unacceptably fast proton decay with $\tau_0 \sim 10^{29\pm2}$ yr, care must be exercised in realistic model building (the situation is exacerbated if $\tan\beta$ is large). This problem is easily avoided in the anomalous $\mathcal{U}(1)$ mediated SUSY breaking scenario. Note that in second term of (37) the I depends on mass of $\tilde{e}^k$ state and even if the latter belongs to the third family, it can have mass in the 10 TeV range if $\tan\beta$ is either of intermediate ($\sim 10-15$) or low value (this is required for preserving the desired gauge hierarchy). The contributions coming from first term of (37) could be adequately suppressed due to CKM matrix elements [note that first term in (37) contains extra multiplier $(V^+)_{n1}$]. In sect. 3, for $SU(5)$ GUT example we will precisely show this. Due to same reasons the contributions from terms of (38) are suppressed. Since these terms would induce $p \to K\mu$ decays, the additional inhensiment factor (of the order of 10) in proton lifetime come from the hadronic matrix element, which correspond to proton decay with emission of charged lepton.

As far as the right handed $d=5$ operators $u^{c\,i}u^{c\,j}d^{cm}e^{cn}$ are concerned, the dominant contribution from them comes through higgsino dressings. Due to antisymmetry in respect of color in this $d=5$ operator, u^c states should be taken from the different generations. So, either c^c or t^c will appear. Thay must not be appeared in an external line after dressing, otherwise the diagram would not be relevant for nucleon decay. Since first two generation sparticles and also all $\tilde{d^c}^i$ states are in 10 TeV range, relevant diagram would be that one which inside the loop contain $\tilde{t}^c$ and $\tilde{\tau}^c$ states (the latter is neccessarily light within $SU(5)$ GUT, because it comes from 10_3 plet). Due to suppressed mixings of third generation states with first and second generations and also due to small Yuakawas (in small $\tan\beta$ regime) appearing due to higgsino dressing, the suppression of nucleon decay can be guaranteed also for this case [this is shown in sect. 3 on an $SU(5)$ example].

In general, within this scenario universality does not hold and nucleon decays through gluino dressings would occur. However, heavy sparticles will still play a crucial role in suppression of nucleon decay. Which contribution is dominant depends on the details of the scenario and in sect. 3 we study this issue within a realistic $SU(5)$ model.

Thus, thanks to the anomalous $\mathcal{U}(1)$ symmetry, in addition to avoiding dangerous FCNC, one can also obtain adequate suppression of nucleon decay. Interestingly, this disfavors the large $\tan\beta$ regime which could be a characteristic feature of this class of models.

2.3 Possible neutrino oscillation scenarios

As we previously mentioned, MSSM does not fix the values of n, k, p in (18), (24). Because of this, the solar neutrino oscillation scenario is not specified. According to (23) both LMA and LOW are possible solutions. Namely, for $n = 3$ we have $\Delta m_{12}^2 \sim 10^{-5}$ eV^2, which corresponds to LMA. $n = 6$ gives $\Delta m_{12}^2 \sim 10^{-7}$ eV^2, which is the scale for the LOW solution

In SUSY $SU(5)$ GUT, due to unified 10, $\bar{5}$ multiplets:

$$Q_q = Q_{e^c} = Q_{u^c} = Q_{10} \ , \qquad Q_l = Q_{d^c} = Q_{\bar{5}} \ . \tag{43}$$

Hierarchies of the CKM matrix elements in (5) dictate the relative $\mathcal{U}(1)$ charges of the 10-plets

$$Q_{10_3} = 0 \ , \quad Q_{10_2} = 2 \ , \quad Q_{10_1} = 3 \ , \tag{44}$$

while the Yukawa hierarchies (2)-(4), together with (44), require that

$$Q_{\bar{5}_3} = Q_{\bar{5}_2} = k \ , \quad Q_{\bar{5}_1} = k + 2 \ . \tag{45}$$

Comparing (43)-(45) with (18), (24) we see that the minimal $SU(5)$ GUT fixes n and p as

$$n = 2 \ , \qquad p = 0 \ . \tag{46}$$

The mass squared splitting in (23) then equals $\Delta m_{12}^2 \sim 10^{-4}$eV^2, which is a reasonable scale for LMA scenario. Therefore, realisation of our bi-maximal mixing scenario in the framework of $SU(5)$ GUT dictates that the LMA scenario is responsible for the solar neutrino deficit (more detailed discussion of $SU(5)$ GUT will be presented in the next section). The same conclusion can be reached for $SO(10)$ GUT where we have three 16-plets of chiral supermultiplets which unify the quark-lepton superfields. We do not present the details here but refer the reader to [35], where an explicit $SO(10)$ model with anomalous $\mathcal{U}(1)$ flavor symmetry is considered for explanations of fermion masses, their mixings, as well as neutrino anomalies.

For models in which an anomalous flavor $\mathcal{U}(1)$ also provides SUSY breaking soft masses, from (31) we should require $Q_{\tilde{q},\tilde{l}} > 0$. On the other hand, from (23), the realization of LMA and LOW respectively require $n = 3$ and $n = 6$ (if $p \gtrsim 1$), both of which guarantee $Q_{e^c} \gtrsim 0$. Therefore, a scenario in which an anomalous flavor $\mathcal{U}(1)$ mediates SUSY breaking permits LMA and LOW oscillations for solar neutrinos (the large angle vacuum oscillation would not be realized within this scenario, since it requires $n = 10$ giving negative Q_{e^c}).

3 An $SU(5)$ Example

Let us now consider in detail a SUSY $SU(5)$ GUT and show how things discussed in the previous section work out in practice. We also present here the possibility for avoiding the problematic asymptotic mass relations $\hat{M}_d^0 = \hat{M}_e^0$ (for first and second families).

The three families of matter in $(10 + \bar{5})$ representations have the transformation properties as given in (44), (45), while the scalar superfields $\bar{H}(\bar{5}) + H(5)$ have $Q_{\bar{H}} = Q_H = 0$. The couplings relevant for the generation of up, down quark and charged lepton masses respectively are given by

$$\begin{array}{c} \\ 10_1 \\ 10_2 \\ 10_3 \end{array} \begin{array}{c} \begin{array}{ccc} 10_1 & 10_2 & 10_3 \end{array} \\ \left(\begin{array}{ccc} \epsilon^6 & \epsilon^5 & \epsilon^3 \\ \epsilon^5 & \epsilon^4 & \epsilon^2 \\ \epsilon^3 & \epsilon^2 & 1 \end{array} \right) \end{array} H , \tag{47}$$

$$\begin{array}{c} \\ 10_1 \\ 10_2 \\ 10_3 \end{array} \begin{array}{c} \begin{array}{ccc} \bar{5}_1 & \bar{5}_2 & \bar{5}_3 \end{array} \\ \left(\begin{array}{ccc} \epsilon^5 & \epsilon^3 & \epsilon^3 \\ \epsilon^4 & \epsilon^2 & \epsilon^2 \\ \epsilon^2 & 1 & 1 \end{array} \right) \end{array} \epsilon^k \bar{H} . \tag{48}$$

Upon diagonalization of (47), (48) we obtain the desirable hierarchies (2)-(5) and $\lambda_b \sim \lambda_\tau \sim \epsilon^k$.

The reader will note, however, that (48) implies the asymptotic mass relations $\hat{M}_d^0 = \hat{M}_e^0$, which are unacceptable for the two light families. This is readily avoided through the mechanism suggested in [18], [7] by employing two pairs of $(\overline{15} + 15)_{1,2}$ matter states. Namely, with $\mathcal{U}(1)$ charges

$$Q_{15_1} = -Q_{\overline{15}_1} = 3 \,, \qquad Q_{15_2} = -Q_{\overline{15}_2} = 2 \,, \tag{49}$$

consider the couplings

$$\begin{array}{c} \\ \overline{15}_1 \\ \overline{15}_2 \end{array} \begin{array}{c} \begin{array}{ccc} 10_1 & 10_2 & 10_3 \end{array} \\ \left(\begin{array}{ccc} 1 & 0 & 0 \\ \epsilon & 1 & 0 \end{array} \right) \end{array} \Sigma \,, \quad \begin{array}{c} \\ \overline{15}_1 \\ \overline{15}_2 \end{array} \begin{array}{c} \begin{array}{cc} 15_1 & 15_2 \end{array} \\ \left(\begin{array}{cc} 1 & 0 \\ \epsilon & 1 \end{array} \right) \end{array} M_{15} \,, \tag{50}$$

where Σ is the scalar 24-plet whose VEV breaks $SU(5)$ down to $SU(3)_c \times SU(2)_L \times U(1)_Y$. For $M_{15} \sim \langle\Sigma\rangle$, we see that the 'light' $q_{1,2}$ states reside both in $10_{1,2}$ and $15_{1,2}$ states with similar 'weights'. At the same time, the other light states from 10-plets (u^c and e^c) will not be affected because the 15-plets do not contain fragments with the relevant quantum numbers. Thus, the relations $m_s^0 = m_\mu^0$ and $m_d^0 = m_e^0$ are avoided, while $m_b^0 = m_\tau^0$ still holds since the terms in (50) do not affect 10_3.

As far as the sparticle spectrum is concerned, since the superfields $10_3, \bar{H}, H$ have zero $\mathcal{U}(1)$ charges, the soft masses of their scalar components will be in the 1 TeV range,

$$m_{\tilde{10}_3} \sim m_{\bar{H}} \sim m_H \sim m_{3/2} = 1 \text{ TeV} , \qquad (51)$$

while for $10_{1,2}$ and $\bar{5}_1$ we have

$$m_{\tilde{10}_1} \sim m_{\tilde{10}_2} \sim m_{\tilde{\bar{5}}_1} \sim m \sim 10 \text{ TeV} . \qquad (52)$$

The soft masses of the scalar fragments from $\bar{5}_{2,3}$ depend on the value of k, and for $k \neq 0$, they also will be in the 10 TeV range, which is preferred for proton stability. As far as neutrino oscillations are concerned, as already pointed out in sect. 1.3, due to $SU(5)$ and $\mathcal{U}(1)$ charge prescriptions, the LMA solution is preferred with the texture in (21).

3.1 Nucleon Decay in $SU(5)$

Turning to the issue of nucleon decay in $SU(5)$, we will take $k \neq 0$ in (45), which provides soft masses for $\bar{5}_{2,3}$ states in the 10 TeV range. Let us first make sure that this ensures proton stability. As pointed out in sect. 2, diagrams with slepton inside the loop [2nd term of (37) and 1st term of (38)] are appropriately suppressed.

The diagrams corresponding to the first terms in (37) induce nucleon decay with the dominant mode $p \to K\nu_{\mu,\tau}$ with $\tilde{t}$ and $\tilde{b}$ running in the loop. For our scenario, taking into account the couplings (47), (48), (50), the dominant contribution accompanying I (from first term in (37)) is of order order of $\epsilon^{k+2}\lambda_t V_{td} V_{ub} \sim \lambda_s \lambda_t V_{td} V_{ub}$. The corresponding partial lifetime in units of $\tau_0 \equiv \tau$[min. SUSY $SU(5)$] will be $\tau(p \to K\nu_{\mu,\tau}) = \tau_0 \left(\frac{\lambda_c \sin^2\theta_c}{\lambda_t V_{td} V_{ub}} \right)^2$. For the CKM matrix elements we have

$$V_{td} = 0.004 - 0.014 , \qquad V_{ub} = 0.0025 - 0.0048 , \qquad (53)$$

and for their central values we have $\tau(p \to K\nu_{\mu,\tau}) = 217\tau_0$, which presumably is still compatible with the available limits (of course, for lower values in (53) we can have much longer lived nucleon $\tau(p \to K\nu_{\mu,\tau}) \sim 2 \cdot 10^3 \tau_0$).

The second term in (38) induces the decay $p \to K\mu$. It can be verified that the suppression discussed above occurs in this case too. Furthermore, additional inhancement of the partial lifetime by a factor of order 10 occurs from the hadronic matrix element corresponding to proton decay with emission of charged lepton.

As far as the right handed $u^c u^c d^c e^c$ type $d = 5$ operators are concerned, due to arguments presented at the end of sect. 2, the relevant $d = 5$ operator will be $\tilde{\tau}^c \tilde{t}^c u^c s^c$ which in our $SU(5)$ model, according to (47), (48), (50) will appear as $\frac{1}{M_T}\lambda_t\lambda_b\epsilon^3 \cdot \tilde{\tau}^c\tilde{t}^c u^c s^c$. After higgsino dressing relevant $d = 6$ operator will have the form

$$\frac{\kappa}{M_T}\lambda_t^2\lambda_b^2\epsilon^3 V_{td} \cdot (\overline{u^c}\ \overline{s^c})(d\nu_{\mu,\tau}) \ . \tag{54}$$

κ is factor which coincides with those appearing in minimal SUSY $SU(5)$ (we assume that $M_{\tilde{W}} \simeq \mu$-term). Operator (54) leads to $p \to K\nu_{\mu,\tau}$ and corresponding partial lifetime is $\tau(p \to \nu_{\mu,\tau}) = \left(\frac{g_2^2\lambda_s\lambda_c \sin^2\theta_c}{\lambda_b^2\lambda_t^2\epsilon^3 V_{td}}\right)^2 \tau_0$, which for $\lambda_b \sim 10^{-2}$ ($\tan\beta \sim 1$) and $V_{td} = 0.004$ gives $\tau(p \to K\nu_{\mu,\tau}) \sim 400\tau_0$, compatible with experimental data.

In considered $SU(5)$ GUT due to different $\mathcal{U}(1)$ charges the universality of soft mass^2 is violated. Because of this nucleon decay through gluino dressings will not be canceled (as is the case in minimal $N = 1$ SUGRA). So, these diagrams are important since the proton decay amplitude will be increased by the factor α_3/α_2. For our $SU(5)$ scenario the dominant contributions come from diagrams inside which run the third generation sparticles (namely $\tilde{t}$, $\tilde{b}$ from q_3). The relevant $d = 5$ operators, obtained after integration of colored triplet higgs fields, will be

$$\frac{1}{M_T}\lambda_t\lambda_b\epsilon^3 \cdot \tilde{t}\tilde{b}d\nu_{\mu,\tau} \ , \quad \frac{1}{M_T}\lambda_t\lambda_b\epsilon^2 \cdot \tilde{t}\tilde{s}d\nu_{\mu,\tau} \ , \tag{55}$$

$$\frac{1}{M_T}\lambda_t\lambda_b\epsilon \cdot \tilde{t}\tilde{b}u\mu \ , \tag{56}$$

which, after gluino dressings, lead to the following four-fermion operators:

$$\frac{\kappa}{M_T}\alpha_3\lambda_t\lambda_b\epsilon^2 L_{13}^u \left[\epsilon L_{23}^d \cdot (d\nu_{\mu,\tau})(us) + L_{13}^d \cdot (s\nu_{\mu,\tau})(ud)\right] \ , \tag{57}$$

$$\frac{\kappa}{M_T}\alpha_3\lambda_t\lambda_\mu\epsilon L_{13}^u L_{23}^d \cdot (u\mu)(us) \ . \tag{58}$$

$L^{u,d}$ are unitary matrices that rotate the left-handed up and down quark flavor states. Operators (57) and (58) respectively lead to the decays $p \to K\nu_{\mu,\tau}$, $p \to K\mu$, and the corresponding lifetimes are

$$\tau(p \to K\nu_{\mu,\tau}) = \left[\frac{\alpha_2}{\alpha_3}\frac{\lambda_s\lambda_c\sin^2\theta_c}{\lambda_b\lambda_t\epsilon^3 L_{13}^u L_{23}^d}\right]^2 \tau_0 \ , \tag{59}$$

$$\tau(p \to K\mu) = 10\left[\frac{\alpha_2}{\alpha_3}\frac{\lambda_s\lambda_c\sin^2\theta_c}{\lambda_\mu\lambda_t\epsilon L_{13}^u L_{23}^d}\right]^2 \tau_0 \ . \tag{60}$$

For V_{cb} we have

$$V_{cb} = 0.036 - 0.046 \ , \tag{61}$$

which together with (53) can dictate that the values of $L^{u,d}$ can vary,

$$L_{12}^u \ , \ L_{13}^d = (1.8 - 3.5)\cdot 10^{-3} \ , \quad L_{23}^d = (2 - 4)\cdot 10^{-2} \ . \tag{62}$$

Taking the lowest values from (62), the nucleon lifetimes are

$$\tau(p \to K\nu_{\mu,\tau}) \sim 177\tau_0 \ , \qquad \tau(p \to K\mu) \sim 2 \cdot 10^3\tau_0 \ , \tag{63}$$

which are still compatible with the experimental bounds[b] , with the dominant decay mode being $p \to K\nu_{\mu,\tau}$.

Before concluding, let us note that the Planck scale suppressed baryon number violating $d = 5$ operator $\frac{1}{M_P} q_1 q_1 q_2 l_{2,3}$, which could cause unacceptably fast proton decay, is also suppressed, since it emerges from the coupling

$$\frac{1}{M_P}\left(\frac{X}{M_P}\right)^{8+k} 10_1 10_1 10_2 \bar{5}_{2,3} \ , \tag{64}$$

with the suppression guaranteed by the $\mathcal{U}(1)$ symmetry.

4 Conclusions

It is quite remarkable that the introduction of an anomalous $\mathcal{U}(1)$ symmetry within a supersymmetric setting can have several far reaching consequences. In particular, the flavor and dimension five proton decay problems encountered in SUSY models can be overcome. The atmospheric and solar neutrino puzzles can be nicely explained via neutrino oscillations. Although we have emphasized the $SU(5)$ model here, the discussion can be extended to other realistic models such as those based on $SO(10)$.

Acknowledgements

We are very grateful to the organizers of the NOON2001 workshop, especially Professor M. Bando, for creating a very stimulating environment at the conference and for their generous hospitality. We also thank the organizers of the post-NOON workshop in Kyoto.

1. F. Gabbiani, E. Gabrielli, A. Masiero, L. Silvestrini, Nucl.Phys. B477 (1996) 321; G. Eyal, A. Masiero, Y. Nir, L. Silvestrini, hep-ph/9908382.
2. S. Weinberg, Phys. Rev. D 26 (1982) 287; N. Sakai, T. Yanagida, Nucl. Phys. B 197 (1982) 533.
3. P. Nath, R. Arnowitt, Phys. Rev. D 38 (1988) 1479; J. Hisano, H. Murayama, T. Yanagida, Nucl. Phys. B 402 (1993) 46.
4. V. Ben-Hamo, Y. Nir, Phys. Lett. B 339 (1994) 77; L. Hall, H. Murayama, Phys. Rev. Lett. 75 (1995) 3985; Z. Berezhiani, hep-ph/9602325; C. Carone, et al., Phys. Rev. D 54 (1996) 2328; Z. Berezhiani, Z. Tavartkiladze, M. Vysotsky, hep-ph/9809301.

[b] The lifetimes can be further increased if say L^d_{23} is more suppressed and in the appropriate entry of CKM matrix the main contribution comes from L^u.

5. Particle Data Group, Phys. Rev. D 66 (2002) 010001.
6. G. Dvali, Phys. Lett. B 287 (1992) 101; K. Babu, S. Barr, Phys. Rev. D 48 (1993) 5354; J. Hisano, et al., Phys. Lett. B 342 (1995) 138; J. Pati, Phys. Lett. B 288 (1996) 532; I. Gogoladze, A. Kobakhidze, hep-ph/9610389; Y. Achiman, C. Merten, hep-ph/9911314; Z. Chacko, R. Mohapatra, Phys. Rev. D 59 (1999) 011702; Q. Shafi, Z. Tavartkiladze, Nucl. Phys. B 573 (2000) 40.
7. Q. Shafi, Z. Tavartkiladze, Phys. Lett. B 459 (1999) 563.
8. L. Ibanez, G. Ross, Nucl. Phys. B 368 (1992) 3.
9. I. Antoniadis, hep-th/0102202.
10. I. Antoniadis, et al., Phys. Lett. B 194 (1987) 231; G. Lazarides, C. Panagiotakopoulos, Q. Shafi, Phys. Lett. B 315 (1993) 325; G. Dvali, Q. Shafi, Phys. Lett. B 403 (1997) 65; Q. Shafi, Z. Tavartkiladze, Nucl. Phys. B 552 (1999) 67.
11. Q. Shafi, Z. Tavartkiladze, Nucl. Phys. B 549 (1999) 3.
12. M. Kakizaki, M. Yamaguchi, hep-ph/0203192.
13. C.A. Lee, Q. Shafi, Z. Tavartkiladze, hep-ph/0206258.
14. S. Fukuda et al. [Super-Kamiokande Collaboration], Phys. Rev. Lett. 88 (2000) 3999; N. Fornengo et al., Nucl. Phys. B 580 (2000) 58.
15. S. Fukuda et al. [Super-Kamiokande Collaboration], Phys. Lett. B 539 (2002) 179; J. Bahcall, P. Krastev, A. Smirnov, hep-ph/0006078; M.C. Gonzalez-Garcia et al., Nucl. Phys. B 573 (2000) 3.
16. M. Gell-Mann, P. Ramond and R. Slansky, in: Supergravity, eds. P. van Nieuwenhuizen and D. Freedman (North-Holland, Amsterdam, 1979) p. 315; T. Yanagida, Prog. Th. Phys. B 135 (1979) 66; R. Mohapatra and G. Senjanovic, Phys. Rev. Lett. 44 (1980) 912.
17. H. Gorgi, C. Jarlskog, Phys. Lett. B 88 (1979) 279; J. Harvey, P. Ramond, D. Reiss, Phys. Lett. B 92 (1980) 309; S. Dimopoulos, L.J. Hall, S. Raby, Phys. Rev. Lett. 68 (1992) 1984, Phys. Rev. D 45 (1992) 4192; V. Barger et. al., Phys. Rev. Lett. 68 (1992) 3394; H. Arason et. al., Phys. Rev. D 47 (1993) 232.
18. Q. Shafi, Z. Tavartkiladze, Phys. Lett. B 451 (1999) 129.
19. J. Ellis, M. Gaillard, Phys. Lett. B 88 (1979) 315; R. Barbieri, D. Nanopoulos, Phys. Lett. B 95 (1980) 43; Z. Berezhiano, hep-ph/9602325; G. Altarelli, F. Feruglio, hep-ph/9812475; see also references therein; Q. Shafi, Z. Tavartkiladze, Phys. Lett. B 522 (2001) 102.
20. S. Weinberg, Phys. Rev. D 13 (1976) 974, Phys. Rev. D 19 (1979) 1277; L. Susskind, Phys. Rev. D 20 (1979) 2619; G. 't Hooft, Proceedings of the NATO Advanced Summer Institute, Cargese, 1979, edited by G. 't Hooft et. al (Plenum, New York, 1980) p. 135.
21. G. Dvali and A. Pomarol, Phys. Rev. Lett. 77 (1996) 3738; P.

Binetruy and E. Dudas, Phys. Lett. B 389 (1996) 503.
22. Q. Shafi, Z. Tavartkiladze, Phys. Lett. B 473 (2000) 272.
23. M. Dine, A. Kagan, S. Samuel, Phys. Lett. B 243 (1990) 250; L. Everett et al., Phys. Lett. B 477 (2000) 233; J. Feng, K. Matchev, hep-ph/0011356; J. Hisano, K. Kurosawa, Y. Nomura, hep-ph/0002286; See also references therein.
24. S. Dimopoulos, G. Giudice, Phys. Lett. B 357 (1995) 537; N. Arkani-Hamed, H. Murayama, hep-ph/9703259.
25. J. Hisano, K. Kurosawa, Y. Nomura, Phys. Lett. B 445 (1999) 316.
26. A. Nelson, D. Wright, Phys.Rev. D 56 (1997) 1598.
27. M. Green and J. Schwarz, Phys. Lett. B 149 (1984) 117.
28. M. Dine, N. Seiberg and E. Witten, Nucl. Phys. B 289 (1987) 584; J. Atick, L. Dixon and A. Sen, Nucl. Phys. B 292 (1987) 109; M. Dine, I. Ichinose and N. Seiberg, Nucl. Phys. B 293 (1987) 253.
29. P. Ramond, hep-ph/9808489, and references therein; J. Sato and T. Yanagida, hep-ph/9809307; See also F. Vissani, hep-ph/9810435.
30. D. Suematsu, Phys. Lett. B 392 (1997) 413; S. Davidson, S.F. King, hep-ph/9808296.
31. P. Binetruy et al., Nucl. Phys. B 496 (1997) 3; V. Barger, T.J. Weiler and K. Whisnant, Phys. Lett. B 427 (1998) 97; Y. Grossman, Y. Nir and Y. Shadmi, hep-ph/9808355; Q. Shafi and Z. Tavartkiladze, Phys. Lett. B 448 (1999) 46, hep-ph/9811463. In the context of realistic $SU(4)_c \times SU(2)_L \times SU(2)_R$ model with maximal $\nu_\mu - \nu_\tau$ mixing, it was also discussed in ref. [11].
32. P. Binetruy , S. Lavignac, S. Petcov, Nucl. Phys. B 496 (1997) 3; Z. Berezhiani, Z. Tavartkiladze, Phys. Lett. B 396 (1997) 150; Phys. Lett. B 409 (1997) 220; Y. Grossman, Y. Nir, Y. Shadmi , JHEP 9810 (1998) 007; M. Fukugita, et al., hep-ph/9809554; M. Gomez et al., Phys. Rev. D 59 (1999) 116009; C. Froggat, M. Gibson, H. Nielsen, hep-ph/9811265; S. Kang, C. Kim, hep-ph/9811379; J. Feng, Y. Nir, hep-ph/9911370; J.M. Mira et al, Phys. Lett. B 492 (2000) 81; G. Altarelli et al., hep-ph/0007254; R. Mohapatra, hep-ph/0008232; M. Berger, K. Siyeon, hep-ph/0010245; I. Gogoladze, A. Perez-Lorenzana, hep-ph/0112034; see also references therein.
33. J. Elwood, N. Irges, P. Ramond, hep-ph/9807228; F. Vissani, hep-ph/9810435; Z. Berezhiani, A. Rossi, hep-ph/9811447; R. Barbieri, L. Hall, G. Kane, G. Ross, hep-ph/9901228; S. Barr, I. Dorsner, hep-ph/0003058; J. Chkareuli, C. Froggatt, H. Nielsen, hep-ph/0109156; T. Kitabayashi, M. Yasue, hep-ph/0112287; F. Feruglio, A. Strumia, F. Vissani, hep-ph/0201291; H.J. He, D.A. Dicus, J.N. Ng, hep-ph/0203237; N.N. Singh, M. Patgiri, hep-ph/0204021; S.F. King, hep-ph/0204360; see also references therein.

34. Q. Shafi, Z. Tavartkiladze, Phys. Lett. B 482 (2000) 145, hep-ph/0101350.
35. Q. Shafi, Z. Tavartkiladze, Phys. Lett. B 487 (2000) 145.
36. Q. Shafi, Z. Tavartkiladze, Phys. Lett. B 448 (1999) 46.
37. R. Barbieri at el., Hep-ph/9807235; A.S. Joshipura, S.D. Rindani, hep-ph/9811252; R.N. Mohapatra, A. Perez-Lorenzana, C.A. de S. Pires, hep-ph/9911395; T. Kitabayashi, M. Yasue, hep-ph/0006014; K.S. Babu, R.N. Mohapatra, hep-ph/0201176.

THE GUT? NEUTRINO BI-LARGE MIXING AND PROTON DECAY

NOBUHIRO MAEKAWA

Department of Physics, Kyoto University, Kyoto 606-8502, Japan
E-mail: maekawa@gauge.scphys.kyoto-u.ac.jp

In this talk, we introduce a new scenario of grand unified theory (GUT) with anomalous $U(1)_A$ gauge symmetry, which can explain doublet-triplet splitting, quark and lepton masses and mixing angles. In neutrino sector, the scenario realizes LMA solution for solar neutrino problem and large $U_{e3} = O(0.1)$. Moreover, the scenario predicts that the main decay mode of proton is from dimension 6 operators and the lifetime of proton must be near the present limit. The realization of gauge coupling unification requires that the cutoff scale of the scenario must be around the usual GUT scale $\Lambda_G \sim 2 \times 10^{16}$ GeV, which is smaller than the Planck scale. It may suggest the extra dimension in which gauge fields in visible sector do not propagate. This talk is based on the papers.[1,2,3,4,5]

1. Introduction

The most of people regard the standard model as the real theory which describe our world but does not satisfy the model as the final theory, because there are a lot of things which are not explained by the model; unstability of the weak scale due to quadratic divergent loop correction to the Higgs mass term: the miracle anomaly cancellation between quark and leptons: the origin of hierarchies of gauge and Yukawa couplings: the origin of small mixings in quark sector and large mixings in lepton sector: the charge quantization: no gravity, etc. The idea of grand unified theories (GUT)[6] not only explain the hierarchy of three gauge couplings in the standard model, anomaly cancellation and charge quantization, but also gives a natural unification of quark and leptons in a few multiplets in a simple gauge group. Since supersymmetry (SUSY) can stabilize the weak scale, SUSY GUT is one of the most promising model beyond the standard model. Unfortunately, it is not so easy to obtain the realistic SUSY GUT, because it is difficult to solve the doublet-triplet splitting problem [7,8,9] with stable proton and to obtain realistic quark and lepton mass matrices. On the other hand, it is known [10,11,12] that the hierarchy of Yukawa couplings in the supersymmetric standard model are realized by introducing anomalous

$U(1)$ gauge symmetry, [14] whose anomaly is cancelled by the Green-Schwarz mechanism. [15] Of course, it is not so straightforward to extend the argument in the SUSY standard model into in the GUT scenario, especially with large neutrino mixing angles, but it is important to examine the GUT scenario with anomalous $U(1)_A$ gauge symmetry.[16,17,13]

Recently, in a series of papers,[1,2,3,4,5] the interesting GUT scenario with anomalous $U(1)_A$ gauge symmetry has been proposed with $SO(10)$ unified group[1] and with E_6 unified group.[4,5] In the scenario, the anomalous $U(1)_A$ gauge symmetry plays an important role not only in obtaining realistic quark and lepton mass matrices, including bi-large neutrino mixings (LMA for solar neutrino problem) but also in solving doublet-triplet splitting problem. Moreover, since generic interactions are allowed to be introduced, it predicts the mass spectrum of superheavy fields and (GUT) symmetry breaking scales once the symmetry of the theory is fixed. It is surprising that the success of coupling unification in the minimal SUSY standard model (MSSM) can be naturally explained in the scenario, though the mass spectrum of superheavy fields does not respect $SU(5)$ symmetry. [3] It is shown that the gauge coupling unification requires the cutoff scale must be around the usual GUT scale $\Lambda_G = 2 \times 10^{16}$ GeV and the unification scale is just below the usual GUT scale if all the fields except those of MSSM have superheavy masses. [3] It is interesting that this result is independent of the detail of Higgs sector. The GUT with anomalous $U(1)_A$ gauge symmetry with a simple unified gauge group predicts the above result. Therefore, proton decay via dimension 6 operators may be seen in near future experiments. Moreover, once SUSY breaking parameters are introduced, the μ problem is naturally solved [2] and natural suppression of flavor changing neutral current (FCNC) is realized in E_6 GUT.[4]

As introduced in the above, the scenario predicts the proton decay via dimension 6 operators even though low energy SUSY is required. Moreover, bi-large neutrino mixings are obtained (especially, LMA solution for the solar neutrino problem is predicted). The scenario predicts large $U_{e3} \sim O(0.1)$ and small $\tan\beta$.

It is interesting that all the above solutions are realized in non-trivial ways once only several anomalous $U(1)_A$ charges are determined. Actually, the input parameters are only 8 integer anomalous $U(1)_A$ charges (+3 for singlet Higgs) for the Higgs sector and 3(or 4) (half) integer charges for the matter sector in E_6 (or $SO(10)$) GUT. In this talk, we will explain some of them.

2. Doublet-triplet splitting

One of the most interesting feature of anomalous $U(1)_A$ gauge theory is that the vacuum expectation values (VEV) are determined by anomalous $U(1)_A$ charges as

$$\langle Z^+ \rangle = 0, \tag{2.1}$$

$$\langle Z^- \rangle \sim \lambda^{-z^-}, \tag{2.2}$$

where $Z^\pm$ are singlet operators with the charges $z^+ > 0$ and $z^- < 0$, and $\lambda = \langle \Theta \rangle / \Lambda$. Here Θ is a Froggatt-Nielsen field. [18] Through this paper, we use unit in which the cutoff $\Lambda = 1$ and denote all the superfields by uppercase letters and their anomalous $U(1)_A$ charges by the corresponding lowercase letters. Such VEVs do not change the order of the coefficients obtained by the Froggatt-Nielsen mechanism:

$$W = \left(\frac{\Theta}{\Lambda}\right)^{x+y+z} XYZ \to \lambda^{x+y+z} XYZ, \tag{2.3}$$

if the total charge $x+y+z$ of the operator XYZ is positive. Note that even if the operator $\frac{Z^-}{\Lambda}$ is used instead of $\left(\frac{\Theta}{\Lambda}\right)^{z^-}$ in the interactions, the order of the coefficients does not change. This feature is critically different from the naive expectation that the contribution from the higher dimensional operators is more suppressed. If the total charge $x + y + z$ is negative, such interaction is not allowed by the anomalous $U(1)_A$ gauge symmetry because only negatively charged fields have non-vanishing VEVs. This is called SUSY zero mechanism. Note that this mechanism leads to the finite number of non-renormalizable interactions, and therefore we can control the generic superpotential.

Actually, under the vacua (2.1), the generic superpotential to determine the VEVs of Z^- can be written as

$$W = \sum_i^{n_+} W_{Z_i^+}, \tag{2.4}$$

where W_X denotes the terms linear in the X field. This is because the F-flatness conditions of negatively charged fields are automatically satisfied and the terms with more than two positively charged fields do not contribute in the F-flatness condition of positively charged fields.

Let us discuss an $SO(10)$ GUT model with anomalous $U(1)_A$ gauge symmetry in which doublet-triplet splitting is naturally realized. The Higgs content is listed in Table I. Here the symbols $\pm$ denote the Z_2 parity.

Table I. Typical values of anomalous $U(1)_A$ charges.

	non-vanishing VEV	vanishing VEV
45	$A(a=-1,-)$	$A'(a'=3,-)$
16	$C(c=-4,+)$	$C'(c'=3,-)$
$\overline{\mathbf{16}}$	$\bar{C}(\bar{c}=-1,+)$	$\bar{C}'(\bar{c}'=6,-)$
10	$H(h=-3,+)$	$H'(h'=4,-)$
1	$\Theta(\theta=-1,+)$,$Z(z=-2,-)$, $\bar{Z}(\bar{z}=-2,-)$	$S(s=5,+)$

The VEVs of the negatively charged Higgs fields are determined by the superpotential

$$W = W_{A'} + W_{C'} + W_{\bar{C}'} + W_{H'} + W_S. \tag{2.5}$$

We do not have spaces enough to explain the vacuum structure in detail, so we here point out only one good feature in realizing the doublet-triplet splitting.

If $-3a \le a' < -5a$, the superpotential $W_{A'}$ is in general written

$$W_{A'} = \lambda^{a'+a}\alpha A'A + \lambda^{a'+3a}(\beta(A'A)_{\mathbf{1}}(A^2)_{\mathbf{1}} + \gamma(A'A)_{\mathbf{54}}(A^2)_{\mathbf{54}}), \tag{2.6}$$

where the suffices **1** and **54** indicate the representation of the composite operators under the $SO(10)$ gauge symmetry, and α, β and γ are parameters of order 1. Here we assume $a+a'+c+\bar{c}<0$ to forbid the term $\bar{C}A'AC$, which destabilizes the DW form of the VEV $\langle A\rangle$. The D-flatness condition requires the VEV $\langle A\rangle = i\tau_2 \times \mathrm{diag}(x_1, x_2, x_3, x_4, x_5)$, and the F-flatness conditions of the A' field requires $x_i(-\alpha\lambda^{-2a} + (2\beta - \frac{\gamma}{5})(\sum_j x_j^2) + \gamma x_i^2) = 0$. This allows only two solutions, $x_i^2 = 0$ and $x_i^2 = \frac{\alpha}{(1-\frac{N}{5})\gamma + 2N\beta}\lambda^{-2a} \equiv v^2$. Here $N = 0-5$ is the number of $x_i = v$ solutions. When $N=3$, the vacuum becomes $\langle A(\mathbf{45})\rangle_{B-L} = \tau_2 \times \mathrm{diag}(v,v,v,0,0)$, which breaks $SO(10)$ into $SU(3)_C \times SU(2)_L \times SU(2)_R \times U(1)_{B-L}$ at the scale $\Lambda_A \equiv \langle A\rangle \sim \lambda^{-a}$. This Dimopoulos-Wilczek form of the VEV plays an important role in solving the DT splitting problem. Actually through the interaction $W = H'AH$, the DW type of the VEV gives superheavy masses only to the triplet Higgs, and therefore the doublet Higgs remains massless. Taking account of the mass term H'^2, only one pair of Higgs doublets becomes massless.

Note that the higher terms $A'A^{2L+1}$ $(L>1)$ are forbidden by the SUSY zero mechanism. If they were allowed, the number of possible VEVs other than the DW form would become larger, and thus it would become less natural to obtain the DW form. This is a critical point of this mechanism, and the anomalous $U(1)_A$ gauge symmetry plays an essential role in forbidding the undesired terms.

The spinor Higgs fields C and $\bar{C}$ break $SU(2)_R \times U(1)_{B-L}$ into $U(1)_Y$ by developing $\langle C \rangle (= \langle \bar{C} \rangle \equiv \Lambda_C \sim \lambda^{-(c+\bar{c})/2})$. Then this model becomes MSSM at a low energy scale.

3. Quark and lepton mass matrices

One of the most attractive features of grand unified theory is to unify the quark and lepton into fewer multiplets. For example, in $SO(10)$ GUT scenario, a **16** representation field contains one family quark and lepton fields including right-handed neutrino field. However, this attractive feature directly leads to unrealistic Yukawa relations. For example, if we introduce $3 \times \mathbf{16}\ \Psi_i (i = 1, 2, 3)$ for 3 family quark and leptons, the Yukawa couplings are obtained from the interaction

$$W = Y_{ij} \Psi_i \Psi_j H \tag{3.1}$$

as $Y_u = Y_d$ and $Y_d = Y_e$, which lead to unrealistic mass relation. We have to pick up the VEV $\langle C \rangle$ in the Yukawa matrices to avoid the former unrealistic relation $Y_u = Y_d$, and the VEV $\langle A \rangle$ to avoid the latter unrealistic relation $Y_d = Y_e$. In our scenario, we introduce an additional matter field $T(\mathbf{10})$. Then after breaking the GUT gauge group into the standard model gauge group, one pair of vector-like fields $\mathbf{5}$ and $\bar{\mathbf{5}}$ of $SU(5)$ becomes massive. The mass matrix is obtained from the interaction

$$W = \lambda^{\psi_i + t + c} \Psi_i T C + \lambda^{2t} T^2 \tag{3.2}$$

as

$$\mathbf{5}_T (\lambda^{t+\psi_1+(c-\bar{c})/2}, \lambda^{t+\psi_2+(c-\bar{c})/2}, \lambda^{t+\psi_3+(c-\bar{c})/2}, \lambda^{2t}) \begin{pmatrix} \bar{\mathbf{5}}_{\Psi 1} \\ \bar{\mathbf{5}}_{\Psi 2} \\ \bar{\mathbf{5}}_{\Psi 3} \\ \bar{\mathbf{5}}_{\mathrm{T}} \end{pmatrix}, \tag{3.3}$$

where actually the VEV $\langle \bar{C} \rangle = \langle C \rangle \sim \lambda^{-(c+\bar{c})/2}$ appear in the mass matrix. Since $\psi_3 < \psi_2 < \psi_1$, the massive mode $\bar{\mathbf{5}}_M$, the partner of $\mathbf{5}_T$, must be either $\bar{\mathbf{5}}_{\Psi 3} (\Delta \equiv 2t - (t + \psi_3 + (c - \bar{c})/2) > 0)$ or $\bar{\mathbf{5}}_T (\Delta < 0)$. The former case is interesting, and in this case, the three massless modes $(\bar{\mathbf{5}}_1, \bar{\mathbf{5}}_2, \bar{\mathbf{5}}_3)$ can be written $(\bar{\mathbf{5}}_{\Psi 1} + \lambda^{\psi_1 - \psi_3} \bar{\mathbf{5}}_{\Psi 3}, \bar{\mathbf{5}}_T + \lambda^{\Delta} \bar{\mathbf{5}}_{\Psi 3}, \bar{\mathbf{5}}_{\Psi 2} + \lambda^{\psi_2 - \psi_3} \bar{\mathbf{5}}_{\Psi 3})$. If we adopt their charges $(\psi_1, \psi_2, \psi_3, t) = (9/2, 7/2, 3/2, 5/2)$ in addition to the charges of Higgs fields, then we can estimate quark and lepton mass matrices as

$$M_u = \begin{pmatrix} \lambda^6 & \lambda^5 & \lambda^3 \\ \lambda^5 & \lambda^4 & \lambda^2 \\ \lambda^3 & \lambda^2 & 1 \end{pmatrix} \langle H_u \rangle, \quad M_d = M_e = \lambda^2 \begin{pmatrix} \lambda^4 & \lambda^{3.5} & \lambda^3 \\ \lambda^3 & \lambda^{2.5} & \lambda^2 \\ \lambda^1 & \lambda^{0.5} & 1 \end{pmatrix} \langle H_d \rangle . \tag{3.4}$$

And for the neutrino sector, we take into account the interaction

$$\lambda^{\psi_i+\psi_j+2\bar{c}}\Psi_i\Psi_j\bar{C}\bar{C}, \tag{3.5}$$

which lead to the right-handed neutrino masses

$$M_R = \lambda^{\psi_i+\psi_j+2\bar{c}}\left\langle\bar{C}\right\rangle^2 = \lambda^{2n+\bar{c}-c}\begin{pmatrix}\lambda^6 & \lambda^5 & \lambda^3\\ \lambda^5 & \lambda^4 & \lambda^2\\ \lambda^3 & \lambda^2 & 1\end{pmatrix}. \tag{3.6}$$

Since the Dirac neutrino mass is given by

$$M_{\nu_D} = \lambda^2\begin{pmatrix}\lambda^4 & \lambda^3 & \lambda\\ \lambda^{3.5} & \lambda^{2.5} & \lambda^{0.5}\\ \lambda^3 & \lambda^2 & 1\end{pmatrix}\langle H_u\rangle\,\eta, \tag{3.7}$$

the neutrino mass matrix is obtained by the seesaw mechanism as

$$M_\nu = M_{\nu_D}M_R^{-1}M_{\nu_D}^T = \lambda^{4-2n+c-\bar{c}}\begin{pmatrix}\lambda^2 & \lambda^{1.5} & \lambda\\ \lambda^{1.5} & \lambda & \lambda^{0.5}\\ \lambda & \lambda^{0.5} & 1\end{pmatrix}\langle H_u\rangle^2\eta^2. \tag{3.8}$$

Note that the ratio $\frac{m_{\nu_\mu}}{m_{\nu_\tau}} \sim \lambda$ is realized, that predicts LMA solution for the solar neutrino problem. It is interesting that we obtain the small mixing angles for the Cabibbo-Kobayashi-Maskawa matrix

$$U_{\rm CKM} = \begin{pmatrix}1 & \lambda & \lambda^3\\ \lambda & 1 & \lambda^2\\ \lambda^3 & \lambda^2 & 1\end{pmatrix}, \tag{3.9}$$

and the large mixing angles for the Maki-Nakagawa-Sakata matrix

$$U_{\rm MNS} = \begin{pmatrix}1 & \lambda^{0.5} & \lambda\\ \lambda^{0.5} & 1 & \lambda^{0.5}\\ \lambda & \lambda^{0.5} & 1\end{pmatrix}. \tag{3.10}$$

Since we use a rule that $\lambda^{0.5}+\lambda^{0.5} \sim \lambda^{0.5}$ in calculating the MNS matrix and $\lambda^{0.5} \sim 0.5$, this model gives large mixing angles [19,20] for the atmospheric neutrino problem and for the solar neutrino problem. And $U_{e3} \sim \lambda$ is predicted, which is around the present upper limit given by CHOOZ.[21] At this stage, the unrealistic GUT relation $Y_d = Y_e$ still remains. However, in our scenario, the same amount of the Yukawa couplings are given by the higher dimensional interactions

$$W = \lambda^{\psi_i+\psi_j+na+h}\Psi_i A^n \Psi_j H \tag{3.11}$$

by developing the VEV $\langle A\rangle \sim \lambda^{-a}$. It is critical that the Yukawa couplings from the higher dimensional interactions have not kept the unrealistic GUT relation. Usually, the corrections from such higher dimensional interactions are suppressed by the factor $\frac{\langle A \rangle}{\Lambda}$. But in our scenario, the suppression factor $\frac{\langle A \rangle}{\Lambda}$ is just cancelled by the enhancement factor λ^a in the coefficients, and therefore we can obtain the same order coefficients as from the tree interaction. This is an attractive feature in our scenario, and the realistic mass matrices are naturally obtained.

4. Gauge coupling unification

First, we show the relation of the determinants of the mass spectrum of superheavy fields in terms of their anomalous $U(1)_A$ charges. If we use the notation of the fields $Q(\mathbf{3},\mathbf{2})_{\frac{1}{6}}$, $U^c(\bar{\mathbf{3}},\mathbf{1})_{-\frac{2}{3}}$, $D^c(\bar{\mathbf{3}},\mathbf{1})_{\frac{1}{3}}$, $L(\mathbf{1},\mathbf{2})_{-\frac{1}{2}}$, $E^c(\mathbf{1},\mathbf{1})_1, N^c(\mathbf{1},\mathbf{1})_0$, $X(\mathbf{3},\mathbf{2})_{-\frac{5}{6}}$ and their conjugate fields, and $G(\mathbf{8},\mathbf{1})_0$ and $W(\mathbf{1},\mathbf{3})_0$ with the standard gauge symmetry, under $SO(10) \supset SU(5) \supset SU(3)_C \times SU(2)_L \times U(1)_Y$, the spinor **16**, vector **10** and the adjoint **45** of $SO(10)$ are divided as

$$\mathbf{16} \to \underbrace{[Q + U^c + E^c]}_{\mathbf{10}} + \underbrace{[D^c + L]}_{\bar{\mathbf{5}}} + \underbrace{N^c}_{\mathbf{1}}, \tag{4.1}$$

$$\mathbf{10} \to \underbrace{[D^c + L]}_{\bar{\mathbf{5}}} + \underbrace{[\bar{D}^c + \bar{L}]}_{\mathbf{5}}, \tag{4.2}$$

$$\mathbf{45} \to \underbrace{[G + W + X + \bar{X} + N^c]}_{\mathbf{24}} + \underbrace{[Q + U^c + E^c]}_{\mathbf{10}} + \underbrace{[\bar{Q} + \bar{U}^c + \bar{E}^c]}_{\overline{\mathbf{10}}} + \underbrace{N^c}_{\mathbf{1}} \tag{4.3}$$

Then the determinants of the mass matrices $\bar{M}_I$ of superheavy fields $I = Q,U^c,E^c,D^c,L,G,W,X$ are estimated as

$$\det \bar{M}_I = \lambda^{\sum_i c_i}, \tag{4.4}$$

where c_i are anomalous $U(1)_A$ charges of superheavy fields.

Secondly, the conditions of the gauge coupling unification in using one loop renormalization group equations

$$\alpha_3(\Lambda_A) = \alpha_2(\Lambda_A) = \frac{5}{3}\alpha_Y(\Lambda_A) \equiv \alpha_1(\Lambda_A), \tag{4.5}$$

where $\alpha_1^{-1}(\mu > \Lambda_C) \equiv \frac{3}{5}\alpha_R^{-1}(\mu > \Lambda_C) + \frac{2}{5}\alpha_{B-L}^{-1}(\mu > \Lambda_C)$, are rewritten by the determinants of the mass matrices of the superheavy fields. Here $\alpha_X = \frac{g_X^2}{4\pi}$, and the parameters $g_X (X = 3, 2, R, B-L, Y)$ are the gauge couplings of $SU(3)_C$, $SU(2)_L$, $SU(2)_R$, $U(1)_{B-L}$ and $U(1)_Y$, respectively.

The gauge couplings at the scale Λ_A are roughly given by

$$\alpha_1^{-1}(\Lambda_A) = \alpha_1^{-1}(M_{SB}) + \frac{1}{2\pi}\left(b_1 \ln\left(\frac{M_{SB}}{\Lambda_A}\right) + \sum_i \Delta b_{1i} \ln\left(\frac{m_i}{\Lambda_A}\right) - \frac{12}{5}\ln\left(\frac{\Lambda_C}{\Lambda_A}\right)\right), \tag{4.6}$$

$$\alpha_2^{-1}(\Lambda_A) = \alpha_2^{-1}(M_{SB}) + \frac{1}{2\pi}\left(b_2 \ln\left(\frac{M_{SB}}{\Lambda_A}\right) + \sum_i \Delta b_{2i} \ln\left(\frac{m_i}{\Lambda_A}\right)\right), \quad (4.7)$$

$$\alpha_3^{-1}(\Lambda_A) = \alpha_3^{-1}(M_{SB}) + \frac{1}{2\pi}\left(b_3 \ln\left(\frac{M_{SB}}{\Lambda_A}\right) + \sum_i \Delta b_{3i} \ln\left(\frac{m_i}{\Lambda_A}\right)\right), \quad (4.8)$$

where M_{SB} is a SUSY breaking scale, $(b_1, b_2, b_3) = (33/5, 1, -3)$ are the renormalization group coefficients for the minimal SUSY standard model (MSSM), and Δb_{ai} $(a = 1, 2, 3)$ are the corrections to the coefficients from the massive fields with mass m_i. The last term in Eq. (4.6) is from the breaking $SU(2)_R \times U(1)_{B-L} \to U(1)_Y$ caused by the VEV $\langle C \rangle$. Since the gauge couplings at the SUSY breaking scale M_{SB} are given by

$$\alpha_i^{-1}(M_{SB}) = \alpha_G^{-1}(\Lambda_G) + \frac{1}{2\pi}\left(b_i \ln\left(\frac{\Lambda_G}{M_{SB}}\right)\right) \quad , \quad (i = 1, 2, 3) \quad (4.9)$$

where $\alpha_G^{-1}(\Lambda_G) \sim 25$ and $\Lambda_G \sim 2 \times 10^{16}$ GeV, the unification conditions $\alpha_1(\Lambda_A) = \alpha_2(\Lambda_A)$, $\alpha_1(\Lambda_A) = \alpha_3(\Lambda_A)$ and $\alpha_2(\Lambda_A) = \alpha_3(\Lambda_A)$ can be rewritten

$$\left(\frac{\Lambda_A}{\Lambda_G}\right)^{14}\left(\frac{\Lambda_C}{\Lambda_A}\right)^{6}\left(\frac{\det \bar{M}_L}{\det \bar{M}_{D^c}}\right)\left(\frac{\det \bar{M}_Q}{\det \bar{M}_U}\right)^{4}\left(\frac{\det \bar{M}_Q}{\det \bar{M}_{E^c}}\right)^{3}\left(\frac{\det \bar{M}_W}{\det \bar{M}_X}\right)^{5}$$
$$= \Lambda_A^{-\bar{r}_{D^c}+\bar{r}_L-4\bar{r}_{U^c}-3\bar{r}_{E^c}+7\bar{r}_Q-5\bar{r}_X+5\bar{r}_W}, \quad (4.10)$$

$$\left(\frac{\Lambda_A}{\Lambda_G}\right)^{16}\left(\frac{\Lambda_C}{\Lambda_A}\right)^{4}\left(\frac{\det \bar{M}_{D^c}}{\det \bar{M}_L}\right)\left(\frac{\det \bar{M}_Q}{\det \bar{M}_U}\right)\left(\frac{\det \bar{M}_Q}{\det \bar{M}_{E^c}}\right)^{2}\left(\frac{\det \bar{M}_G}{\det \bar{M}_X}\right)^{5}$$
$$= \Lambda_A^{-\bar{r}_L+\bar{r}_{D^c}-\bar{r}_{U^c}-2\bar{r}_{E^c}+3\bar{r}_Q-5\bar{r}_X+5\bar{r}_G}, \quad (4.11)$$

$$\left(\frac{\Lambda_A}{\Lambda_G}\right)^{4}\left(\frac{\det \bar{M}_{D^c}}{\det \bar{M}_L}\right)\left(\frac{\det \bar{M}_U}{\det \bar{M}_Q}\right)\left(\frac{\det \bar{M}_G}{\det \bar{M}_W}\right)^{2}\left(\frac{\det \bar{M}_G}{\det \bar{M}_X}\right)$$
$$= \Lambda_A^{-\bar{r}_L+\bar{r}_{D^c}-\bar{r}_Q+\bar{r}_U-2\bar{r}_W-\bar{r}_X+3\bar{r}_G}. \quad (4.12)$$

Here $\bar{r}_I$ are the ranks of the mass matrices of superheavy fields $\bar{M}_I$. Note that the above conditions are dependent only on the ratio of the determinants of mass matrices which are included in the same multiplet of $SU(5)$ and on the symmetry breaking scales Λ_A, Λ_C. If all the component fields in a multiplet have been superheavy, the above ratios would be of order one, because the determinants are given by $\det \bar{M} = \lambda^{\sum_i c_i}$. However, since part of the component fields (massless Higgs doublets or Nambu-Goldstone modes) do not appear in the mass matrices, the above ratios are dependent only on the charges of these massless modes. If all the other fields than in

MSSM become superheavy, the above ratios are easily estimated as

$$\frac{\det \bar{M}_L}{\det \bar{M}_{D^c}} \sim \lambda^{-2h} \tag{4.13}$$

$$\frac{\det \bar{M}_Q}{\det \bar{M}_{E^c}} \sim \frac{\det \bar{M}_{U^c}}{\det \bar{M}_{E^c}} \sim \lambda^{c+\bar{c}-2a} \tag{4.14}$$

$$\frac{\det \bar{M}_G}{\det \bar{M}_X} \sim \frac{\det \bar{M}_W}{\det \bar{M}_X} \sim \lambda^{-2a}. \tag{4.15}$$

Then the conditions for the coupling unification becomes

$$\Lambda \sim \lambda^{\frac{h}{7}} \Lambda_G,, \quad \Lambda \sim \lambda^{-\frac{h}{8}} \Lambda_G,, \quad \Lambda \sim \lambda^{-\frac{h}{2}} \Lambda_G. \tag{4.16}$$

So the unification conditions become $h \sim 0$, and thus the cutoff scale must be taken as $\Lambda \sim \Lambda_G$. It is obvious that if the cutoff scale have been another scale (for example, the Planck scale), in MSSM three gauge couplings would meet at the scale. This means that in this scenario it is not accidental that three gauge couplings meet at a scale in MSSM, even though the unification scale in our scenario is different from the usual unification scale. Note that the above results are independent of the detail of the Higgs sector, because the requirement that all the other fields than those in MSSM become superheavy determines the field content of the massless fields, whose charges are important to examine whether gauge couplings meet at the unification scale Λ_A or not. The above argument can be applied also to the scenario of E_6 unification, though instead of usual doublet Higgs charge h we have to use effective Higgs charges $h_{eff} \equiv h + \frac{1}{4}(\phi - \bar{\phi})$, where E_6 is broken into $SO(10)$ by the VEV $|\langle \Phi \rangle| = |\langle \bar{\Phi} \rangle| \sim \lambda^{-\frac{1}{2}(\phi+\bar{\phi})}$.

Note that the condition $h \sim 0$ does not mean $h = 0$, because there is an ambiguity involving order 1 coefficients and we have used only one loop RGEs. However, the above analysis is fairly useful to provide a rough picture of the behavior.

5. Proton decay

The proton decay via dimension 5 operators[22] is suppressed in our scenario, because effective colored Higgs mass is given by $m_H^c \sim \lambda^{2h}$ and the dimension 5 operators are suppressed due to anomalous $U(1)_A$ gauge symmetry. On the other hand, proton decay via dimension 6 operators is reachable, because the cutoff scale must be around the usual GUT scale $\Lambda_G \sim 2 \times 10^{16}$ GeV and the unification scale is given by λ^{-a}. If we adopt $a = -1$, then the lifetime of the proton is roughly estimated as

$$\tau_p(p \to e\pi^0) \sim 3 \times 10^{33} \left(\frac{\Lambda_A}{5 \times 10^{15}\ \text{GeV}} \right)^4 \left(\frac{0.015(\text{GeV})^3}{\alpha} \right)^2 \text{years}, \tag{5.1}$$

which is near the present experimental lower bound.[23]. Here α is the hadron matrix element parameter. Here we use the formula[24] and the value α given by lattice calculation. [25] Though the prediction is strongly dependent on the actual unification scale which is dependent on the order one coefficients, this rough estimation gives a strong motivation for the future experiments for proton decay search.

References

1. N. Maekawa, *Prog. Theor. Phys.* **106**, 401 (2001); hep-ph/0110276.
2. N. Maekawa, *Phys. Lett.* **B521**, 42 (2001).
3. N. Maekawa, *Prog. Theor. Phys.* **107**, 597 (2002).
4. M. Bando and M. Maekawa, *Prog. Theor. Phys.* **106**, 1255 (2001).
5. N. Maekawa and T. Yamashita, hep-ph/0202050 (to appear in *Prog. Theor. Phys.*).
6. H. Georgi and S.L. Glashow, *Phys. Rev. Lett.* **32**, 438 (1974).
7. E. Witten, *Phys. Lett.* **B105**, 267 (1981);
   A. Masiero, D.V. Nanopoulos, K. Tamvakis and T.Yanagida, *Phys. Lett.* **115**, 380 (1982);
   B. Grinstein, *Nucl. Phys.* **B206**, 387 (1982);
   K. Inoue, A. Kakuto and T. Takano, *Prog. Theor. Phys.* **75**, 664 (1986);
   E. Witten, *Nucl. Phys.* **B258**,75 (1985);
   T. Yanagida, *Phys. Lett.* **B344**, 211 (1995);
   Y. Kawamura, *Prog. Theor. Phys.* **105**, 691 (2001);*ibid* **105**, 999 (2001);
   L. Hall and Y. Nomura, *Phys. Rev.* **D64**, 055003 (2001).
8. S. Dimopoulos and F. Wilczek, NSF-ITP-82-07;
   M. Srednicki, *Nucl. Phys.* **B202**, 327 (1982).
9. S.M. Barr and S. Raby, *Phys. Rev. Lett.* **79**, 4748 (1997).
10. L. Ibáñez and G.G. Ross, *Phys. Lett.* **B332**, 100 (1994);
    P. Binétruy and P. Ramond, *Phys. Lett.* **B350**, 49 (1995);
    E. Dudas, S. Pokorski and C.A. Savoy, *Phys. Lett.* **B356**, 45 (1995);
    P. Binétruy, S. Lavignac and P. Ramond, *Nucl. Phys.* **B477**, 353 (1996).
11. P. Binétruy, S. Lavignac, S. Petcov and P. Ramond, *Nucl. Phys.* **B496**, 3 (1997).
12. H. Dreiner, G.K. Leontaris, S. Lola, G.G. Ross and C. Scheich, *Nucl. Phys.* **B436**, 461 (1995).
13. Q. Shafi and Z. Tavartkiladze, *Phys. Lett.* **B487**, 145 (2000).
14. E. Witten, *Phys. Lett.* **B149**, 351 (1984);
    M. Dine, N. Seiberg and E. Witten, *Nucl. Phys.* **B289**, 589 (1987);
    J.J. Atick, L.J. Dixon and A. Sen, *Nucl. Phys.* **B292**, 109 (1987);
    M. Dine, I. Ichinose and N. Seiberg, *Nucl. Phys.* **B293**, 253 (1987).
15. M. Green and J. Schwarz, *Phys. Lett.* **B149**, 117 (1984).
16. M. Bando and T. Kugo, *Prog. Theor. Phys.* **101**, 1313 (1999);
    M. Bando, T. Kugo and K. Yoshioka, *Prog. Theor. Phys.* **104**, 211 (2000).
17. K.-I. Izawa, K. Kurosawa, Y. Nomura, T. Yanagida *Phys. Rev.* **D60**, 115016 (1999).

18. C.D. Froggatt and H.B. Nielsen, *Nucl. Phys.* **B147**, 277 (1979).
19. Y. Fukuda et al.(The Super-Kamiokande Collaboration), *Phys. Lett.* **B436**, 33 (1998); *Phys. Rev. Lett.* **81**, 1562 (1998); *Phys. Rev. Lett.* **86**, 5656 (2001).
20. The SNO Collaboration, hep-ex/0204008;hep-ex/0204009.
21. The CHOOZ Collaboration, *Phys. Lett.* **B420**, 397 (1998).
22. N. Sakai and T. Yanagida, *Nucl. Phys.* **B197**, 533 (1982).
23. Super-Kamiokande Collaboration, *Phys. Rev. Lett.* **81**, 3319 (1998); *ibid* **83**, 1529 (1999).
24. J. Hisano, H. Murayama and T. Yanagida, *Nucl. Phys.* **B402**, 46 (1993).
25. JLQCD Collaboration, S. Aoki et al., *Phys. Rev.* **D62**, 014506 (2000).

NEUTRINO BIMAXIMAL TEXTURE AND LEPTON FLAVOR VIOLATION

NORIYUKI SHIMOYAMA

Department of Physics, Niigata University, Ikarashi 2-8050, 950-2181 Niigata, JAPAN

E-mail: simoyama@muse.hep.sc.niigata-u.ac.jp

We investigate the lepton flavor violation in the framework of the MSSM with right-handed neutrinos taking the large mixing angle MSW solution in the quasi-degenerate and the inverse-hierarchical neutrino masses. We predict the branching ratio of $\mu \to e + \gamma$ processe assuming the degenerate right-handed Majorana neutrino masses. We find that the branching ratio in the quasi-degenerate neutrino mass spectrum is 100 times smaller than the one in the inverse-hierarchical and the hierarchical neutrino spectra. We emphasize that the magnitude of U_{e3} is one of important ingredients to predict $\mathrm{BR}(\mu \to e + \gamma)$. Furtheremore, we examine the Shafi-Tavartkiladze model, which gives the inverse-hierarchical neutrino masses. Predicted branching ratios of $\mu \to e + \gamma$ is smaller than the experimantal bound.

1 Introduction

If neutrinos are massive and mixed in the SM, there exists a source of the lepton flavor violation (LFV) through the off-diagonal elements of the neutrino Yukawa coupling matrix. However, due to the smallness of the neutrino masses, the predicted branching ratios for these processes are tiny. On the other hand, in the supersymmetric framework the situation is quite different. Many authors have already studied the LFV in the minimal supersymmetric standard model (MSSM) with right-handed neutrinos assuming the relevant neutrino mass matrix [1,2]. In the MSSM with soft breaking terms, there exist lepton flavor violating terms such as off-diagonal elements of slepton mass matrices $(\mathbf{m}^2_{\tilde{\mathrm{L}}})_{ij}$, $(\mathbf{m}^2_{\tilde{\mathrm{e}}_\mathrm{R}})_{ij}$ and trilinear couplings $\mathbf{A}^\mathrm{e}_{ij}$. Strong bounds on these matrix elements come from requiring branching ratios for LFV processes to be below observed ratios. For the present, the most stringent bound comes from the $\mu \to e + \gamma$ decay ($\mathrm{BR}(\mu \to \mathrm{e} + \gamma) < 1.2 \times 10^{-11}$) [4]. However, if the LFV occurs at tree level in the soft breaking terms, the branching ratio of this process exceeds the experimental bound too much. Therefore one assumes that the LFV does not occur at tree level in the soft parameters. This is realized by taking the assumption that soft parameters such as $(\mathbf{m}^2_{\tilde{\mathrm{L}}})_{ij}$, $(\mathbf{m}^2_{\tilde{\mathrm{e}}_\mathrm{R}})_{ij}$, $\mathbf{A}^\mathrm{e}_{ij}$, are universal *i.e.* proportional to the unit matrix. This assumption follows from the m-SUGRA. However, even though there are no flavor violation at tree level, it is generated by the effect of the renormalization group equations

(RGE's) via neutrino Yukawa couplings. Suppose that neutrino masses are produced by the see-saw mechanism, there are the right-handed neutrinos above a scale M_R ($\gtrsim 10^{12}$GeV). Then neutrinos have the Yukawa coupling matrix $\mathbf{Y}_\nu$ with off-diagonal entries in the basis of the diagonal charged-lepton Yukawa couplings. The off-diagonal elements of $\mathbf{Y}_\nu$ drive off-diagonal ones in the $\left(\mathbf{m}_{\tilde{L}}^2\right)_{ij}$ and $\mathbf{A}^{\mathrm{e}}_{ij}$ matrices through the RGE's running [5].

One can construct $\mathbf{Y}_\nu$ by the recent data of neutrino oscillations. Assuming that oscillations need only accounting for the solar and the atmospheric neutrino data, we take the LMA-MSW solution, which gives the lepton mixing matrix (MNS) [6] and the neutrino mass scales as seen in ref.7 . Since the data of neutrino oscillations only indicate the differences of the mass square Δm^2_{ij}, the neutrinos have three possible mass spectra: the hierarchical spectrum $m_{\nu 3} \gg m_{\nu 2} \gg m_{\nu 1}$, the quasi-degenerate one $m_{\nu 1} \simeq m_{\nu 2} \simeq m_{\nu 3}$ and the inverse-hierarchical one $m_{\nu 1} \simeq m_{\nu 2} \gg m_{\nu 3}$.

This paper is organized as follows. In section 2, we give the general form of $\mathbf{Y}_\nu$ and $\mathbf{Y}_\nu^\dagger \mathbf{Y}_\nu$, which play a crucial role in generating the LFV through the RGE's running. In section 3, we calculate the branching ratio of the processes $\mu \to e+\gamma$ in the three different neutrino mass spectra. In section 4, we examine the Shafi-Tavartkiladze model, which gives the inverse-hierarchical neutrino masses. In section 5, we summarize our results and give discussions.

2 LFV in the MSSM with Right-handed Neutrinos

2.1 LFV in Yukawa Couplings

In this section, we introduce the general expression of the neutrino Yukawa coupling $\mathbf{Y}_\nu$, which is useful in the following arguments, and investigate the LFV triggered by the neutrino Yukawa couplings. The superpotential of the lepton sector is described as follows:

$$W_{\text{lepton}} = \mathbf{Y_e} L H_d e^c_R + \mathbf{Y}_\nu L H_u \nu^c_R + \frac{1}{2} \nu^{c\,T}_R \mathbf{M_R} \nu^c_R \ , \tag{1}$$

where H_u, H_d are chiral superfields for Higgs doublets, L is the left-handed lepton doublet, e_R and ν_R are the right-handed charged lepton and the neutrino superfields, respectively. The $\mathbf{Y_e}$ and $\mathbf{Y}_\nu$ are Yukawa coupling matrices, $\mathbf{M_R}$ is Majorana mass matrix of the right-handed neutrinos. We take the charged lepton Yukawa couplings and the Majorana neutrino mass matrix to be diagonal. It is well-known that the neutrino mass matrix is given as

$$\mathbf{m}_\nu = (\mathbf{Y}_\nu v_u)^{\mathrm{T}} \mathbf{M_R}^{-1} (\mathbf{Y}_\nu v_u) \ , \tag{2}$$

via the see-saw mechanism, where v_u is the vacuum expectation value (VEV) of Higgs H_u. The neutrino mass matrix $\mathbf{m}_\nu$ is diagonalized by a single unitary matrix $\mathbf{m}_\nu^{\rm diag} \equiv \mathbf{U}_{\rm MNS}^{\rm T}\mathbf{m}_\nu\mathbf{U}_{\rm MNS}$, where $\mathbf{U}_{\rm MNS}$ is the MNS matrix. Following the expression in ref.[3], we write the neutrino Yukawa coupling as

$$\mathbf{Y}_\nu = \frac{1}{v_u}\sqrt{\mathbf{M}_{\mathbf{R}}^{\rm diag}}\ \mathbf{R}\ \sqrt{\mathbf{m}_\nu^{\rm diag}}\ \mathbf{U}_{\rm MNS}^{\rm T}\ , \tag{3}$$

where $\mathbf{R}$ is a 3×3 complex orthogonal matrix, which depends on models.

At first, let us take the degenerate right-handed Majorana masses $M_{R1} = M_{R2} = M_{R3} \equiv M_R$. This assumption is reasonable for the case of the degenerate neutrino masses. Otherwise a big conspiracy would be needed between $\mathbf{Y}_\nu$ and $\mathbf{M_R}$.

As mentioned in the previous section, there are three possible neutrino mass spectra. The hierarchical type ($m_{\nu1} \ll m_{\nu2} \ll m_{\nu3}$) gives the neutrino mass spectrum as

$$m_{\nu1} \sim 0\ , \qquad m_{\nu2} = \sqrt{\Delta m_\odot^2}\ , \qquad m_{\nu3} = \sqrt{\Delta m_{\rm atm}^2}\ , \tag{4}$$

the quasi-degenerate type ($m_{\nu1} \sim m_{\nu2} \sim m_{\nu3}$) gives

$$m_{\nu1} \equiv m_\nu\ , \qquad m_{\nu2} = m_\nu + \frac{1}{2m_\nu}\Delta m_\odot^2\ , \qquad m_{\nu3} = m_\nu + \frac{1}{2m_\nu}\Delta m_{\rm atm}^2\ , \tag{5}$$

and the inverse-hierarchical ($m_{\nu1} \sim m_{\nu2} \gg m_{\nu3}$) type gives

$$m_{\nu2} \equiv \sqrt{\Delta m_{\rm atm}^2}\ , \quad m_{\nu1} = m_{\nu2} - \frac{1}{2m_{\nu2}}\Delta m_\odot^2\ , \quad m_{\nu3} \simeq 0\ . \tag{6}$$

We take the typical values $\Delta m_{\rm atm}^2 = 3 \times 10^{-3}{\rm eV}^2$ and $\Delta m_\odot^2 = 7 \times 10^{-5}{\rm eV}^2$ in our calculation of the LFV.

2.2 LFV in Slepton Mass

We consider the MSSM with the right handed neutrinos, which have the Majorana mass M_R. Since SUSY is spontaneously broken at the low energy, we have to add the soft SUSY breaking terms:

$$\begin{aligned} -\mathcal{L}_{\rm soft} = &+(\mathbf{m}_{\tilde{L}}^2)_{ij}\tilde{l}_{Li}^\dagger\tilde{l}_{Lj} + (\mathbf{m}_{\tilde{e}}^2)_{ij}\tilde{e}_{Ri}^*\tilde{e}_{Rj} + (\mathbf{m}_{\tilde{\nu}}^2)_{ij}\tilde{\nu}_{Ri}^*\tilde{\nu}_{Rj} \\ &+\mathbf{A}_{lij}H_d\tilde{e}_{Ri}^*\tilde{l}_{Lj} + \mathbf{A}_{\nu ij}H_u\tilde{\nu}_{Ri}^*\tilde{l}_{Lj}\ + ... \end{aligned} \tag{7}$$

where $\mathbf{m}_{\tilde{L}}^2, \mathbf{m}_{\tilde{e}}^2$, and $\mathbf{m}_{\tilde{\nu}}^2$ are the left-handed slepton masses, the right-handed charged lepton and sneutrino masses, respectively.

Note that the lepton flavor violating processes come from the diagram including non-zero off-diagonal elements in the soft parameter because the off-diagonal elements mix among three generations.

In this paper we assume the m-SUGRA, therefore we put the assumption of universality for soft SUSY breaking terms at the unification scale:

$$
\begin{aligned}
(\mathbf{m}^2_{\tilde{L}})_{ij} &= (\mathbf{m}^2_{\tilde{e}})_{ij} = (\mathbf{m}^2_{\tilde{\nu}})_{ij} = \cdots = \delta_{ij} m_0^2 \ , \\
\mathbf{A}^{ij}_\nu &= \mathbf{Y}^{ij}_\nu a_0 m_0, \ \mathbf{A}^{ij}_e = \mathbf{Y}^{ij}_e a_0 m_0 \ ,
\end{aligned} \tag{8}
$$

where m_0 and a_0 stand for the universal scalar mass and the universal A-parameter, respectively. Because of the universality, the LFV is not caused at the unification scale.

To estimate values of the soft parameters at the low energy, we need to know the effect of radiative corrections. As a result, the lepton flavor conservation is violated at the low energy.

The RGE's for the left-handed slepton soft mass are given by

$$
\begin{aligned}
\mu \frac{d}{d\mu} (\mathbf{m}^2_{\tilde{L}})_{ij} = (\mu \frac{d}{d\mu} (\mathbf{m}^2_{\tilde{L}})_{ij})_{\mathrm{MSSM}} + \frac{1}{16\pi^2} \left[(\mathbf{m}^2_{\tilde{L}} \mathbf{Y}^\dagger_\nu \mathbf{Y}_\nu + \mathbf{Y}^\dagger_\nu \mathbf{Y}_\nu \mathbf{m}^2_{\tilde{L}})_{ij} \right. \\
\left. + 2(\mathbf{Y}^\dagger_\nu \mathbf{m}_{\tilde{\nu}} \mathbf{Y}_\nu + \tilde{m}^2_{h2} \mathbf{Y}^\dagger_\nu \mathbf{Y}_\nu + \mathbf{A}^\dagger_\nu \mathbf{A}_\nu)_{ij} \right] ,
\end{aligned} \tag{9}
$$

while the first term in the right hand side is the normal MSSM term which has no LFV, and the second one is a source of the LFV through the off-diagonal elements of neutrino Yukawa couplings.

3 Numerical Analyses of Branching Ratio

Let us calculate the branching ratio of $e_i \to e_j + \gamma$ $(j < i)$. The decay rate can be calculated using amplitudes $A^{L,R}$

$$
\Gamma(e_i \to e_j + \gamma) = \frac{e^2}{16\pi} m^5_{e_i} (|A^L|^2 + |A^R|^2) \ . \tag{10}
$$

Since we know the relation $m^2_{e_i} \gg m^2_{e_j}$, then we can expect $|A^R| \gg |A^L|$. The $A^{L,R}$ contain the contribution of the neutralino loop and the chargino loop as seen in fig.1. In order to clarify parameter dependence, let us present an approximate estimation. The decay amplitude is approximated as

$$
|A^R|^2 \simeq \frac{\alpha_2^2}{16\pi^2} \frac{|(\Delta \mathbf{m}^2_{\tilde{L}})_{ij}|^2}{m_S^8} \tan^2\beta \ , \tag{11}
$$

where α_2 is the couplng constant of $SU(2)_W$, respectively, m_S is a common SUSY particle mass. RGE's develop the off-diagonal elements of the slepton mass matrix and A-term. These terms at the low energy are approximated as

$$
(\Delta \mathbf{m}^2_{\tilde{L}})_{ij} \simeq -\frac{(6 + 2a_0^2) m_0^2}{16\pi^2} (\mathbf{Y}^\dagger_\nu \mathbf{Y}_\nu)_{ij} \ln \frac{M_{\mathrm{X}}}{M_{\mathrm{R}}} \ , \tag{12}
$$

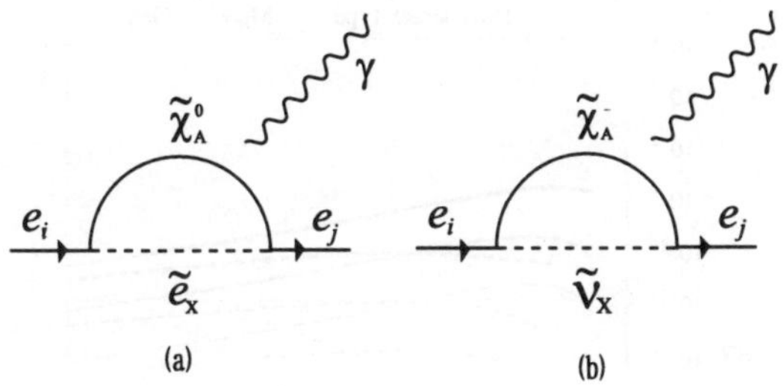

Figure 1. Feynman diagrams which contribute to the branching ratio of $e_i \to e_j + \gamma$. There are two types of diagrams, (a) neutralino-slepton loop and (b) chargino-sneutrino loop.

where M_X is the GUT scale. Therefore, off-diagonal elements of $(\mathbf{Y}_\nu^\dagger \mathbf{Y}_\nu)_{ij}$ are the crucial quantity to estimate the branching ratio.

We present a qualitative discussion on $(\mathbf{Y}_\nu^\dagger \mathbf{Y}_\nu)_{21}$ before predicting the branching ratio $\mathrm{BR}(\mu \to e + \gamma)$. This is given in terms of neutrino masses and mixings at the electroweak scale as follows:

$$(\mathbf{Y}_\nu^\dagger \mathbf{Y}_\nu)_{21} = \frac{M_R}{v_u^2}\left[U_{\mu 2}U_{e2}^*(m_{\nu 2} - m_{\nu 1}) + U_{\mu 3}U_{e3}^*(m_{\nu 3} - m_{\nu 1})\right] , \tag{13}$$

where $v_u \equiv v \sin\beta$ with $v = 174\mathrm{GeV}$ is taken as an usual notation and the unitarity condition of the MNS matrix elements is used. Taking the three cases of the neutrino mass spectra: the degenerate, the inverse-hierarchical and the normal hierarchical masses, one obtains the following forms, repectively,

$$\begin{aligned} (\mathbf{Y}_\nu^\dagger \mathbf{Y}_\nu)_{21} &\simeq \frac{M_R}{\sqrt{2}v_u^2}\frac{\Delta m_{\mathrm{atm}}^2}{2m_\nu}\left[\frac{1}{2}\frac{\Delta m_\odot^2}{\Delta m_{\mathrm{atm}}^2} + U_{e3}^*\right] , \quad \text{(Degenerate)} \\ &\simeq \frac{M_R}{\sqrt{2}v_u^2}\sqrt{\Delta m_{\mathrm{atm}}^2}\left[\frac{1}{4}\frac{\Delta m_\odot^2}{\Delta m_{\mathrm{atm}}^2} - U_{e3}^*\right] , \quad \text{(Inverse)} \\ &\simeq \frac{M_R}{\sqrt{2}v_u^2}\sqrt{\Delta m_{\mathrm{atm}}^2}\left[\frac{1}{2}\sqrt{\frac{\Delta m_\odot^2}{\Delta m_{\mathrm{atm}}^2}} + U_{e3}^*\right] , \quad \text{(Hierarchy)} \end{aligned} \tag{14}$$

where we take the bi-maximal mixing. For the case of the degenerate neutrino masses, $(\mathbf{Y}_\nu^\dagger \mathbf{Y}_\nu)_{21}$ depends on the unknown neutrino mass scale m_ν. As one takes the smaller m_ν, one predicts the larger branching ratio. In our calculation, we take $m_\nu = 0.3\mathrm{eV}$, which is close to the upper bound from the neutrinoless double beta decay experiment, and also leads to the smallest branching ratio.

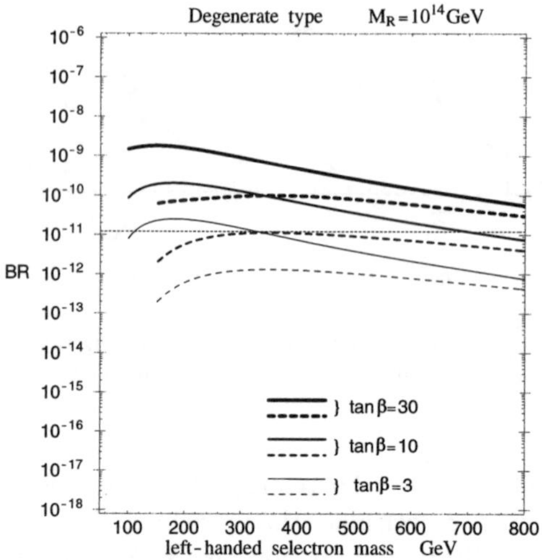

Figure 2. Predicted branching ratio $BR(\mu \to e + \gamma)$ versus the left-handed selectron mass for $\tan\beta = 3,\ 10,\ 30$ in the case of the degenerate neutrino masses. Here $M_R = 10^{14}$GeV and $U_{e3} = 0.2$ are taken. The solid curves correspond to $M_2 = 150$GeV and the dashed ones to $M_2 = 300$GeV. A horizontal dotted line denotes the experimental upper bound.

We also note that the degenerate case gives the smallest branching ratio $BR(\mu \to e+\gamma)$ among the three cases as seen in eqs.(14) owing to the scale of m_ν. It is easy to see the fact that the second terms in eqs.(14) are dominant as far as $U_{e3} \gtrsim 0.01$(degenerate), 0.01(inverse) and 0.07(hierarchy), respectively.

At first, we present numerical results in the case of the degenerate neutrino masses assuming $\mathbf{M_R} = M_R\mathbf{I}$. We take a universal scalar mass (m_0) for all scalars and $a_0 = 0$ as a universal A-term at the GUT scale ($M_X = 2 \times 10^{16}$ GeV). The branching ratio of $\mu \to e + \gamma$ is given versus the left-handed selectron mass $m_{\tilde{e}_L}$ for each $\tan\beta = 3,\ 10,\ 30$ and a fixed wino mass M_2 at the electroweak scale. In fig.2, the branching ratios are shown for $M_2 = 150,\ 300$ GeV in the case of $U_{e3} = 0.2$ with $M_R = 10^{14}$GeV and $m_\nu = 0.3$eV, in which the solid curves correspond to $M_2 = 150$GeV and the dashed ones to $M_2 = 300$GeV. As the $\tan\beta$ increases, the branching ratio increases because the decay amplitude from the SUSY diagrams is approximately proportional to $\tan\beta$ [1]. It is found that the branching ratio is almost larger than the experimental upper bound in the case of $M_2 = 150$GeV. On the other hand, the predicted values are smaller than the experimental bound except for $\tan\beta = 30$ in the case of $M_2 = 300$GeV.

Next we show results in the case of the inverse-hierarchical neutrino masses. As expected in eq.(14), the branching ratio is much larger than the one in the degenerate case. In fig.3, the branching ratio is shown for

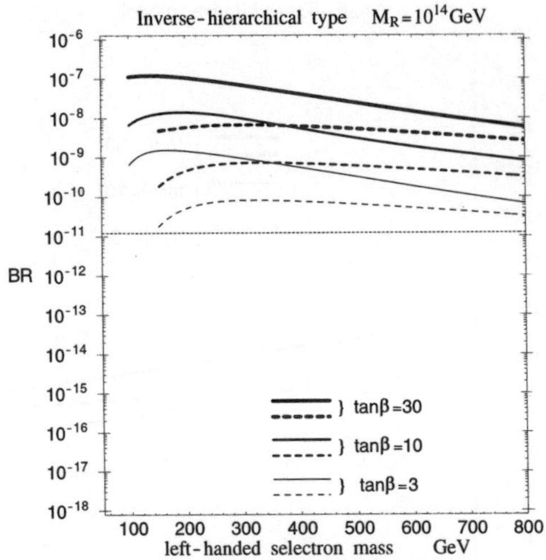

Figure 3. Predicted branching ratio $\mathrm{BR}(\mu \to e + \gamma)$ versus the left-handed selectron mass for $\tan\beta = 3,\ 10,\ 30$ in the case of the inverse-hierarchical neutrino masses. Here $M_R = 10^{14}$GeV and $U_{e3} = 0.2$ are taken. The solid curves correspond to $M_2 = 150$GeV and the dashed ones to $M_2 = 300$GeV.

$M_2 = 150,\ 300$ GeV in the case of $U_{e3} = 0.2$ with $M_R = 10^{14}$GeV. The M_R dependence is similar to the case of the degenerate neutrino masses. The predictions almost exceed the experimental bound as far as $U_{e3} \geq 0.05$, $\tan\beta \geq 10$ and $M_R \simeq 10^{14}$GeV. This result is based on the assumption $\mathbf{M_R} = M_R\mathbf{I}$, which is not guaranteed in the case of the inverse-hierarchical neutrino masses. Therefore we will examine a typical model [9], which gives $\mathbf{M_R} \neq M_R\mathbf{I}$ in section 4.

4 The Shafi-Tavartkiladze model - Inverse-hierarchical type

The model with the inverse-hierarchical neutrino masses is the Zee model [10], in which the right-handed neutrinos do not exist. However, one can also consider a Yukawa texture which leads to the inverse-hierarchical masses through the see-saw mechanism, namely the Shafi-Tavartkiladze model [9].

Shafi and Tavartkiladze used the anomalous $U(1)$ flavor symmetry . In this model, due to the Froggatt-Nielsen mechanism, the Yukawa interaction term in the effective theory is given as

$$\overline{e}_{Ri} L_j H_d \left(\frac{S}{M_{\rm pl}}\right)^{m_{ij}}, \tag{15}$$

where $\overline{e}_{R_i}$ and L_j are the right-handed charged lepton and the left-handed

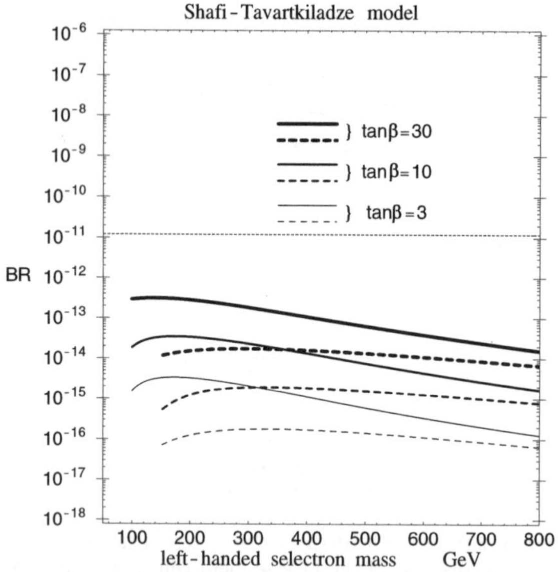

Figure 4. Predicted branching ratio BR($\mu \to e+\gamma$) versus the left-handed selectron mass for $\tan\beta = 3,\ 10,\ 30$ in the case of the Shafi-Tavartkiladze model. The solid curves correspond to $M_2 = 150\text{GeV}$ and the dashed ones to $M_2 = 300\text{GeV}$.

charged lepton doublet, respectively, H_d is Higgs doublet, and S is singlet field. The Yukawa couplings are given in term of $\lambda \equiv \langle S \rangle / M_{\rm pl} \simeq 0.2$.

The neutrino mass matrix is given in Appendix D. Fixing the $U(1)$ flavor charges k, k', n as $k = 0, n = 2, k' = 2$, which is consistent with neutrino mass data, the Yukawa coupling and the right-handed neutrino Majorana mass matrix are given as

$$\mathbf{Y}_\nu = \begin{pmatrix} \lambda^4 & \lambda^2 & \lambda^2 \\ 1 & 0 & 0 \end{pmatrix} , \ \mathbf{M}_R = M_R \begin{pmatrix} \lambda^4 & 1 \\ 1 & 0 \end{pmatrix} . \tag{16}$$

In eq.(16), components 2-2 and 2-3 in Y_ν must be zero for the sake of holomorphy of superpotential, it is so-called SUSY zero. The neutrino mass matrix is given by the see-saw mechanism as

$$\mathbf{m}_\nu = \mathbf{Y}_\nu^T \mathbf{M}_R^{-1} \mathbf{Y}_\nu v_u^2 = \frac{\lambda^2 v_u^2}{M_R} \begin{pmatrix} \lambda^2 & 1 & 1 \\ 1 & 0 & 0 \\ 1 & 0 & 0 \end{pmatrix} , \tag{17}$$

where the order one coefficient in front of each entry is neglected. This mass matrix gives the inverse-hierarchical neutrino masses. The $\mathbf{Y}_\nu^\dagger \mathbf{Y}_\nu$ is given as

$$\mathbf{Y}_\nu^\dagger \mathbf{Y}_\nu \sim \begin{pmatrix} 1 & \lambda^6 & \lambda^6 \\ \lambda^6 & \lambda^4 & \lambda^4 \\ \lambda^6 & \lambda^4 & \lambda^4 \end{pmatrix} . \tag{18}$$

It is noticed that the component $(\mathbf{Y}_\nu^\dagger \mathbf{Y}_\nu)_{21}$ is suppressed as

$$(\mathbf{Y}_\nu^\dagger \mathbf{Y}_\nu)_{21} \sim \lambda^6 \sim \mathcal{O}(10^{-5}) . \tag{19}$$

Then, we expect that the branching ratio of $\mu \to e + \gamma$ in this model is much smaller than the one in the case of $(\mathbf{M}_R)_{3\times3} = M_R(\mathbf{I})_{3\times3}$ in section 3.

In fig.4, the branching ratio is shown for $M_2 = 150,\ 300$ GeV. The predictions are given by taking $\lambda = 0.2$ and all order one coefficients in the Yukawa couplings are fixed to be one. The predicted value is much smaller than the one in the inverse-hierarchical case discussed in the section 3. Since the $(\mathbf{Y}_\nu^\dagger \mathbf{Y}_\nu)_{21}$ is proportional to λ^6, the smallness of the branching ratio is understandable.

5 Summary and Discussions

We investigate the lepton flavor violating processes $\mu \to e+\gamma$ in the framework of the MSSM with the right-handed neutrinos. Even if we impose the universal condition for the soft scalar masses and A-terms at the GUT scale, off-diagonal elements of the left-handed slepton mass matrix is generated through the RGE's running effects from the GUT scale to the right-handed neutrino mass scale M_R. We take the LMA-MSW solution for the neutrino masses and mixings.

The branching ratios of $\mu \to e + \gamma$ and processe are proportional to $(\mathbf{Y}_\nu^\dagger \mathbf{Y}_\nu)_{ij}^2$. Since $(\mathbf{Y}_\nu^\dagger \mathbf{Y}_\nu)_{ij}$ depends on the mass spectrum of neutrinos, we can compare the branching ratio of three cases of neutrino mass spectra: the degenerate, the inverse-hierarchical and the hierarchical case.

First, we study the three types in the case of $\mu \to e + \gamma$, in which we take $\mathbf{M}_R = M_R \mathbf{I}$. For the case of the degenerate neutrino masses, the branching ratio depends on the unknown neutrino mass m_ν. We take $m_\nu = 0.3$eV, which gives us the largest branching ratio. It is emphasized that the magnitude of U_{e3} is one of important ingredients to predict BR($\mu \to e + \gamma$). The branching ratio of the inverse-hierarchical case almost exceeds the experimental upper bound and is much larger than the degenetate case for $M_2 = 150,\ 300$GeV. In general, we expect the relation BR(degenerate)$\ll$BR(inverse-hierarchical)$<$ BR(hierarchical).

Finally, we investigate the branching ratio of $\mu \to e + \gamma$ in the typical model of the inverse-hierarchical case. The Shafi-Tavartkiladze model, which is a typical one of the inverse-hierarchical case, predicts the very small branching ratio. Thus, the models can be tested by the $\mu \to e + \gamma$ process.

The branching ratio of $\mu \to e + \gamma$ will be improved to the lebel 10^{-14} in the PSI. Therefore, future experiments can probe the framework for the neutrino masses.

I would like to thank my collaborators S. Kaneko, A. Kageyama and M. Tanimoto.

References

1. J. Hisano, T. Moroi, K. Tobe, M. Yamaguchi and T. Yanagida, Phys. Lett. **B357** (1995) 579; J. Hisano, T. Moroi, K. Tobe and M. Yamaguchi, Phys. Rev. **D53** (1996) 2442.
2. J. Hisano et al., Phys. Lett. **B437** (1998) 351; J. Hisano et al., Phys. Rev. **D59** (1999) 116005; M.E. Gomez et al., Phys. Rev. **D59** (1999) 116009; W. Buchmüller et al., Phys. Lett. **B459** (1999)171; W. Buchmüller et al., Nucl. Phys. **B576** (2000) 445; J. Ellis et al., Eur. Phys. J. **C14** (2000) 319; J. L. Feng et al., Phys. Rev. **D61** (2000) 113005; S. Baeket al., Phys. Rev. **D63** (2001) 051701; J. Sato, K. Tobe, and T. Yanagida, Phys. Lett. **B498** (2001) 189; J. Sato and K. Tobe, Phys. Rev. **D63** (2001) 116010; S. Lavignac et al., hep-ph/0106245;
3. J. A. Casas et al., Nucl. Phys. **B618** (2001) 171.
4. MEGA Collab., M. L. Brooks et al., Phys. Rev. Lett. **83** (1999) 1521.
5. F. Borzumati and A. Masiero, Phys. Rev. Lett. **57** (1986) 961.
6. Z. Maki, M. Nakagawa and S. Sakata, Prog. Theor. Phys. **28** (1962) 870.
7. M. Fukugita and M. Tanimoto, Phys. Lett. **B515** (2001) 30.
8. A. Kageyama, S. Kaneko, N. Simoyama and M. Tanimoto, Phys. Lett. **B527** (2002) 206; A. Kageyama, S. Kaneko, N. Simoyama and M. Tanimoto, hep-ph/0112359.
9. Q. Shafi and Z. Tavartkiladze, Phys. Lett. **B482** (2000) 145.
10. A. Zee, Phys. Lett. **B93** (1980) 389; **B161** (1985) 141; L. Wolfenstein, Nucl. Phys. **B175** (1980) 92; S. T. Petcov, Phys. Lett. **B115** (1982) 401; C. Jarlskog, M. Matsuda, S. Skadhauge and M. Tanimoto, Phys. Lett. **B449** (1999) 240; P. H. Frampton and S. Glashow, Phys. Lett. **B461** (1999) 95.

$\tau \to \mu\gamma$: STATUS AND PROSPECTS

TAKAYOSHI OHSHIMA
Department of Physics, Nagoya University
Chikusa, Nagoya 484-9802, Japan
E-mail: ohshima@hepl.phys.nagoya-u.ac.jp

Based on a previous Belle data analysis, from which the highest sensitivity of $\mathrm{Br}(\tau \to \mu\gamma) < 1 \times 10^{-6}$ at the 90% CL was attained last Summer in 2001, we here discuss possible ways to enhance our sensitivity on this decay mode in the coming years by the KEKB-Belle experiment and also by a proposed Super-KEKB facility.

1 Introduction

The lepton flavor violating (LFV) $\tau \to \mu\gamma$ process is a counterpart of a flavor-changing neutral current, $b \to s\gamma$, decay in the quark sector. While the latter transition has been observed at 10^{-5}, the former is forbidden in the Standard Model. Even neutrinos have flavor mixing among them, as indicated by recent neutrino-oscillation experiments; its branching fraction (Br) would be far too small for detection, for example, $\mathrm{Br} \propto \sin^2 2\theta \cdot (\Delta m_\nu^2/m_W^2)^2 \sim 10^{-54}$, due to the smallness of the neutrino masses. Even though heavy right-handed neutrinos are supposed to exist to give rise to non-zero neutrino masses by means of a seesaw mechanism, the anticipated branching ratio is still undetectably small due to the heaviness of the right-handed neutrinos[1]: $\mathrm{Br} \propto 1/(M_{\nu_R})^n$. When the SUSY world opens with its breaking at the electroweak scale, a large left-right mixing of heavy slepton masses could possibly enhance the branching fraction to detectable magnitudes, for instance[2] $\mathrm{Br}{\sim}10^{-(7-8)}$. Thus, $\tau \to \mu\gamma$ is an attractive and unique process to employ in looking for SUSY phenomena and physics beyond the SM.

Figure 1 shows the experimental results on $\mathrm{Br}(\tau \to \mu\gamma)$ so far measured[3], where the best upper limit was obtained by CLEO[4] and Belle[5] as $\mathrm{Br} = 1.1 \times 10^{-6}$ and 1.0×10^{-6} at the 90% CL, respectively. Some theoretical models which predict a relatively large $\mathrm{Br}(\tau \to \mu\gamma)$, for instance, by SUSY models are also indicated.

Since only the B-factory currently provides an opportunity to look for this decay mode with a higher sensitivity than so far achieved, this paper focuses on my personal view concerning the status and perspective of $\tau \to \mu\gamma$ research by the B-factory experiment, special focus is placed on a Belle experiment which just reported a preliminary result as an upper limit of $\mathrm{Br}{<}1.0\mathrm{x}10^{-6}$

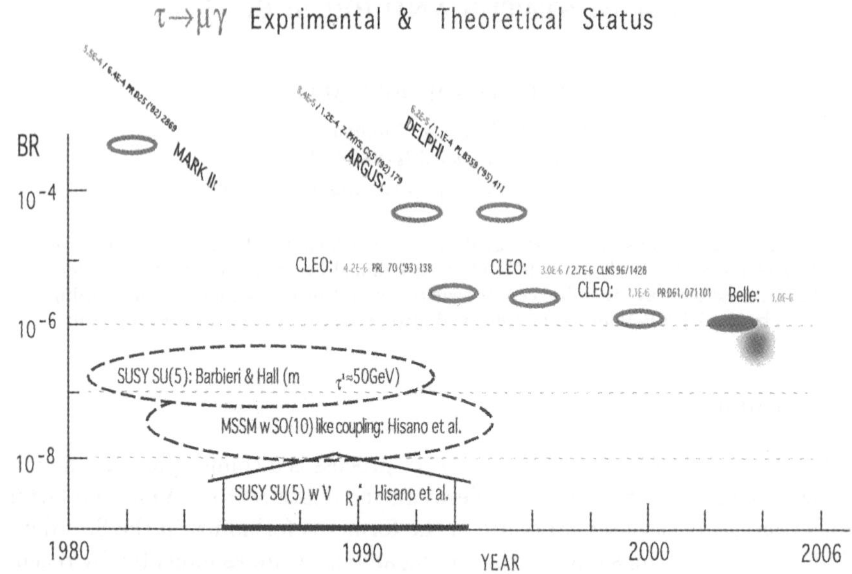

Figure 1. Experimental data, Br($\tau \to \mu\gamma$), so far achieved plotted by colored ellipses. Some predicted ranges on Br($\tau \to \mu\gamma$) by certain theoretical models are illustrated.

at the 90% CL. from the first 21 fb^{-1} data the last Summer, 2001[5]. The achievable sensitivity in the coming years and at a higher intensity B-factory, the so-called Super-KEKB, which is being proposed, is also examined.

2 KEKB/Belle Experiment

KEKB[6] is a high-luminosity e^+e^- asymmetric collider machine of 3.5 GeV positrons and 8.0 GeV electrons built as a B-factory for CP violation studies in B-meson decays to scrutinize the Kobayashi-Maskawa theory. It provides a huge number of τ-pairs as well as $B\overline{B}$ pairs: $\sigma(\tau\tau) \approx 0.9$ nb and $\sigma(B\overline{B}) \approx 1.0$ nb. The B-factory can also act as a Tau-factory. Belle is a general-purpose type of spectrometer[7] placed on an interaction region in the KEKB machine, as illustracted in Figure 2. Among many detectors, the Central Drift Chamber (CDC) and EM Calorimeter (ECL) are especially essential for tracking charged particles and detecting photons, respectively, in this type of physics; an Aerogel Cherenkov Counter (ACC), a K_L/μ detector (KLM) and a Time-Of-Flight counter (TOF) and a CDC in terms of dE/dX play

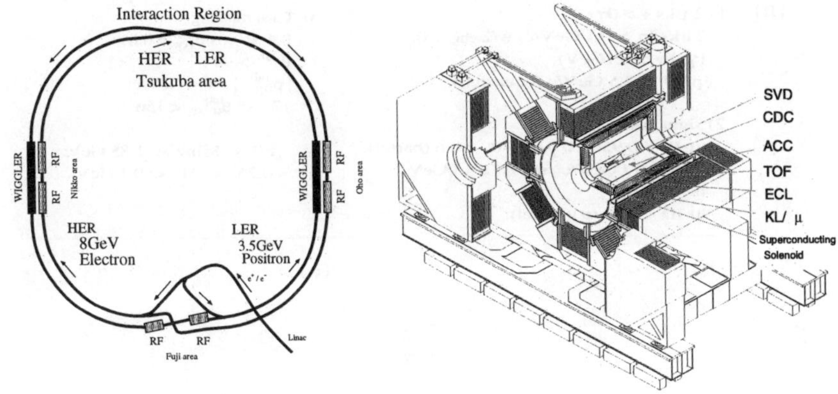

Figure 2. Schematic drawing of the KEKB collider (left) and Belle spectrometer (right).

important roles in particle identification.

3 Previous Belle Analysis

The data used were the first recorded 21fb^{-1} that corresponds to 19.3M τ-pair productions at $\Upsilon(4S)$. We searched for the event

$$(\tau \to \mu\gamma) + (\tau \to \mu^{not} + \geq 0\gamma + X(\text{missing})),$$

where "μ^{not}" means a charged particle, but not a muon. Both the tracks and photon are required to be detected by the barrel part of the detectors, and certain conditions are imposed on the momentum and polar-angle of the missing quantities in order to confirm that the missing particle(s) are neutrino(s). The applied selection criteria are listed in Figure 3.

Figure 3 also shows the remaining data events and Monte Calro in terms of M_{inv} (an invariant mass of the $\mu\gamma$ system) and ΔE (energy difference between the $\mu\gamma$ system and the beam at the CM system) distributions and their scatter plots, where the central square in (d) and (e) is the signal-box assigned in this analysis.

In a way of analysis we located the origins of the backgrounds (BG). The main BG is attributed to radiative $\gamma\tau\tau$: The MC result is shown in Fig.4(a). Such events distribute only in the ΔE <0 region, while some remaining data events seen in Fig.3(d) also scatter at the ΔE >0 region. It was identified as $\gamma\mu\mu$ events, but contaminated as $\gamma\mu$+μ^{not} events due to an inefficiency of η ≈7% in the KLM muon identification. Figure 4(b) gives the real data

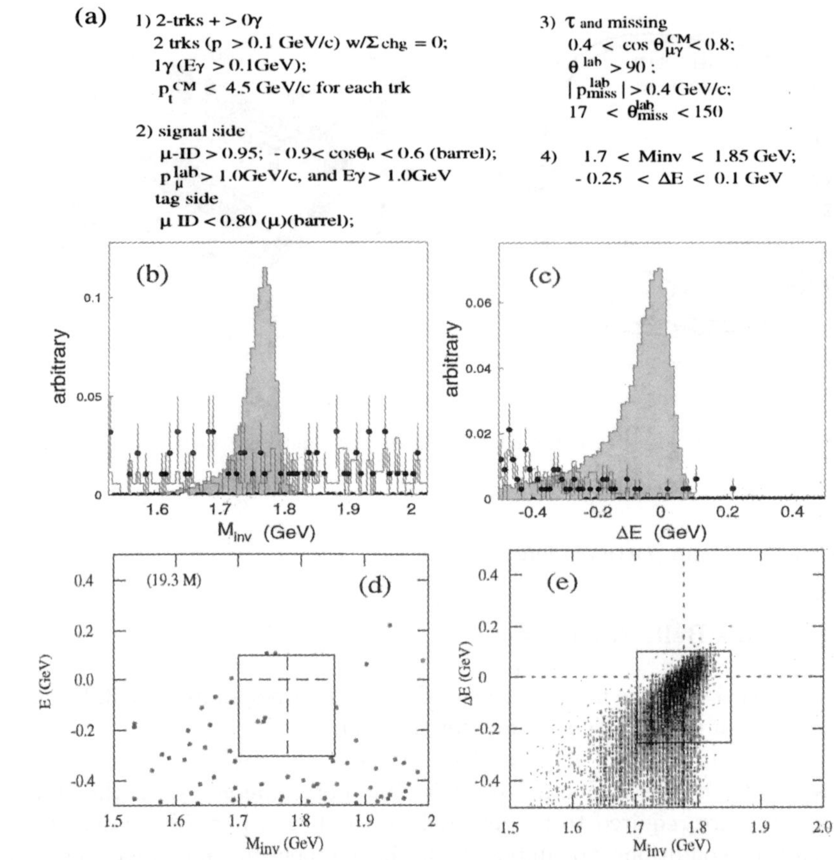

Figure 3. The selection criteria are listed in (a). Data events (closed circles) which survived from the selection and MC signal samples (histogram) are shown in (b) M_{inv} and (c) ΔE distributions. (d) and (e) are their scatter plot on M_{inv}-vs-ΔE for data events (19.3 M) and MC samples, respectively.

events that passed all selections, but whose two tracks were required to be muons (denoted as "$\mu\mu$"): A rather dense distribution at $\Delta E > 0$ is seen. Since generic $\tau\tau$, $B\overline{B}$, continuum, Bhabha, and most of two-photon processes do not yield this sort of BG, according to MC simulations, the origin might be understood as radiative muon-pair processes, $\mu\mu\gamma(\gamma)$. The BG was therefore

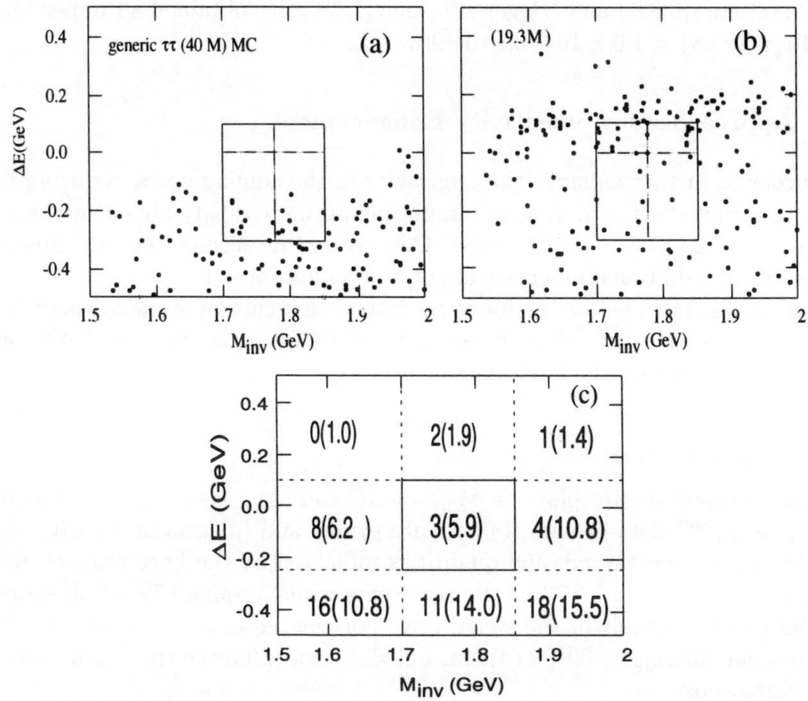

Figure 4. M_{inv}-vs-ΔE scatter plots for (a) generic $\tau\tau$ MC samples (generated 40 M) and (b) $\mu\mu$ data events (19.3 M). (c) Comparison between the resulted number of data events and the expected BGs. The central square is the signal-box.

evaluated to be the sum of these contributions as

$$N_{BG} = N_{\tau\tau}^{MC} + N_{\mu\mu}^{data} \times \eta, \tag{1}$$

where the first term is generic $\tau\tau$ events (mostly $\gamma\tau\tau$) estimated by MC and the second term is the $\mu\mu$ event just discussed above.

The thus-estimated BG rates and the survived data event numbers are compared, and a well-matched agreement is found between them, as shown in Figure 4(c). The expected BG and candidate events in the signal box are 5.9 and 3, respectively, and the upper limit at the 90% confidence level was evaluated on the number of signals as 3.5 events. Taking into account the systematic uncertainties, such as track(2.3%) and photon(5.3%) reconstraction efficiencies, cut selection(1.5%), luminosity(1.4%), muon identification(4%),

MC statistics(0.8%) and trigger efficiency(1.7%), we obtained an upper limit of $Br(\tau \to \mu\gamma) = 1.0 \times 10^{-6}$ at the 90% CL.

4 Approaches for Sensitivity Enhancement

In order to further enhance the sensitivity in the coming years, reducing the BG contamination as well as accumulating an increasingly higher integrated luminosity are two required tasks. Concerning the former, we are thinking to apply an additional kinematical cut and an appropriate signal-area assignment, as discussed below. As for to the latter, an inclusion of endcap detectors would enlarge the detection efficiency to ~12% from the previous 9.4%, and would then improve the sensitivity.

4.1 M^2_{miss} vs p_{miss}

Figure 5 shows scatter plots on $M_{\rm inv}$-vs-ΔE and $M^2_{\rm miss}$-vs-$p_{\rm miss}$ for (a) signal MC, (b) $\mu\mu^{not}$ data events, (c) $\mu\mu$ data events and (d) generic $\tau\tau$ MC. The additional cut on the missing quantities indicated by the lines removes 68% of the $\gamma\tau\tau$ events and 87% of the $\mu\mu$ events while keeping 87% of all signals or 95% of the signals in the signal-box. For the remaining data events, this additional cut rejects 60% of them, but does not influence the candidates in the signal-box.

As a result, because the BG distribution becomes quite flat on both the ΔE and $M_{\rm inv}$ (BGs show a steep distribution on ΔE without the cut), the evaluation on the BG rate in the signal-box can be carried out reliably based on eq.(1).

4.2 Photon resolution and signal-area

As can be seen in Figures 3(b), (c) and (e), the signal MC events exhibit long low-energy tails in both the $M_{\rm inv}$ and ΔE distributions. Figure 6 is a scatter plot of the MC signals. The tail along the ΔE-axis is due to initial radiation, while the tail across the plane is mostly due to photon-energy leakage out of the ECL detector. The Belle detector has a good energy resolution for photons, for instance, $\sigma_E/E \sim$1.8 % @1 GeV, and a good momentum resolution for charged particles, as $\sigma_p/p \sim$0.35 % @1 GeV/c. Thus, the intrinsic resolutions would be $\sigma_{M_{\rm inv}}$=18 MeV and $\sigma_{\Delta E}$=48 MeV compared to the CLEO[4] case of 25 MeV and 53 MeV, respectively. In spite of this advantage, due to these tails the signal-box cannot be assigned correspondingly smaller than the CLEO's one to ensure that it will capture as many signals as possible.

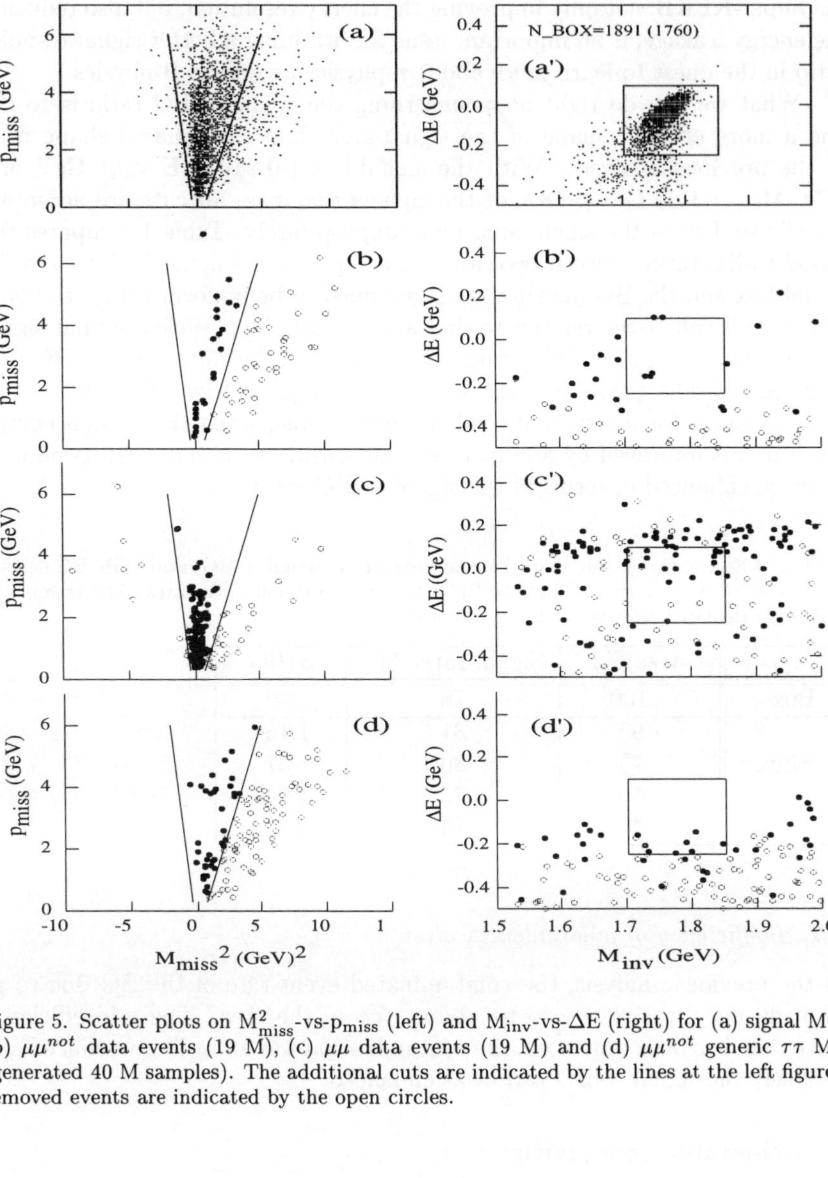

Figure 5. Scatter plots on M^2_{miss}-vs-p_{miss} (left) and M_{inv}-vs-ΔE (right) for (a) signal MC, (b) $\mu\mu^{not}$ data events (19 M), (c) $\mu\mu$ data events (19 M) and (d) $\mu\mu^{not}$ generic $\tau\tau$ MC (generated 40 M samples). The additional cuts are indicated by the lines at the left figures; removed events are indicated by the open circles.

To improve the current ECL energy resolution, we should examine a possible effect on reducing the leakage tail by extending the CsI crystal region used to measure the energy deposite. Once we have built a new ECL detector for the Super-KEKB, not only improving the energy resolution, but also reducing the energy leakage, is an important issue for attaining a better signal-to-noise ratio in the quest to learn more about τ-physics as well as B-physics.

What we can do right now concerning the signal-to-BG ratio is to define a more efficient shape of the signal-area than the squared shape used in the previous analysis. With the signal-box (-0.25< ΔE <0.1 GeV and 1.7< M_{inv} <1.85 GeV), 78% of the signals that pass all cuts are accepted. An ellipse defines the signal area more appropriately. Table 1 compares the signal-to-BG ratio between two cases. An ellipse area is defined relative to the signal-box and the BG distribution is assumed to be uniform within the box. The ratio is calculated relative to that at the signal-box. With a similar signal acceptance rate of 77% (76%) with that of the box, the ellipse area or BG rate decreases to 60% (48%) and its ratio is then improved by 1.64 (2.03). This is a big factor for a sensitivity enhancement; thus, when the photon energy resolution is improved by any method, the sensitivity would correspondingly be more enhanced in terms of the signal-to-BG ratio.

Table 1. Effect of the elliptic shaped signal-area on the signal-to-BG ratio. THe BG density is assumed to be uniform, so that the BG rate is proportional to the area. The area of the ellipse is calculated relative to that of the signal-box.

	Area(%)	Signal-rate(%)	S/BG
Box	100	78	1.00
Ellipse	90	81	1.15
	75	80	1.37
	60	77	1.64
	48	76	2.03

4.3 Inefficiency of muon identification

In the previous analysis, the comtaminated event rate of the BG due to μ-id inefficiency was shown to be about 25% in the area. The μ-id efficiency since then has been improved, the contamination was accordingly reduced and inversely the signal-to-BG ratio was enhanced.

5 Achievable Sensitivities

Let us summarize the improvement factors discussed above.

<u>BG</u>:

The additional criteria on M^2_{miss}-vs-p_{miss} would remove the BG by 60%. When the μ-id efficiency is improved by any amount by any method, $\mu\mu^{not}$ contamination, 25% of BG, is suppressed by an equivalent rate.

Signal – to – BGratio:
Taking a proper shape for defining the signal-area, for instance, an ellipse would enhance the ratio by a factor of 1.5-2.0 while not largely decreasing the signal acceptance rate.

Statistics:
Including the endcap part of detectors would increase the detection efficiency from 9.5% to ~12%.

Without any of the above-mentioned improvements, an achievable upper limit at the 90% CL under the present Belle condition can be approximated as

$$\mathrm{Br} \sim \frac{1.28\sqrt{(6\ \mathrm{ev}/20\ \mathrm{M}) \times \mathrm{N}_{\tau\tau}(\mathrm{M})}}{2 \times \varepsilon \times \mathrm{N}_{\tau\tau}(\mathrm{M})} \tag{2}$$

$$\sim \frac{4}{\sqrt{\mathrm{N}_{\tau\tau}(\mathrm{M})}} \times 10^{-6}, \tag{3}$$

where $N_{\tau\tau}$ is the number of produced τ-pair events in a unit of millions. We here assume, based on the previous analysis, that an expected BG rate is 6 events in the signal-box for 20 M τ-pair productions with a detection efficiency of $\varepsilon \approx$9%. Figure 6 indicates the thus-calculated upper-limit as a function of the integrated luminosity.

On the other hand, when we assume improvements to be possibly attainable in the next analysis, that is, with an additional criterion on M^2_{miss}-vs-p_{miss} and endcap inclusion, the expected sensitivity can be enhanced by a factor of 1.5-1.8: $1/\sqrt{0.4}$ ~1.5 by the former and ~1.2 by the latter,

$$\mathrm{Br} \sim \frac{2.5}{\sqrt{\mathrm{N}_{\tau\tau}(\mathrm{M})}} \times 10^{-6}. \tag{4}$$

Within the next few years, because the Belle experiment is expected to accumulate 300 fb^{-1} of data, an upper limit of $\mathrm{Br}(\tau \to \mu\gamma) \approx 1 \times 10^{-7}$ will be attained.

Super-KEKB, a super-high luminosity B-factory, has been attempting to push CP violation physics further. With 3-4 years running under $\approx 10^{36}/\mathrm{cm}^2/\mathrm{sec}$ luminosity, because 10,000 M τ-pairs could be produced, the expected sensitivity would be around 2×10^{-8}. In the case of the Super-KEKB era, a better than hoped for sensitivity is highly anticipated.

6 Summary

We have discussed ways to further develop our sensitivity in the search for $\tau \to \mu\gamma$ decay. Proofs of some of these proposals can be demonstrated in an analysis being carried out using 30 fb^{-1} data and 80 fb^{-1} data to be obtained in the coming Summer, 2002.

An expected sensitivity with 10,000 M tau-pairs production at the proposed Super-KEKB is estimated to be around a few $\times 10^{-8}$.

It is my personal evaluation that the sensitivity seems to be rather limited at the $\mathbf{O}(10^{-8})$ level and not feasible to reach $\mathbf{O}(10^{-9})$ or better. A new path up the mountain to $\tau \to \mu\gamma$ is waiting to be explored at any time now.

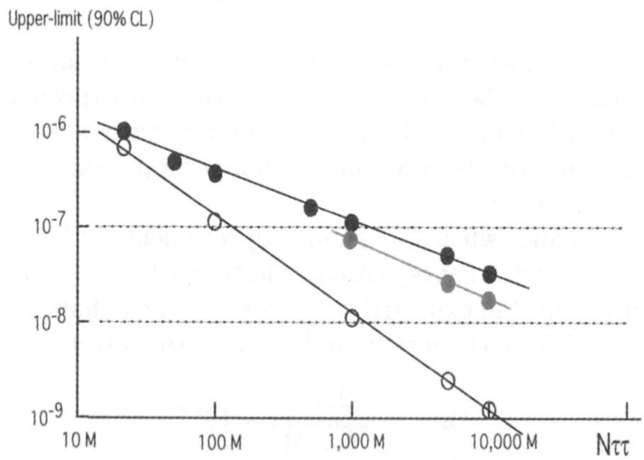

Figure 6. Expected achievable sensitivity by the KEKB-Belle experiment and a proposed Super-KEKB. The upper line is an achievable sensitivity, eq.(3), evaluated under the previous analysis condition at the Belle experiment. The second upper line, eq.(4), is with the attainable improvements by the M^2_{miss}-vs-p_{miss} additional cut and the inclusion of the endcap detector. The bottom line is in an extreme case that no candidate is survived. Within a few years, Belle will accumulate 300 M τ-pairs or more. At the Super-KEKB with a luminosity of $\approx 10^{36}/cm^2/sec$, 10,000 M τ-pairs could be produced in 3-4 years running.

References

1. R. Stroynowski, Nucl. Phys. B (Proc. Suppl.) 76 (1999) 185-191.
2. J. Hisano, T. Moroi, K. Tobe, M. Yamaguchi and T. Yanagida, Phys. Lett. B357 (1995) 579-587; J. Hisano, D. Nomura and T. Yanagida, Phys. Lett. B437 (1998) 351-358; R. Barbieri and L.J. Hall, Phys. Lett. B338 (1994) 212-218.
3. Particle Data Group, Review of Particle Physics, Eur. Phys. J. C 3 (1998) 1-794.
4. S. Ahmed et al., Phys. Rev. D61 (2000) 071101(R); K.W. Edwards, et al., Phys. Rev. D55 (1997) R3919-3923.
5. BELLE-CONF-0118, "Search for $\tau \to \mu\gamma$ decay at Belle"; Proceedings of EPS2001 Conf., to be published.
6. KEKB accelerator group, KEKB B Factory Design Report, KEK Report 95-7, 1995; K. Akai et al., Proc. 1999 Particle Accelerator Conference, New York (1999); Y. Funakoshi et al., Proc. 2000 European Particle Accelerator Conference, Vienna (2000).
7. Belle Collaboration, *The Belle Detector*, Nucl. Instrum. Meth., A479 (2002) 117-232.

NEW EXPERIMENT TO SEARCH FOR $\mu^+ \rightarrow e^+\gamma$ AT PSI

W. OOTANI

International Center for Elementary Particle Physics, University of Tokyo, 7-3-1 Hongo, Bunkyo-ku, Tokyo, 113-0033, Japan
E-mail: wataru@icepp.s.u-tokyo.ac.jp

A new experiment to search for the lepton flavor violating process, $\mu^+ \rightarrow e^+\gamma$ is planned at Paul Scherer Institut(PSI) where the most intensive DC muon beam in the world is available. The experiment is designed to have a sensitivity down to 10^{-14} branching ratio which is below the present experimental bound by three orders of magnitude and could reveal the nature of new physics beyond the standard model such as SUSY-GUT. The current status of the preparation for the experiment and prospects are described in this article.

1 Introduction

A new experiment is planned to search for the rare decay of muon into electron and gamma which violates lepton flavor conservation [1]. We aim at searching with a sensitivity down to 10^{-14} branching ratio by using the intensive DC muon beam at Paul Scherer Institut(PSI). This sensitivity is below the current experimental limit by three orders of magnitude. Many promising theories beyond the standard model such as SUSY-GUT naturally predict the branching ratio of $\mu \rightarrow e\gamma$ decays somewhere above 10^{-14}. Therefore, this experiment could give us a chance to explore such new physics [2]. Sizable decay rate is also predicted in the SUSY model with the seesaw mechanism which can explain the finite mass of neutrino. The branching ratio above 10^{-14} can be expected for the 'large mixing angle solution' in this model which is favored by the recent results from the solar neutrino measurements at Super Kamiokande [3].

Figure 1 shows the current design of the detector for the $\mu^+ \rightarrow e^+\gamma$ search experiment at PSI. The positron and gamma-ray from the $\mu^+ \rightarrow e^+\gamma$ decay is detected by a positron spectrometer and gamma-ray detector placed just outside the positron spectrometer, respectively. The design of each detector component and current status of the preparation are described in this article.

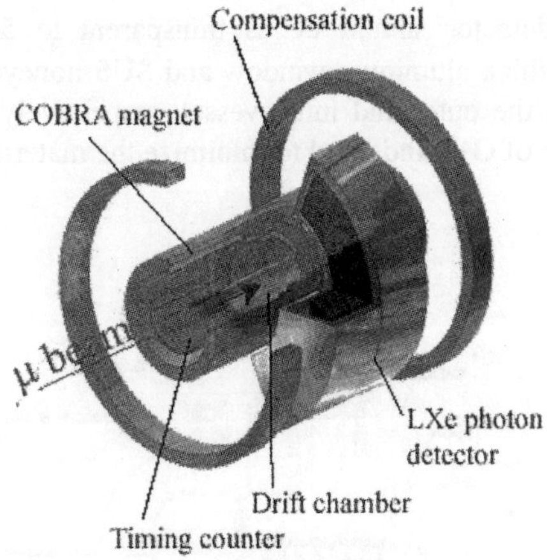

Figure 1 Detector for new μ->eγ search experiment at PSI.

2 Detector

2.1 Liquid xenon gamma detector

We have been developing a new type of liquid xenon scintillation detector for gamma-ray detection in this experiment. A gamma-ray detector with good energy-, position-, and timing resolutions is indispensable to minimize accidental overlaps which could be a major background source in this experiment. The gamma-ray detector utilizes scintillation light from the liquid xenon with a volume of 800 liters. The scintillation light is collected by an array of 800 photomultipliers(PMTs) which is immersed in the liquid xenon so that the scintillation light can be collected efficiently. The fast response and large light yield of the xenon serve to realize the good energy-, timing-, and position- resolutions.

In the previous study, the response to gamma-rays up to 2MeV was measured using a small prototype detector with an active volume of 2.3 liters viewed by 32 PMTs[4]. The naive extrapolation form the results shows the detector could have resolutions for 52.8MeV gamma-rays from the μ->eγ of 0.7% for energy, less than a few mm for position, and less than 50psec for timing where the resolutions are measured in sigma. In order to prove this estimation and test various elements to be used in the full-scale detector such as a refrigerator, cables, feedthrough, etc., we constructed a larger prototype with liquid xenon of 69 liters active volume so that the high energy gamma-ray event can be fully contained within the active volume.

Figure 2 shows the schematic view of the constructed prototype detector. A total of 228 PMTs arrayed in rectangular shape is installed in the liquid xenon. The front face of the detector should be as transparent to 50MeV gamma-rays as possible. A 100μm thick aluminum window and SUS honeycomb window are used on the front face in the outer and inner vessel, respectively. The front face of the PMT holder is made of G10 and acryl to minimize the material thickness.

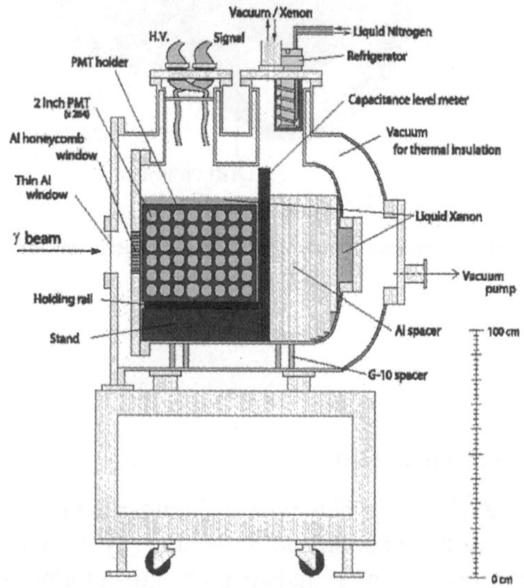

Figure 2 Schematic view of the constructed prototype of the liquid xenon gamma-ray detector.

Powerful and reliable cryogenics is required to liquefy such large amount of xenon and to keep it stable after the liquefaction over a long period. The liquid nitrogen is used to liquefy xenon. After the completion of the liquefaction, the liquid is maintained by a pulse-tube refrigerator. The pulse-tube refrigerator is easy to handle and the vibration noise is much lower compared with the conventional GM refrigerator because there is no mechanical moving part in the low temperature stage. The heat load into the liquid xenon is dominated by heat generation by the PMTs (18W) and heat inflow through the cables(10W). The cooling power at liquid xenon temperature (165K) is about 70W which is large enough to cover the heat load of the prototype detector.

In the first liquefaction test, the liquid xenon with total amount of 100 liters was liquefied in two days with heat exchanger pipe cooled by liquid nitrogen. After the completion of the liquefaction, the liquid is kept at stable temperature over a week with the pulse tube refrigerator. High voltage is applied to all PMTs during the liquefaction test.

We can believe that the performance of the cryogenics based on the pulse tube refrigerator is sufficient for our purpose and similar cryogenics with larger cooling power is applicable to the full-scale detector.

To demonstrate the performance of the liquid xenon detector for the 52.8MeV gamma-rays from the $\mu \rightarrow e\gamma$ decays, we have started to study with the prototype detector using high energy gamma-rays. We are using the high energy gamma-rays up to 40MeV from laser Compton backscattering at the TERAS electron storage ring of National Institute of Advanced Industrial Science and Technology (AIST) in Tsukuba, Japan (Figure 3).

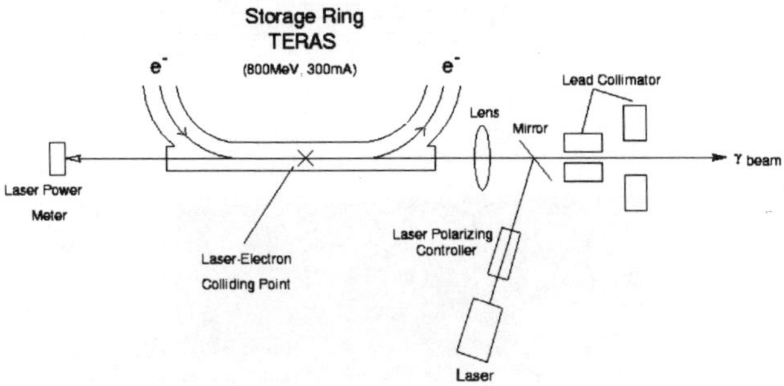

Figure 3 Laser Compton backscattering facility at TERAS of AIST.

Energy resolution can be evaluated by the spread of the Comptopn edge. Position resolution can be estimated by analyzing distribution of the PMT outputs with collimated gamma-ray beam. Timing resolution can be estimated by averaging the arrival time after proper ADC slewing collections. First gamma-ray beam test at AIST was performed in June 2001 and the analysis is now underway.

2.2 *Positron spectrometer*

The positron spectrometer which is called COnstant Bending RAdius(COBRA) spectrometer will consist of a special solenoidal magnet with a gradient magnetic field, drift chamber system for the positron tracking, and scintillation counters for the timing measurement.

2.2.1 COBRA magnet

COBRA magnet is a thin superconducting solenoidal magnet specially designed to produce a gradient magnetic field. Under the gradient field arranged properly, positron emitted from the target curls with a constant projected bending radius independently of the emission angle and then is swept out quickly. Therefore, a sharp momentum window for the signal positrons can be defined and pile-up problem due to the positrons from the Michel decays in the drift chamber system can be avoided.

COBRA magnet consists of five coils with three different diameters to realize the gradient field (Figure 4). The field strength is 1.26T at the center and 0.49T at the end of the magnet.

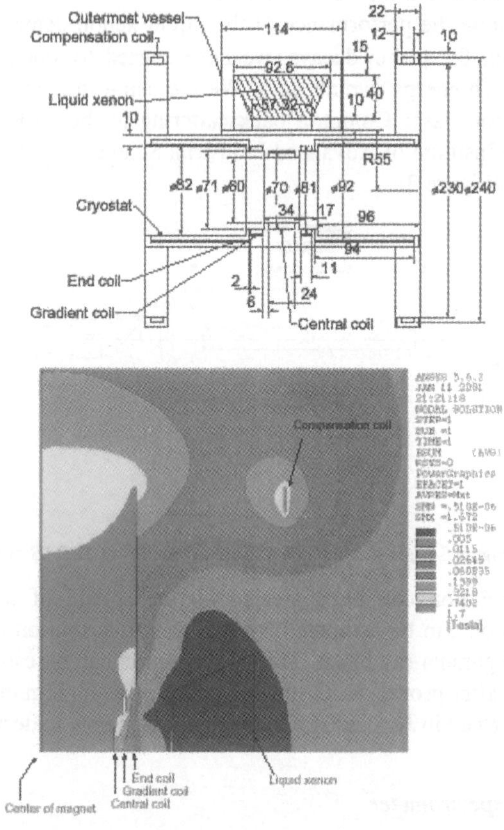

Figure 4 Layout of the COBRA magnet(top) and residual magnetic field around the liquid xenon calorimeter(bottom). Note that the intervals of the contour are not linear.

A pair of compensation coils with a diameter of 240cm is installed around the COBRA magnet to reduce the residual field around the liquid xenon calorimeter for a proper operation of the photomultipliers (PMT) of the calorimeter. The magnetic field around the calorimeter can be suppressed down to around 50Gauss which is acceptable for the PMTs to be used in the calorimeter, while giving a negligible effect inside the main solenoid where the positron trajectories are measured. It was also confirmed that the essential features such as the constant bending radius and fast sweeping out of the positrons from the Michel decay is preserved in this design.

High strength superconducting cable with aluminum stabilizer specially designed for this magnet was already developed and the construction of the magnet has been started.

2.2.2 Drift chamber

Drift chamber system for the positron tracking consists of seventeen chamber sectors aligned radially with 10 degrees intervals. Each chamber sector has two staggered arrays of drift cells. Such cell configuration allows us to determine the r-coordinate and the time of the positron track. Position of positron hit point along the beam direction is roughly estimated with an accuracy of ~ 1cm by the charge ratio observed at both ends of the sense wire and then determined with an accuracy of ~ 300μm by the readout from the aluminum pad deposit on the chamber wall with a vernier pattern.

Two prototypes have been constructed for optimizing the detector parameters such as gas mixture, high voltage and for studying the vernier patterns. Study on the operation of the chamber under the magnetic field is also being tried using the prototype.

2.2.3 Timing counter

Positron timing counters of cylindrical hodoscope arrays of plastic scintillators are placed at both sides in the COBRA magnet. Test facility of the timing counter prototypes by using cosmic ray muons was constructed as shown in Figure 5. A telescope of Microstrip Gas Chambers is used to reconstruct the muon track. It allows us to determine the hit point of the muon on the timing counter prototype with a resolution better than 1mm and to correct for the position dependence of the timing resolution.

BC404 scintillator bars (cross section 5 cm × 1 cm), wrapped with aluminized Mylar of 50μm thick and coupled to PMTs through the light guides at both ends were used for the prototype counters. We used Philips XP2020 UR and Hamamatsu R5946 fine mesh PMTs for the prototype.

Figure 5 shows the measured timing resolutions as a function of the hit position of the muon. The timing resolution obtained by weighted average of the resolutions measured by each PMT of the counter is about 60psec independently of the hit position. We also observed the timing resolution is improved with a dependence on the square root of the number of photoelectrons by tilting the counter, thus increasing the muon path in the counter.

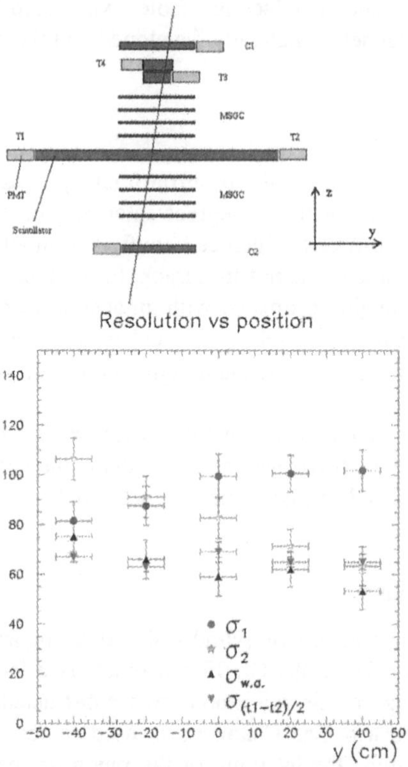

Figure 5 Test facility of the timing counter prototype(top) and obtained timing resolutions(bottom).

2.3 *Trigger electronics*

The trigger rate due to the accidental overlap background which is considered to be the most serious background source in our experiment was evaluated to be about 20Hz after some simple selection.

We are designing the trigger system with high density FPGAs coupled with fast FADCs of 100MHz which is programmed to evaluate the physical quantities used for the trigger such as the energy, direction, and timing of the photon and positron.

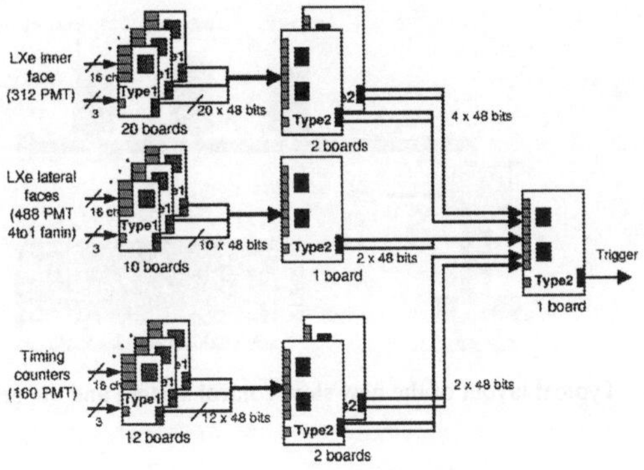

Figure 6 Structure of the trigger system. Two types of trigger electronics board are used.

The trigger system will have two different electronics boards which are arranged in a three structure as shown in Figure 6. A first board (type1) digitizes the signal from the calorimeter and positron spectrometer and transfer information to a second board (type2) through LVDS connections where the different physics quantities are evaluated.

The design work on the FPGA configurations for the type1 boards is almost finished and all the trigger requirements for the calorimter board are found to be acceptable. The printed circuit design of the type1 board is in progress.

2.4 *Slow control system*

A new slow control system is now under development at PSI to control the detector by monitoring the parameters such as temperature, pressure, and high voltage [5]. Reliable control system is indispensable for stable operation of the detector over a long period of time.

Figure 7 shows the scheme of the new slow control system. Each slow control node has microcontrollers equipped with both analog-to-digital and digital-to-analog converters. The system communicates via a RS485 based network. Several prototypes have been already built and tested. A prototype has been implemented and successfully tested at a drift chamber system. The final system design is now fixed and under production.

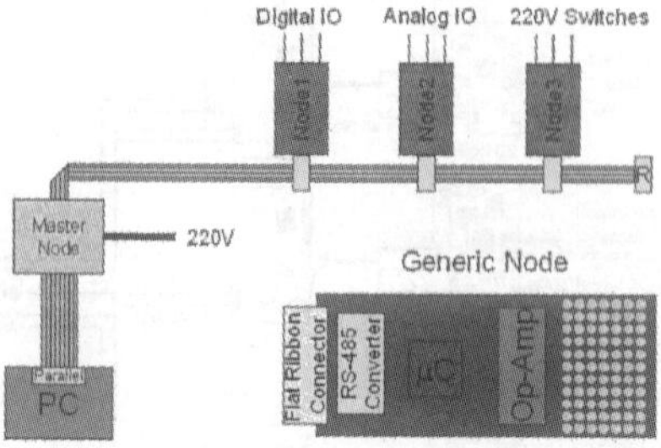

Figure 7 Typical layout of the new slow control system and the generic node.

3 Summary and Prospects

A new μ->eγ search experiment at Paul Scherer Institut(PSI) is planned. The experiment is designed to have a sensitivity of 10^{-14} branching ratio by using a high performance liquid xenon calorimeter and positron spectrometer with excellent features of the constant bending radius and fast sweeping out of the positrons from the Michel decays.

The preparation for each detector component is going on as well as the development on the data acquisition system and study on the πE5 beam line at PSI. The liquid xenon calorimeter is a key item of this experiment. We constructed the large prototype of the xenon calorimeter and have just started a series of beam tests at the laser Compton backscattering facility at TERAS of AIST to study the performance of the detector for gamma-rays of around 40MeV. The results from the beam tests will be reflected in the design of the full-scale detector and the construction will be started soon.

Acknowledgement

This work is in part supported by Grant-in-Aid for Scientific Research on Priority Areas (A) provided by the Ministry of Education, Culture, Sports, Science and Technology of Japan. We would like to thank the staff at TERAS including Dr. H. Toyokawa for their support.

References

1. T. Mori et al., "Search for μ->eγ down to 10^{-14} branching ratio", Research Proposal to PSI, May 1999
2. R. Barbieri and L.J.Hall, *Phys. Lett.* B338(1994)212; R. Barbieri, L. J. Hall, and A. Strumia, *Nucl, Phys.* B445(1995)219

3. J. Hisano and D. Nomura, *Phys. Rev.* D59(1999)116005; J. Hisano, T. Moroi, K. Tobe, and M. Yamaguchi, *Phys. Lett.* B357(1995)579; J. Hisano, D. Nomura, T. Yanagida, *Phys. Lett.* B437(1998)351
4. S. Mihara et al., gDevelopment of a Liquid Xe Photon Detector for mu->e gamma Decay Search Experiment at PSI", Proc. IEEE2000 - Nuclear Science Symposium and Medical Imaging Conference, 2000.10.15-20, Lyon, France
5. http://midas.psi.ch/mscb

PROBING PHYSICS BEYOND THE STANDARD MODEL FROM LEPTON SECTOR

J. HISANO
ICRR, University of Tokyo, Kashiwa 277-8582, Japan
Theory Group, KEK, Tsukuba 305-0801, Japan
E-mail: hisano@icrr.u-tokyo.ac.jp

In this review we discuss physics of the lepton sector, the anomalous dipole moment of muon, the charged lepton-flavor violation, and the electric dipole moments of charged leptons, from viewpoints of the minimal supersymmetric standard model and the extensions.

1 Introduction

The Standard Model (SM) is the most successful model to explain physics below the weak scale. The recent results for $\sin 2\beta$ or $\sin 2\phi_1$ by Belle[1] and Babar[2] are converging, and they are consistent with the Kobayashi-Maskawa mechanism. Also, the precision measurements of the electroweak parameters suggest the light SM Higgs boson such as $m_h \lesssim 196$GeV (95%CL)[3]. Now we are waiting for signature of physics beyond the SM.

Nowadays the minimal supersymmetric standard model (MSSM) is one the most promising extension of standard model (SM), and many experiments are searching for the possible evidence of the low-energy supersymmetry. In this review we discuss the possibilities in physics of charged lepton, paying attention to the dipole-moment operators;

$$\mathcal{L} = e\frac{m_{l_j}}{2}\bar{l}_i\sigma_{\mu\nu}F^{\mu\nu}(L_{ij}P_L + R_{ij}P_R)l_j \tag{1}$$

where $P_{R/L} = (1 \pm \gamma_5)/2$, and i, j are for the generation. These operators are sensitive to physics beyond the SM. The real diagonal parts of L_{ij} and R_{ij} contributes to the anomalous magnetic moments of charged leptons, $a_{l_i}(\equiv (g_{l_i} - 2)/2)) = m_{l_i}^2(R_{ii} + L_{ii})$. If non-vanishing L_{ij} or R_{ij} $(i \neq j)$ exits, the charged lepton-flavor violating (LFV) processes, such as $\mu \to e\gamma$, are predicted; $Br(\mu \to e\gamma) \propto (|R_{\mu e}|^2 + |L_{\mu e}|^2)$. When diagonal parts of L_{ij} and/or R_{ij} have imaginary part, CP is violating and the electric dipole moments (EDM) are predicted; $d_{l_i} = em_{l_i}\mathrm{Im}(R_{ii} - L_{ii})$.

We organize this review as follows. In the next section we summarize the current status of the muon $(g-2)$. In section 2 we discuss dependence of the charged LFV processes on the SUSY breaking models, and show the branching ratios of the charged LFV processes in the supersymmetric seesaw model, using the neutrino oscillation data. Section 3 is for the EDMs of charged leptons. Section 4 is summary.

2 Muon anomalous magnetic moment

The latest result for the anomalous magnetic moment of muon[4] is $a_\mu^{\rm exp} = (116\,592\,023 \pm 151) \times 10^{-11}$, while the SM prediction is $a_\mu^{\rm SM} = (116\,592\,768 \pm 65) \times 10^{-11}$. The deviation of the measurement from the SM prediction is $a_\mu^{\rm NP} (\equiv a_\mu^{\rm exp} - a_\mu^{\rm SM}) = 255 \pm 164 \times 10^{-11}$, and it is 1.6σ away. At present the significance of the deviation is small. The experimental error is expected to be improved by a factor 2 in BNL'00 data, and the ultimate goal may be $\sim 40 \times 10^{-11}$.

First, we review the error in the SM prediction. The largest ambiguities in the SM prediction come from the leading hadronic vacuum polarization contribution $a_\mu^{\rm Had}({\rm VP1})$ and the hadronic light-by-light (LbyL) scattering contribution $a_\mu^{\rm Had}({\rm LbyL})$. The $a_\mu^{\rm Had}({\rm VP1})$ in the above SM prediction is derived by M. Davier and A. Hocker[5] from the e^+e^- hadronic cross section and the hadronic τ decay data, including perturbative QCD calculation in the high q^2 part. This estimation will be further improved by high quality data for the e^+e^- hadronic cross section by CMD2 in Novosibirsk, KLOE in Fascati, and BES in Beijing. The Babar may contribute to it by measurement via the initial state radiation of hard photon. The CLEO and LEP data for the tau decay are also important.

On the other hand, the estimate of the LbyL scattering contribution relies on the model calculation. The average value for the latest results[6,7] ($a_\mu^{\rm Had}({\rm LbyL}) = (89.6 \pm 15.4) \times 10^{-11}$ and $(83 \pm 32) \times 10^{-11}$) is used in the above SM prediction. The dominant contribution in $a_\mu^{\rm Had}({\rm LbyL})$ comes from the pion-pole diagram. This diagram was reevaluated by several groups[8,6,7], and the sign problem has been fixed now. However, still they rely on the model calculation since the diagram is divergent. They are based on the chiral perturbation or the ENJL model. The vector-meson dominance is assumed and the phenomenological parametrization of the pion form factor $\pi\gamma^*\gamma^*$ is introduced in order to regularize the divergence.

In Ref. [9] the pion-pole contribution is evaluated in a model-dependent way, based on the chiral perturbation theory. The result is following;

$$a_\mu^{\rm Had}({\rm LbyL})|_{\pi^0 {\rm pole}} = \frac{3\alpha^3}{16\pi^5} \left(\frac{m_\mu}{F_\pi}\right)^2 \left\{ \log^2 \frac{\Lambda}{m_\mu} + \left(\frac{1}{6}\chi - 0.17\right) \log \frac{\Lambda}{m_\mu} + \tilde{C} \right\},$$

where Λ is the ultraviolet cutoff ($\Lambda \sim 4\pi F_\pi$). The largest term proportional to

$\log^2 \Lambda/m_\mu$ is fixed by the gauge invariance and chiral anomaly. χ is a counter term to regularize the two-loop diagram. While it can be determined by the leptonic decay of the psuedescalar mesons, the sensitivity is low at present. Furthermore, $\tilde{C}$, which is a piece not enhanced by log, cannot be evaluated without explicit models. The uncertainty due to $\tilde{C}$ is $\delta a_\mu = 31 \times 10^{-11}\tilde{C}$.

While the above two model-dependent calculations seem to be converged, we do not have a strategy to derive the pion-pole contribution precisely enough in a model-dependent way. Also, we have a subtle problem in the light-by-light contribution, whether the inclusion of the quark loop is double-counting or not. Thus, the calculation of the light-by-light contribution on base of QCD is strongly desired.

If the hadronic contribution is well-controlled, the muon $g-2$ is so sensitive to physics beyond the SM. Before closing this section, we discuss it from a viewpoint of the MSSM. The nature of two Higgs doublet model in the MSSM can enhance the contribution, and the contribution proportional to $\tan\beta$ is given as

$$a_\mu^{\rm SUSY} \simeq \frac{5\alpha_2 + \alpha_Y}{48\pi}\frac{m_\mu^2}{m_{\rm S}^2}\tan\beta \simeq 1.3 \times 10^{-9}\left(\frac{100{\rm GeV}}{m_{\rm SUSY}}\right)^2 \tan\beta.$$

Here, all relevant SUSY breaking parameters are assumed to be common to m_S. Thus, it may be larger than the electroweak correction in the SM and the deviation from the SM in the MSSM may reach to 10^{-8}[10].

3 Lepton-flavor violation in the charged-lepton sector

While the lepton-flavor violation is observed in the neutrino oscillation experiments, this does not mean sizable LFV processes in the charged-lepton sector exit. The charged LFV processes induced by the small neutrino masses, expected from the neutrino oscillation results, are suppressed by the GIM mechanism, as $Br(\mu \to e\gamma) \lesssim 10^{-48}(m_\nu/1{\rm eV})^4$, even if the neutrino mixing is maximal. On the other hand, if the SM is supersymmetrized, the situation is changed. The SUSY breaking slepton masses are not necessary aligned to the lepton masses, and it may lead to sizable lepton-flavor violating.

Let us asuume that $(m_{\tilde{L}}^2)_{12}$ in the left-handed slepton mass matrix is non-vanishing. In this case $\mu \to e\gamma$ is generated, and the approximate formula is given as $Br(\mu \to e\gamma) \simeq 3 \times 10^{-5}(a_\mu^{\rm SUSY}/10^{-9})^2((m_{\tilde{L}}^2)_{12}/m_S^2)^2$[11]. Since the muon $g-2$ and $\mu \to e\gamma$ come from the dipole operators, the muon $g-2$ gives the normalization of the branching ratio of $\mu \to e\gamma$.

In Table 2 we summarize the current experimental bounds to the charged LFV processes, the sensitivities in the present activities, and prospects in

Table 1. Current experimental bounds to the charged LFV processes, the sensitivities in the present activities, and prospects in the future experiments.

	Current bound	Present Activities	Future
$\tau \to \mu\gamma$	1.0×10^{-6}[12]	$\sim 10^{-7}$ (Belle)[13]	$10^{-(8-9)}$[13]
$\mu \to e\gamma$	1.2×10^{-11}[14]	2×10^{-14} (PSI)[15]	10^{-15}[16]
$\mu \to 3e$	1.0×10^{-12}[17]		$10^{-(15-16)}$[16]
$\mu^N \to e^- N$	6.1×10^{-13}[18]	10^{-14} (SINDRUM II)[19] 5×10^{-17} (MECO)[20]	10^{-18}[16,21]

the future experiments such as the PRISM project[21] and the front ends of neutrino factories under consideration at CERN[16]. The charged LFV processes are radiative-induced in the MSSM as far as the R party is not broken. Thus, the branching ratio of $\mu \to 3e$ and the μ-e conversion rate in nuclei are approximately given as $Br(\mu \to 3e)/Br(\mu \to e\gamma) \simeq 7 \times 10^{-3}$ and $R(\mu^- \mathrm{Ti(Al)} \to e^- \mathrm{Ti(Al)})/Br(\mu \to e\gamma) \simeq 5(3) \times 10^{-3}$. (See the detailed calculation of the μ-e conversion rate in nuclei is given in Refs. [22].) From these simple formulas, the naive current bound on $(m^2_{\tilde{L}})_{12}/m^2_S$ is $\lesssim 6 \times 10^{-4}(\delta a_\mu^{\mathrm{SUSY}}/10^{-9})^{-1}$, and PSI and MECO/BNL (PRISM and NuFACT) may reach to $\sim 10^{-5}(10^{-6})$. These experiments are stringent tests of the lepton-flavor symmetry in the MSSM.

The charged LFV in the MSSM depends on the origin of the SUSY breaking term in the MSSM and the interaction of physics beyond the MSSM. The SUSY breaking model is classified to two types by degeneracy or non-degeneracy of the sfermion masses. The later may predict the large LFV rates and sometimes the broad parameter region has been excluded already. Here, we will concentrate on the SUSY breaking models where the degeneracy of the sfermion masses is predicted by assuming the hidden sector, such as the gravity- [23], gauge- [24], anomaly- [25], gaugino-mediation [26] models.

The magnitude of the charged LFV processes in these models depends on the scale of the SUSY breaking mediation (M_M) and the scale of the physics with LFV interaction (M_{LFV}). The well-motivated candidates for the physics with LFV interaction are the SUSY GUTs and the see-saw mechanism. If $M_M \gtrsim M_{LFV}$, such as in the gravity-mediation model, the LFV slepton masses are radiatively generated by the LFV interaction and they depend on $\log M_M/M_{LFV}$. The LFV processes may have observable rates in this case[27,28,29].

In the gaugino-mediation model, where M_M is the reduced Planck scale or the GUT scale, the scalar masses at M_M are vanishing, and they are generated through the gaugino loops. The LFV slepton masses are induced at higher order and suppressed. However, the suppression of the LFV processes is at most a factor 10^{11}.

When $M_M \lesssim M_{LFV}$, such as in the low-energy gauge-mediation model, the radiative correction is suppressed by a power of M_M/M_{LFV}, and the effect tends to be invisible. The anomaly-mediation model is exceptional. While M_M is the gravitational scale, the SUSY breaking parameters are determined by only the particle contents and interactions in the MSSM in the original anomaly-mediation model (the UV insensitivity), and the LFV slepton masses are suppressed. On the other hand, the model has the problem of tachyonic sleptons. Then, the LFV slepton masses in this model depends on how to care of the problem. For example, in the minimal anomaly-mediation model[30], where the universal scalar mass m_0 is added to the anomaly-mediation contribution, the LFV slepton masses are generated and proportional to m_0^2.

Next, we will discuss the charged LFV processes in the supersymmetric see-saw model using the the neutrino oscillation results. We assume the gravity-mediation model. The atmospheric neutrino result suggests the large mixing of left-handed stau and smuon, and it may imply the large branching ratio of $\tau \to \mu\gamma$. In Fig. (1) we present $Br(\tau \to \mu\gamma)$ in this model. Here, we use $m_{\nu_\tau}^2 = 2 \times 10^{-3}\text{eV}^2$ and $U_{23} = 1/\sqrt{2}$, and asuume that the large mixing comes from the neutrino Yukawa coupling and that the Yukawa unification of the tau-neutrino and top-quark Yukawa couplings is imposed at the reduced Planck scale. These assumptions make $Br(\tau \to \mu\gamma)$ enhanced. For the SUSY breaking parameters, we take $m_0 < 500$GeV, the $\text{U}(1)_Y$ gaugino mass < 500GeV, and the universal A parameter A_0 zero. While a parameter region is excluded, the branching ratio tends to be smaller than the reach of the KEK Belle. If the large deviation from the SM prediction in the muon $g-2$, such as $\sim 10^{-9}$, is observed, the branching ratio may be larger than 10^{-9}.

Now the large-angle MSW solution is the most favored in the solar neutrino problem, and this may lead to the large branching ratio of $\mu \to e\gamma$. In Fig. (2) we show $\mu \to e\gamma$, using the atmospheric neutrino result and the large-angle MSW solution; $m_{\nu_\mu}^2 = 7.5 \times 10^{-5}\text{eV}^2$ and $U_{12} = 1/\sqrt{2}$. Here we take $m_{\nu_e} = 0$ and assume the canonical generational structure for the right-handed neutrino masses. For the SUSY breaking parameters, the universal gaugino mass $M_{1/2} = 200$GeV, $m_0 = 200$GeV, and $A_0 = 200$GeV. The horizontal line is for the right-handed tau neutrino mass M_{N_τ}. A broad region has been excluded already, and the future experiments may cover almost the region above $M_{N_\tau} \gtrsim 10^{11}$GeV. If $M_{N_\tau} \lesssim \text{O}(10^{11})$GeV, the Dirac mass for the tau neutrino is smaller than O(1)GeV, which is much smaller than the top quark mass.

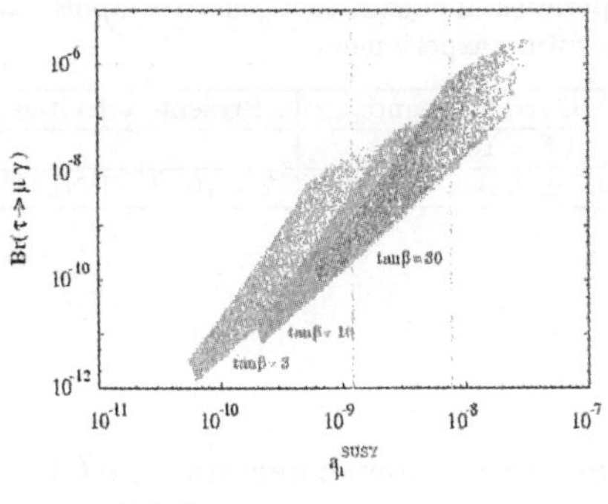

Figure 1. $Br(\tau \to \mu\gamma)$ in the supersymmetric see-saw model, assuming the gravity-mediation model. Here, we use the atmospheric neutrino result. The horizontal line is the MSSM contribution to the muon $g-2$.

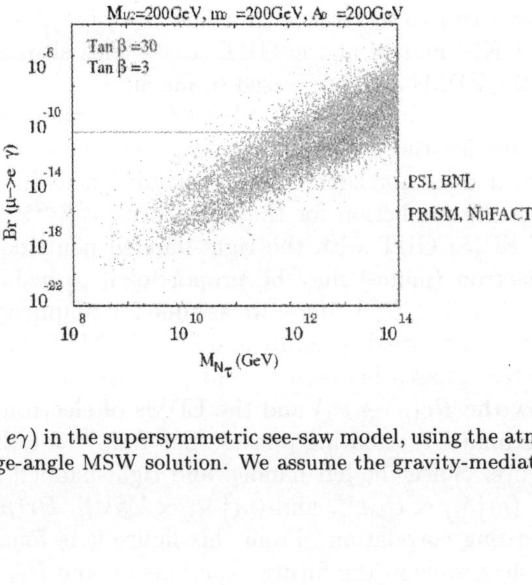

Figure 2. $Br(\mu \to e\gamma)$ in the supersymmetric see-saw model, using the atmospheric neutrino result and the large-angle MSW solution. We assume the gravity-mediation model.

4 EDMs

In this section we discuss the EDMs of charged leptons. The current experimental bounds and the sensitivities of the future experiments are listed in Table 3. While the EDMs are suppressed in the SM as $d_e(d_\mu) < 10^{-40}(10^{-38})e$ cm, they are sensitive to the MSSM. The relative phases of the F-term SUSY breaking parameters, the A and B terms and the gaugino masses, contribute to the EDMs. In this section, we assume for simplicity that

Table 2. The current experimental bounds to the electric dipole moments of charged leptons and the prospects in the future experiments.

	Current bound	Present Activities	Future
$d_e(e\ \mathrm{cm})$	$1.6 \times 10^{-27} e^{31}$		10^{-3332}
$d_\mu(e\ \mathrm{cm})$	$(3.7 \pm 3.4) \times 10^{-1933}$	2×10^{-24} (BNL)[34]	$10^{-26\ 1621}$

the sfermion masses are flavor-independent and the CP-violating phases of the SUSY breaking parameters are zero at the SUSY-breaking mediation scale, and consider the EDMs radiatively-induced in physics beyond the MSSM.

In the minimal SUSY SU(5) GUT, the predicted EDMs are very small. The quark and lepton masses are given by the up-type and down-type quark Yukawa couplings at the GUT scale. As the result, the EDMs of electron and muon are proportional to a Jarskog invariant, $\sim f_b^2 f_c^2 f_t^4\ \mathrm{Im}[V_{11} V_{12}^\star V_{22} V_{21}^\star]$, where V is the CKM matrix at the GUT scale. This situation is similar to the SM. Thus, the EDMs are suppressed so much.

We know that the minimal SUSY SU(5) GUT cannot explain the quark and lepton masses for the first and second generations, and the extension is needed. Also, it does not have the right-handed neutrinos. These extension may change the prediction for the EDM drastically[28]. Let us consider that the SUSY SU(5) GUT with the right-handed neutrinos. In this case, the EDM of electron (muon) may be proportional to a Jarskog invariant, $\sim f_{\nu_\tau}^2 f_t^2 \mathrm{Im}[V_{31(2)} V_{33}^\star U_{1(2)3} U_{33}^\star]$. Here, we assume for simplicity that the right-handed neutrino masses are degenerate and U is the MNS matrix at the GUT scale. The relative phases between U and V contribute to the EDMs. In Fig. (3) we show the $Br(\mu \to e\gamma)$ and the EDMs of electron and muon. We asuume the maximal CP violating phases. See Ref. [35] for the input parameters in this figure. Since the left-handed and right-handed sleptons get the LFV masses as $(m_{\tilde{L}}^2)_{ij} \propto U_{i3} U_{j3}^\star$ and $(m_{\tilde{E}}^2)_{ij} \propto V_{3i} V_{3j}^\star$, $Br(\mu \to e\gamma)$ and the EDMs have a strong correlation. From this figure it is found that the prediction may be accessible in the future experiments, and U_{13} is an important parameter for the electron EDM.

In the supersymmetric see-saw model, if the right-handed neutrino masses are exactly degenerate, the EDMs of charged leptons are also suppressed, similar to the minimal SUSY SU(5) GUT. The non-degeneracy of the right-handed neutrino masses may enhance the EDMs drastically[36], and the muon (electron) EDM can reach to $10^{-26}(10^{-31})e$ cm.

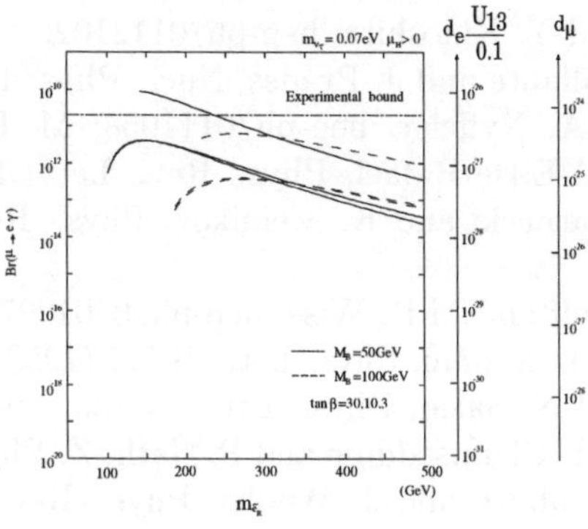

Figure 3. $Br(\mu \to e\gamma)$ and the EDMs of electron and muon in the SUSY SU(5) GUT with the right-handed neutrinos.

5 Summary

In this review we discuss physics of the lepton sector from viewpoints of the minimal supersymmetric standard and the extensions. While the muon $g-2$ is sensitive to the MSSM, the understanding of the systematic error in the the SM prediction, especially the light-by-light contribution, is very serious when the experimental error is reduced furthermore. The charged LFV processes depends on the SUSY breaking models and the LFV interaction beyond the MSSM. The current neutrino data is encouraging. The interesting future experiments may give suggestion for the model discrimination. The EDMs of charged leptons are sensitive to the extension of the SUSY GUTs, and they may be accessible in the future experiments.

References

1. K. Trabelsi, talk given in XXXVII Rencontres de Moriond Electroweak Interactions and Unified Theories, Les Arcs, March 9-16-2002.
2. G. Raven, talk given in XXXVII Rencontres de Moriond Electroweak Interactions and Unified Theories. Les Arcs, March 9-16, 2002.
3. The LEP Electroweak Working Group, http://lepewwg.web.cern.ch/LEPEPWG/.
4. H.N. Brown *et al.* [Muon g-2 Collaboration], Phys. Rev. Lett. 86 (2001) 2227.
5. M. Davier and A. Hocker, Phys. Lett. B 435 (1998) 427.

6. M. Hayakawa and T. Kinoshita, hep-ph/0112102.
7. J. Bijnens, E. Pallante and J. Prades, Nucl. Phys. B 626 (2002) 410.
8. M. Knecht and A. Nyffeler, hep-ph/0111058; M. Knecht, A. Nyffeler, M. Perrottet and E.De Rafael, Phys. Rev. Lett. 88 (2002) 071802; I. Blokland, A. Czarnecki and K. Melnikov, Phys. Rev. Lett. 88 (2002) 071803.
9. M. Ramsey-Musolf and M.B. Wise, hep-ph/0201297.
10. R. Barbieri and L. Maiani, Phys. Lett. B 117 (1982) 203; D.A. Kosower, L.M. Krauss and N. Sakai, Phys. Lett. B 133 (1983) 305; T.C. Yuan, R. Arnowitt, A.H. Chamseddine and P. Nath, Z. Phys. C 26 (1984) 407; C. Arzt, M.B. Einhorn and J. Wudka, Phys. Rev. D 49 (1994) 1370; J.L. Lopez, D.V. Nanopoulos and X. Wang, Phys. Rev. D 49 (1994) 366; U. Chattopadhyay and P. Nath, Phys. Rev. D 53 (1996) 1648; T. Moroi, Phys. Rev. D 53 (1996) 6565 [Erratum-ibid. D 56 (1996) 4424]; M. Carena, G.F. Giudice and C.E. Wagner, Phys. Lett. B 390 (1997) 234.
11. J. Hisano and K. Tobe, Phys. Lett. B 510 (2001) 197.
12. K. Abe *et al.* [BELLE Collaboration], BELLE-CONF-0118.
13. T. Ohoshima, talk at the workshop "Neutrino oscillations and their origin" (NOON2001) (ICRR, Univ. of Tokyo, Kashiwa, Japan, Dec.,2001).
14. M.L. Brooks *et al.* [MEGA Collaboration], Phys. Rev. Lett. 83 (1999) 1521.
15. L.M. Barkov *et al.*, Research Proposal for experiment at PSI (1999).
16. J. Äystö *et al.*, "Physics with Low-Energy Muons at a Neutrino Factory Complex", hep-ph/0109217.
17. D.E. Groom *et al.* [Particle Data Group Collaboration], Eur. Phys. J. C 15 (2000) 1.
18. P. Wintz, in Proceeding of the International Symposium "Lepton and Baryon Number Violation", (eds. H.V. Klapdor-Kleingrothaus and I.V. Krivosheina, Institute of Physics Bristol, 1998), p.534.
19. SINDRUM II Collaboration, Research Proposal for experiment at PSI (1999).
20. M. Bachmann *et al.* [MECO Collaboration], Research Proposal E940 for experiment at BNL (1997).
21. Y. Kuno and Y. Okada, Rev. Mod. Phys. 73 (2001) 151; M. Furusaka *et al.*, JAERI/KEK Joint Project Proposal "The Joint Project for High-Intensity Proton Accelerators", KEK-REPORT-99-4, JAERI-TECH-99-056.
22. A. Czarnecki, W.J. Marciano and K. Melnikov, hep-ph/9801218; R. Kitano, M. Koike and Y. Okada, hep-ph/0203110.

23. R. Barbieri, S. Ferrara and C.A. Savoy, Phys. Lett. B 119 (1982) 343; R. Arnovitt, P. Nath and A.H. Chamseddine, Phys. Rev. Lett. 49 (1982) 970; L.J. Hall, J. Lykken and S. Weinberg, Phys. Rev. D 27 (1983) 2359.
24. M. Dine and A.E Nelson, Phys. Rev. D 48 (1993) 1277; M. Dine, A.E. Nelson and Y. Shirman, Phys. Rev. D 51 (1995) 1362; M. Dine, A.E. Nelson, Y. Nir and Y. Shirman, Phys. Rev. D 53 (1996) 2658; G.F. Giudice and R. Rattazzi, Phys. Rept. 322 (1999) 419.
25. L.J. Randall and R. Sundrum, Nucl. Phys. B 557 (1999) 79; G.F. Giudice, M.A. Luty, H. Murayama and R. Rattazzi, JHEP 9812 (1998) 027.
26. D.E. Kaplan, G.D. Kribs and M. Schmaltz, Phys. Rev. D 62 (2000) 035010; Z. Chacko, M.A. Luty, A.E. Nelson and E. Ponton, JHEP 0001 (2000) 003; M. Schmaltz and W. Skiba, Phys. Rev. D 62 (2000) 095005.
27. L.J. Hall, V.A. Kostelecky and S. Raby, Nucl. Phys. B 267 (1986) 415; R. Barbieri and L.J. Hall, Phys. Lett. B 338 (1994) 212; J. Hisano, T. Moroi, K. Tobe, and M. Yamaguchi, Phys. Lett. B 391 (1997) 341; [Erratum-ibid. B 397 (1997) 357].
28. R. Barbieri, L. Hall, and A. Strumia, Nucl. Phys. B 445 (1995) 219; P. Ciafaloni, A. Romanino and A. Strumia, Nucl. Phys. B 458 (1996) 3; N. Arkani-Hamed, H.C. Cheng, and L.J. Hall, Phys. Rev. D 53 (1996) 413 J. Hisano, D. Nomura, Y. Okada, Y. Shimizu and M. Tanaka, Phys. Rev. D 58 (1998) 116010; A. Romanino and A. Strumia, Nucl. Phys. B 622 (2002) 73.
29. F. Borzumati and A. Masiero, Phys. Rev. Lett. 57 (1986) 961; J. Hisano, T. Moroi, K. Tobe, M. Yamaguchi and T. Yanagida, Phys. Lett. B 357 (1995) 579; J. Hisano, T. Moroi, K. Tobe and M. Yamaguchi, Phys. Rev. D 53 (1996) 2442; J. Hisano and D. Nomura, Phys. Rev. D 59 (1999) 116005.
30. J.L. Feng and T. Moroi, Phys. Rev. D 61 (2000) 095004.
31. B.C. Regan, E.D. Commins, C.J. Schmidt and D. DeMille, Phys. Rev. Lett. 88 (2002) 071805.
32. S.K. Lamoreaux, nucl-ex/0109014.
33. H.N. Brown *et al.* [Muon g-2 Collaboration], Phys. Rev. Lett. 86 (2001) 2227.
34. Y.K. Semertzidis *et al.*, hep-ph/0012087.
35. J. Hisano, D. Nomura and T. Yanagida, Phys. Lett. B 437 (1998) 351.
36. J.R. Ellis, J. Hisano, M. Raidal and Y. Shimizu, Phys. Lett. B 528 (2002) 86.

FOUR PUZZLES OF NEUTRINO MIXING

S.M. BARR
Bartol Research Institute, University of Delaware, Newark DE 19716, US
E-mail: smbarr@bartol.udel.edu

There are four puzzling questions about by the magnitudes of neutrino mixings and mass splittings. A brief sketch is given of the various kinds of models of neutrino masses and how they answer these questions. Special attention is given to so-called "lopsided" models.

1 Comparison of Quarks and Leptons

Over the last three decades theorists have been trying to understand the spectrum of quark and lepton masses. Although no simple model of the many that have been proposed is uniquely compelling, there are certain basic ideas that seem rather probable and are incorporated in most published models. One of these ideas is that there is a direct relation between the mass ratios and the mixing angles of the quarks. Since the charged leptons exhibit a mass hierarchy very similar to those of the quarks, it was widely expected that the lepton mixing angles would also be like those of the quarks. The discovery that the atmospheric neutrino mixing angle θ_{atm} is nearly maximal thus came as a surprise.

In this talk I first review the basic facts about quark masses and mixings and then discuss several features of neutrino mixing that seem at first sight puzzling in light of these facts. I will then show how various types of models explain these puzzling features.

There are two quark mass matrices, M_U and M_D, for the up-type quarks (u, c, t) and down-type quarks (d, s, b) respectively. These are diagonalized by unitary transformations: $V_U^\dagger M_U U_U = \text{diag}(m_u, m_c, m_t)$ and $V_D^\dagger M_D U_D = \text{diag}(m_d, m_s, m_b)$. The mismatch between the unitary transformations of the left-handed quarks gives rise to the CKM matrix, $V_{CKM} = U_U^\dagger U_D$. The CKM angles are $|V_{us}| = \sin\theta_{12}^q \cong 0.2$, $|V_{cb}| = \sin\theta_{23}^q \cong 0.04$, and $|V_{ub}| = \sin\theta_{13}^q \cong 0.003$. The smallness of these angles is presumably due to small ratios of elements in M_U and M_D, and is therefore presumably directly related to the smallness of the mass ratios m_c/m_t, m_u/m_c, m_s/m_b, and m_d/m_s

How mass ratios and mixing angles might be directly related can be seen easily from a 2×2 example.[1] Consider the matrix

$$M = \begin{pmatrix} 0 & \epsilon \\ \epsilon & 1 \end{pmatrix} m. \tag{1}$$

This is diagonalized by $R(\theta)^T M R(\theta)$, where $R(\theta)$ is the 2×2 rotation matrix with $\tan 2\theta = 2\epsilon$, or, for small ϵ, $\theta \cong \epsilon$. The large eigenvalue of M is obviously $m_2 \cong m$, while the fact that $\det M = -\epsilon^2 m$, tells us that other eigenvalue is $m_1 \cong -\epsilon^2 m$. Consequently, one has that $\theta \cong \sqrt{|m_1/m_2|}$. This can be compared to the old and famously successful relation for the Cabibbo angle $\tan\theta_c \cong \sqrt{m_s/m_d}$.

One should note that the matrix in this example is "hierarchical", by which we mean that the entries get smaller *upward* and *to the left* of any diagonal entry. Most realistic models of quark masses and mixings assume such hierarchical mass matrices. For example, a recent model[2] of Babu and Nandi, which fits the data extremely well, has quark matrices of the form

$$M_U \sim \begin{pmatrix} \epsilon^6 & \epsilon^4 & \epsilon^4 \\ \epsilon^4 & \epsilon^2 & \epsilon^2 \\ \epsilon^4 & \epsilon^2 & 1 \end{pmatrix} m, \quad M_D \sim \begin{pmatrix} \epsilon^6 & \epsilon^6 & \epsilon^6 \\ \epsilon^6 & \epsilon^4 & \epsilon^4 \\ \epsilon^6 & \epsilon^4 & \epsilon^2 \end{pmatrix} m, \tag{2}$$

Now let us consider the leptons. Here again there are two mass matrices, M_L for the charged leptons (e^-, μ^-, τ^-) and M_ν for the neutrinos (ν_e, ν_μ, ν_τ). The matrix M_ν is different in two respects from M_U, M_D, and M_L: it has much smaller entries, and it is symmetric, since it is a Majorana matrix connecting left-handed neutrinos to left-handed neutrinos. Nevertheless, as with the quark mass matrices, the lepton mass matrices are diagonalized by unitary transformations that can have a mismatch. That mismatch gives rise to the neutrino mixing matrix sometimes referred to as the MNS matrix: $U_{MNS} = U_L^\dagger U_\nu$. In $(U_{MNS})_{fm}$, $f = e, \mu, \tau$ and $m = 1, 2, 3$. Experimentally one has that $|U_{\mu 3}|(\equiv \sin\theta_{atm} = \sin\theta_{23}^\ell) \cong 0.7$, $|U_{e2}|(\equiv \sin\theta_{sol} = \sin\theta_{12}^\ell) = O(1)$ (probably), and $|U_{e3}|(\equiv \sin\theta_{13}^\ell) \leq 0.15$. There is still a great deal of uncertainty about the solar mixing angle, but the solution with small θ_{sol} (the "SMA" or Small Mixing Angle MSW solution) is disfavored by recent global fits to the data.[3] The best fits are to the "LMA" or Large Mixing Angle MSW solution and the "LOW" solution. The best-fit value for the LMA solution is $\tan^2\theta_{sol} \approx 0.4$.

The mass splittings needed to fit the atmospheric and solar data are $\delta m^2_{atm} = m_3^2 - m_2^2 \approx 3 \times 10^{-3}$ eV^2, and $\delta m^2_{sol} = m_2^2 - m_1^2 \approx 3 \times 10^{-5}$ eV^2 (for LMA, smaller for other solar solutions). The fact that $\delta m^2_{sol} \ll \delta m^2_{atm}$ suggests there is probably a family hierarchy of neutrino masses, although it also possible that the neutrino masses are nearly degenerate and that only their splittings have a hierarchy.

2 Three Puzzles

In the basic facts about neutrino masses and mixings there are three features that appear puzzling in light of the conventional wisdom about quark masses and mixings.

Puzzle 1: *Why are some* $\theta^\ell \sim 1$ *whereas all* $\theta^q \ll 1$ *?* In grand unified theories the quarks and leptons are related, and one expects similar mass ratios and mixing angles for them. In models with flavor symmetry the same flavor symmetries generally control the quark and lepton mass matrices and give them similar structure. Empirically, one indeed sees that the charged leptons have a mass hierarchy qualitatively similar to those of the up-type and down-type quarks. Another similarity is that the 13 mixing angle is by far the smallest in both cases ($|V_{ub}| \ll |V_{us}|, |V_{cb}|$ and $|U_{e3}| \ll |U_{e2}|, |U_{\mu 3}|$).

In light of the expected and actual similarities of quarks and leptons it appears strange that at least one and probably two of the leptonic angles are large, while all the quark angles are very small.

Puzzle 2: *How can there be small lepton mass ratios but large leptonic mixing angles?* As we have seen, for the quarks the smallness of the mixing angles and mass ratios are generally thought to be related. For the charged leptons the mass ratios are certainly small, and for the neutrinos at least the ratios of mass splittings are small, and yet the three families of neutrinos are very strongly mixed.

Puzzle 3: *Why are two leptonic angles large but the third* (θ^ℓ_{13}) *small?* If all the leptonic angles were of order unity it might suggest that all the entries of the neutrino mass matrix M_ν were of the same order, as would typically be the case if it were a "random" matrix, as has indeed been suggested. However, such a matrix would not generally give a hierarchy of neutrino mass splittings, nor would it generally yield a 13 mixing angle much smaller than the others. The smallness of θ^ℓ_{13} and largeness of the other leptonic angles suggests that the leptonic mass matrices have quite special forms. To see what those forms might be let us consider the following product of rotation matrices:

$$
\begin{aligned}
R_{23}(\theta_{atm}) R_{12}(\theta_{sol}) &= \begin{pmatrix} 1 & 0 & 0 \\ 0 & c_{atm} & s_{atm} \\ 0 & -s_{atm} & c_{atm} \end{pmatrix} \begin{pmatrix} c_{sol} & s_{sol} & 0 \\ -s_{sol} & c_{sol} & 0 \\ 0 & 0 & 1 \end{pmatrix} \\
&= \begin{pmatrix} c_{sol} & s_{sol} & 0 \\ -c_{atm} s_{sol} & c_{atm} c_{sol} & s_{atm} \\ s_{atm} s_{sol} & -s_{atm} c_{sol} & c_{atm} \end{pmatrix} .
\end{aligned} \tag{3}
$$

One sees that even if θ_{atm} and θ_{sol} are both large this matrix has the property that $U_{e3} = \sin\theta^\ell_{13}$ vanishes. Thus Puzzle 3 is resolved if one has that $U_{MNS} \equiv$

$U_L^\dagger U_\nu \cong R_{23}(\theta_{atm})R_{12}(\theta_{sol})$. There are three simple possibilities:
Solution A: $U_L \cong I$, $U_\nu \cong R_{23}(\theta_{atm})R_{12}(\theta_{sol})$.
Both large mixing angles come from M_ν, whose diagonalization involves *first* a large 23 rotation and *then* a large 12 rotation.
Solution B: $U_L \cong R_{12}(\theta_{sol})R_{23}(\theta_{atm})$, $U_\nu \cong I$.
Both large mixing angles come from M_L, whose diagonalization involves *first* a large 12 rotation and *then* a large 23 rotation.
Solution C: $U_L \cong R_{23}(\theta_{atm})$, $U_\nu \cong R_{12}(\theta_{sol})$.
The large atmospheric angle comes from M_L, and the large solar angle comes from M_ν.

3 How Non-see-saw Models Resolve the Puzzles

Let us first recall how the see-saw mechanism works. The up quarks, down quarks, and charged leptons all have Dirac masses through the Higgs doublet field (or fields) coupling the left-handed fermions to their right-handed partners. If there are right-handed neutrinos, then an analogous Dirac mass matrix M_ν^{Dirac} can exist for the neutrinos as well. However, there can also exist a Majorana mass matrix M_R connecting the right-handed neutrinos to themselves. These right-handed Majorana masses can be superlarge as they do not break the gauge symmetries of the Standard Model. Integrating out the superheavy right-handed neutrinos leaves behind light left-handed neutrinos with an effective Majorana mass matrix given by the "see-saw" formula $M_\nu = -M_\nu^{Dirac\,T} M_R^{-1} M_\nu^{Dirac}$. In see-saw models, then, the neutrino masses have fundamentally the same origin as the charged lepton and quark masses, namely they come from the existence of both left- and right-handed components coupled together by the Higgs field (or fields).

In non-see-saw models there are no right-handed neutrinos. The masses of the neutrinos therefore have to arise in some other, completely new way *not* directly related to mass generation for the charged leptons and quarks. Many such mechanisms have been proposed.[4] Three popular ones are the Zee mechanism, R-parity violation in SUSY models, and triplet Higgs.

In the Zee mechanism, there exists a singly charged, singlet scalar field h^+, which can couple to a pair of lepton doublets $(h^+ L_i L_j)$ and to a pair of Higgs doublets $(h^+ H_\alpha H_\beta)$. Obviously, with both types of couplings, there is no way consistently to assign lepton number to h^+. Whether one assigns it $L = -2$ or $L = 0$, one of its couplings will violate lepton number by two units, which is what is needed to generate Majorana masses for the left-handed neutrinos. Such masses arise at one-loop.

In theories with low-energy supersymmetry, the neutrinos can acquire mass by coupling to a neutralino that plays the role of right-handed neutrino. The scalar that couples the neutrino to the neutralino is the sneutrino, which is able to obtain a non-zero vacuum expectation value if R-parity is violated. R-parity violation also allows superpotential terms of the type LQD^c and LLE^c, which give one-loop neutrino masses when the sleptons and squarks are integrated out.

Finally, if there is a triplet higgs field T with Standard Model quantum numbers $(1, 3, +1)$, then it can have a renormalizable coupling to a pair of lepton doublets (TL_iL_j) that gives a tree-level neutrino mass.

The great advantage of such non-see-saw mechanisms is that they automatically provide a very plausible answer to Puzzle 1: the lepton mixing angles differ so dramatically from the quark mixing angles simply because M_ν has a very different origin than M_U and M_D. We will now look at specific non-see-saw ideas to see how they resolve the other Puzzles.

Inverted Hierarchy Models. In inverted hierarchy models the neutrino mass matrix has approximately the following form:

$$M_\nu \cong \begin{pmatrix} 0 & A & B \\ A & 0 & 0 \\ B & 0 & 0 \end{pmatrix}, \tag{4}$$

with $A \sim B$. This can arise in various ways. In the Zee model the one-loop mass matrix is symmetric with vanishing diagonal elements. If for some reason the 23 (32) elements are smaller than the others, the form in Eq. (4) results. It can also result from an approximate $L_e - L_\mu - L_\tau$ symmetry.

One can diagonalize the large elements A and B in Eq. (4) by two successive large rotations. First, one can rotate in the "23 plane" by angle $\theta_{23} \cong \tan^{-1}(B/A) = O(1)$ to eliminate the 13 and 31 elements. Then one can rotate in the "12 plane" by $\theta_{12} \cong \pi/4$ to eliminate the 12 and 21 elements:

$$\begin{bmatrix} 0 & A & B \\ A & 0 & 0 \\ B & 0 & 0 \end{bmatrix} \underset{\theta_{23}}{\longrightarrow} \begin{bmatrix} 0 & \sqrt{A^2+B^2} & 0 \\ \sqrt{A^2+B^2} & 0 & 0 \\ 0 & 0 & 0 \end{bmatrix} \underset{\theta_{12}}{\longrightarrow} \begin{bmatrix} \sqrt{A^2+B^2} & 0 & 0 \\ 0 & -\sqrt{A^2+B^2} & 0 \\ 0 & 0 & 0 \end{bmatrix}. \tag{5}$$

Note that this sequence of large rotations is precisely Solution A of Puzzle 3. Even though the hierarchy of neutrino masses is inverted here, in the sense that m_3 is the smallest, the near degeneracy of $|m_1|$ and $|m_2|$ gives the corect hierarchy of splittings, $\delta m^2_{sol} \ll \delta m^2_{atm}$, thus resolving Puzzle 2.

Factorized Mass Matrix Models. In some models M_ν has approximately the form

$$M_\nu \cong \begin{pmatrix} 0 & c & d \\ c & B^2 & AB \\ d & AB & A^2 \end{pmatrix}, \tag{6}$$

where $c, d \ll A \sim B$. One can see that in a sense this form is the opposite of the inverted hierarchy form. (In fact, if M_ν has this form, then M_ν^{-1} has the inverted hierarchy form.) We call this form factorized, because if m_{ij} is the 23 block of M_ν, then $m_{ij} \cong a_i a_j$. This form can arise if the dominant contribution to M_ν comes from integrating out a single heavy fermion N that has Dirac mass $m_i(\nu_i N)$ with the left-handed neutrinos ν_2 and ν_3 (its coupling to ν_1 should be smaller). A notable instance of this occurs in the SUSY models with R-parity violation, where N is the neutralino.

A rotation in the 23 plane by $\theta_{23} = \tan^{-1}(B/A) = O(1)$ eliminates the 23, 32, and 22 elements in Eq. (6). That leaves a matrix whose 12 block has a "pseudo-Dirac" form, with the 12 and 21 elements being larger than the 11 and 22 elements. This block can be diagonalized with a rotation in the 12 plane by $\theta_{12} \cong \pi/4$. The resulting matrix can be diagonalized with only small further rotations. Thus, as in the inverted hierarchy models, one has just the sequence of large rotations corresponding to Solution A of Puzzle 3. The hierarchy of neutrino masses is the "normal" one with $m_1, m_2 \ll m_3$, giving the correct hierarchy of splittings and resolving Puzzle 2.

Flavor Democracy Models. A third kind of model, about which many papers have been written, assumes that the M_L, M_U, and M_D all have approximately the "democratic" form in which all the entries are equal. The matrix M_ν, is assumed to be approximately diagonal. The CKM angles thus end up being small due to cancellation, whereas the large leptonic angles that come from diagonalizing M_L do not get cancelled. The democratic form can be diagonalized by the sequence of rotations

$$\begin{pmatrix} 1 & 1 & 1 \\ 1 & 1 & 1 \\ 1 & 1 & 1 \end{pmatrix} \underset{\theta_{12}}{\longrightarrow} \begin{pmatrix} 0 & 0 & 0 \\ 0 & 2 & \sqrt{2} \\ 0 & \sqrt{2} & 1 \end{pmatrix} \underset{\theta_{23}}{\longrightarrow} \begin{pmatrix} 0 & 0 & 0 \\ 0 & 0 & 0 \\ 0 & 0 & 3 \end{pmatrix}, \tag{7}$$

where $\theta_{12} = \tan^{-1} 1 = \pi/4$ and $\theta_{23} = \tan^{-1}\sqrt{2}$. This sequence of large rotations is just that of Solution B of Puzzle 3. The flavor democracy models also give the "normal" neutrino mass hierarchy, resolving Puzzle 2.

In all three kinds of non-see-saw model we have discussed we see that $\theta_{sol} \cong \pi/4$ (maximal mixing), whereas θ_{atm} is only predicted to be large, but not nearly maximal (though it may be by accident). Curiously, the empirical situation is just the reverse. It is θ_{atm} that is observed to be close to maximal. This is our fourth puzzle:

Puzzle 4: *Why is θ_{atm} so close to maximal?* It is not an accident that many models predict θ_{sol} to be nearly maximal while very few models exist where θ_{atm} is. The reason is essentially the following. The simplest way to arrange that a mixing angle is nearly maximal is to assume that the relevant 2×2 block of the 3×3 mass matrix is pseudo-Dirac. The diagonalization of such a matrix leads to nearly degenerate masses, which is to say, very small δm^2. Thus, models with nearly maximal θ_{sol} tend to give $\delta m^2_{sol} \ll \delta m^2_{atm}$ as observed. However, a model with nearly maximal θ_{atm} will tend to give the wrong result $\delta m^2_{atm} \ll \delta m^2_{sol}$.

4 See-saw Models

See-saw models have three great advantages over non-see-saw models. First, they do not have to invent an exotic mechanism for generating neutrino mass. There is nothing exotic about right-handed neutrinos, which indeed have to exist in most kinds of gauge-unified models ($SU(5)$ being an exception). Indeed, grand unification, which is well motivated on other grounds, naturally leads to see-saw neutrino masses. Second, see-saw/GUT models beautifully explain the scale of neutrino mass. Writing the heaviest neutrino mass as $m_3 = m_t^2/M_R$ (GUTs typically relate the neutrino Dirac masses to the up-type quark masses), and taking $m_3 \cong \sqrt{m_3^2 - m_2^2} = \sqrt{\delta m^2_{atm}} \cong 0.06$ eV, one finds, $M_R \sim 10^{15}$ GeV, which is very close to the GUT scale known from running of the gauge couplings. By contrast, in non-see-saw models the neutrino mass scale depends on many parameters about which virtually nothing is known even as to their order of magnitude. Third, see-saw/GUT models tend to be far more predictive than most non-see-saw schemes.

At first glance, Puzzle 1 seems especially puzzling in the context of see-saw/GUT models, since grand unification closely relates quarks and leptons. And it is certainly true that historically the great majority of GUT models predicted leptonic mixing angles of the same order as the CKM angles. Looking more closely, however, we see that this need not be the case. Indeed, there is a beautiful way to resolve Puzzle 1 in the see-saw/GUT framework.

All grand unified gauge groups contain $SU(5)$ as a subgroup, and $SU(5)$ relates down-type quarks to charged leptons *having the opposite chirality*. The $\overline{\mathbf{5}}$ contains ℓ_L^- and the charge conjugate of d_R, while the $\mathbf{10}$ contains d_L and the charge conjugate of ℓ_R^-. As a consequence, what is related by $SU(5)$ is $\theta_{d_L} \longleftrightarrow \theta_{\ell_R}$ and $\theta_{d_R} \longleftrightarrow \theta_{\ell_L}$. Since the CKM angles are the mixings of *left-handed* quarks, and the MNS angles are the mixings of *left-handed* leptons,

$SU(5)$ does not relate them to each other. Rather, it relates the CKM angles to some unobserved mixing of right-handed leptons, and the MNS angles to some unobserved mixing of right-handed quarks. Consequently, it is perfectly possible for the CKM angles to be small and the "corresponding" MNS angles large if the mass matrices M_L and M_D are highly left-right asymmetric or "lopsided".[5] We can see this in a simple toy model.

Consider an $SU(5)$ model with mass terms for the second and third family given by $\lambda(\mathbf{10}_3\overline{\mathbf{5}}_3+\sigma\mathbf{10}_3\overline{\mathbf{5}}_2+\epsilon\mathbf{10}_2\overline{\mathbf{5}}_3)\langle\overline{\mathbf{5}}_H\rangle$, with $\epsilon \ll \sigma \sim 1$. The mass matrices M_L and M_D appear in the terms $\overline{\ell_{Ri}}(M_L)_{ij}\ell_{Lj} + \overline{d_{Ri}}(M_D)_{ij}d_{Lj}$, so they have the form

$$M_L = \begin{pmatrix} - & - & - \\ - & 0 & \epsilon \\ - & \sigma & 1 \end{pmatrix} m, \quad M_D = \begin{pmatrix} - & - & - \\ - & 0 & \sigma \\ - & \epsilon & 1 \end{pmatrix} m, \tag{8}$$

where the dashes are small entries for the first family coming from other terms. Note the left-right transposition between M_L and M_D, whose origin we have already explained. The 32 entry in M_L is σ, which gives an $O(1)$ contribution to the MNS angle $U_{\mu 3}$. The 32 entry in M_D, on the other hand, is small and gives only a small contribution to the CKM angle V_{ub}. (The 23 entries control mixings of right-handed fermions.) I said that see-saw/GUT models are in general more predictive than non-see-saw models. And, indeed, simple and highly predictive $SO(10)$ models that are very similar (for the second and third families) to the toy model just described have been constructed.[6] In fact, many models based on "lopsided" mass matrices now exist in the literature.[4]

Note the very important point that in such lopsided models the large atmospheric neutrino angle comes from the *charged lepton mass matrix* M_L rather than from M_ν. This shows how such models resolve Puzzle 2. In lopsided models the reason why some of the neutrino mixing angles can be large even though all the neutrino mass ratios are small is that large neutrino mixing angles can be caused by large off-diagonal elements in M_L (here σ) whereas the neutrino mass ratios obviously are determined entirely by M_ν.

How can lopsided models resolve Puzzle 3? There are two interesting and simple possibilities. One possibility is that the large θ_{atm} arises from M_L as just described, but that the large θ_{sol} arises from M_ν. This corresponds to Solution C. Such models are very easy to construct.[7] The other possibility is that both of the large angles θ_{atm} and θ_{sol} come from lopsidedness in M_L.[8] Consider the following matrix

$$M_L = \begin{pmatrix} - & - & - \\ - & - & \epsilon \\ \rho' & \rho & 1 \end{pmatrix} m, \tag{9}$$

where $\epsilon \ll \rho' \sim \rho \sim 1$ and the dashes represent elements yet smaller than ϵ. A rotation in the 12 plane by $\theta_{atm} \cong \tan^{-1}(\rho'/\rho)$ brings the matrix to the form shown in Eq. (8) with $\sigma = \sqrt{\rho^2 + \rho'^2}$. Then a rotation in the 23 plane by angle $\theta_{23} \cong \tan^{-1}\sigma$, as in the toy model, eliminates the large 32 element. The further rotations required to diagonalize M_L are small. This sequence of large rotations in the charged lepton sector gives just Solution B of Puzzle 3. (It should be noted that, as in the toy model, the matrix M_D will have the large elements appearing transposed compared to M_L, so that they only affect mixings of right-handed quarks.)

Very few models in the literature attempt to explain why θ_{atm} is nearly maximal (Puzzle 4). It turns out that within the framework of lopsided models it is not difficult to obtain $\theta_{atm} \cong \pi/4$.[8] Consider a model with M_L having the form in Eq. (9), where some nonabelian symmetry relates μ_L^- and τ_L^- so that $\rho = 1$. That would give the relations $\tan\theta_{sol} \cong \rho'$ and $\tan\theta_{atm} \cong \sqrt{1+\rho'^2}$, which imply the interesting relation $\tan^2\theta_{atm} \cong 1 + \tan^2\theta_{sol}$. Even for the best-fit LMA value $\tan^2\theta_{sol} \approx 0.4$ this gives $\sin^2 2\theta_{atm} \cong 0.97$.

References

1. S. Weinberg, Trans. N.Y. Acad. Sci. **38**, 185 (1977); F. Wilczek and A. Zee, *Phys. Lett.* **B70**, 418 (1977); H. Fritzsch, *Phys. Lett.* **B70**, 436 (1977).
2. K.S. Babu and S. Nandi, *Phys. Rev.* **D62**, 033002 (2000).
3. J.N. Bahcall, M.C. Gonzalez-Garcia, C. Pena-Garay, *JHEP* **0108**, 014 (2001).
4. For a review of models see S.M. Barr and I. Dorsner, *Nucl. Phys.* **B585**, 79 (2000).
5. K.S. Babu and S.M. Barr, *Phys. Lett.* **B381**, 202 (1996); C.H. Albright, K.S. Babu and S.M. Barr, *Phys. Rev. Lett* **81**, 1167 (1998).
6. C.H. Albright and S.M. Barr, *Phys. Lett.* **B452** 287 (1999);
7. C.H. Albright and S.M. Barr, *Phys. Lett.* **B461** 218 (1999).
8. K.S. Babu and S.M. Barr, *Phys. Lett.* **B525**, 289 (2002).

SUPERNOVA NEUTRINOS

J. F. BEACOM

NASA/Fermilab Astrophysics Center, Fermi National Accelerator Laboratory, Batavia, Illinois 60510-0500, USA

E-mail: beacom@fnal.gov

We propose that neutrino-proton elastic scattering, $\nu + p \rightarrow \nu + p$, can be used for the detection of supernova neutrinos. Though the proton recoil kinetic energy spectrum is soft, with $T_p \simeq 2E_\nu^2/M_p$, and the scintillation light output from slow, heavily ionizing protons is quenched, the yield above a realistic threshold is nearly as large as that from $\bar{\nu}_e + p \rightarrow e^+ + n$. In addition, the measured proton spectrum is related to the incident neutrino spectrum, which solves a long-standing problem of how to separately measure the total energy release and temperature of ν_μ, ν_τ, $\bar{\nu}_\mu$, and $\bar{\nu}_\tau$. The ability to detect this signal would give detectors like KamLAND and Borexino a crucial and unique role in the quest to detect supernova neutrinos.

1 Introduction

When the next Galactic supernova comes, will we have enough information to study the supernova neutrino signal in detail? Almost all of the detected events will be charged-current $\bar{\nu}_e + p \rightarrow e^+ + n$, which will be well-measured, both because of the large yield and because the measured positron spectrum is related to the neutrino spectrum. Because of the charged-lepton thresholds, the flavors ν_μ, ν_τ, $\bar{\nu}_\mu$, and $\bar{\nu}_\tau$ can only be detected in neutral-current reactions, of which the total yield is expected to be approximately 10^3 events. However, as will be discussed below, in general one *cannot* measure the neutrino energy in neutral-current reactions. These four flavors are expected to carry away about 2/3 of the supernova binding energy, and are expected to have a higher temperature than ν_e or $\bar{\nu}_e$. However, there is no experimental basis for these statements, and at present, numerical models of supernovae cannot definitively address these issues either. If there is no spectral signature for the detection reactions, then neither the total energy carried by these flavors nor their temperature can be separately determined from the detected number of events.

For example, the total energy is needed to determine the mass of the neutron star, and the temperature is needed for studies of neutrino oscillations. At present, such studies would suffer from the need to make model-dependent assumptions. This problem has long been known, but perhaps not widely enough appreciated. In Ref.[1], we clarify this problem, and provide a realistic solution that can be implemented in two existing detectors, KamLAND and

Borexino. The solution is based on neutrino-proton elastic scattering, which has been observed at accelerators at GeV energies, but has never before been shown to be a realistic detection channel for low-energy neutrinos.

2 Main Ideas

If all terms of order E_ν/M_p or higher powers are neglected, the differential cross section for $\nu + p \to \nu + p$ is very simple:

$$\frac{d\sigma}{dT_p} = \frac{G_F^2 M_p}{\pi}\left[\left(1 - \frac{M_p T_p}{E_\nu^2}\right)c_V^2 + \left(1 + \frac{M_p T_p}{E_\nu^2}\right)c_A^2\right]. \tag{1}$$

The neutral-current coupling constants between the exchanged Z° and the proton are

$$c_V = \frac{1 - 4\sin^2\theta_w}{2} = 0.04 \tag{2}$$

$$c_A = \frac{1.27}{2}, \tag{3}$$

where the factor 1.27 is determined by neutron beta decay and its difference from unity is a consequence of the partially conserved axial current. Since $c_A \gg c_V$, this form makes it clear that the *largest* proton recoils are favored, which is optimal for detection. Note that this is the ***opposite*** behavior compared to neutrino-electron elastic scattering, where both the vector and axial couplings contribute, and where E_ν/m_e is not small. It also means that the neutrino (c_A) and antineutrino ($-c_A$) cross sections are nearly identical. If c_V is neglected and the differential cross section is expressed in terms of $\cos\theta_\nu$, it follows the form $1 - 1/3\cos\theta_\nu$ expected for a non-relativistic axial coupling (i.e., a Gamow-Teller matrix element). The total cross section is

$$\frac{G_F^2 E_\nu^2}{\pi}\left(c_V^2 + 3c_A^2\right). \tag{4}$$

As expected, this is of the same form as the total cross section for the charged-current reaction $\bar{\nu}_e + p \to e^+ + n$. However, note that the vector coupling nearly vanishes in the neutral-current channel, and that the axial coupling is half as large as in the charged-current channel, thus making the total cross section approximately 4 times smaller. This factor of 4 can be immediately obtained by considering the product of the couplings and the propagator factor, and using the definition of θ_W. However, for a supernova, several flavors contribute to the neutral-current signal.

The supernova binding energy release about 3×10^{53} ergs, about 99% of which is carried off by all flavors of neutrinos and antineutrinos over about

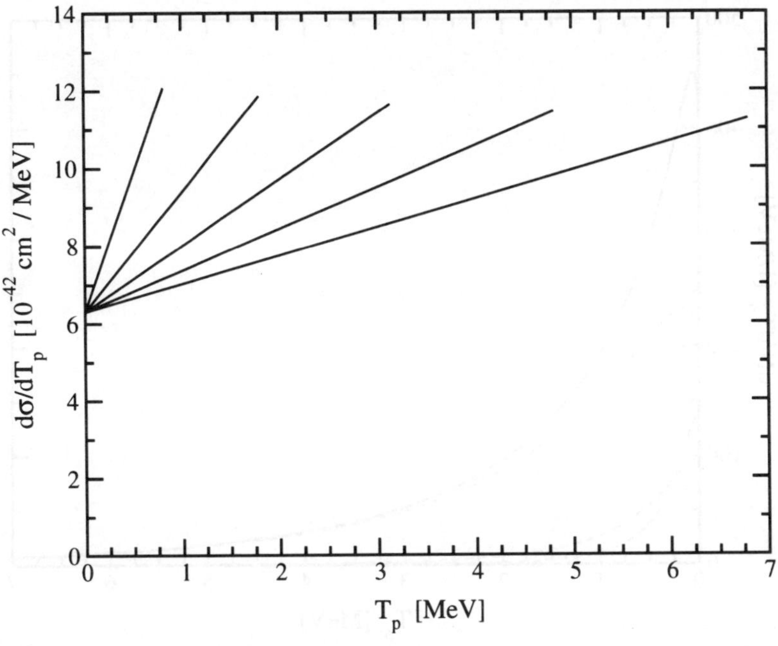

Figure 1. The differential cross section as a function of T_p for fixed E_ν. Note the rise at large T_p, indicating that high recoil energy is preferred. From left to right, the lines are for $E_\nu = 20, 30, 40, 50$, and 60 MeV.

10 s. The emission time is much longer than the light-crossing time of the proto-neutron star because the neutrinos are trapped and must diffuse out, eventually escaping with approximately Fermi-Dirac spectra characteristic of the surface of last scattering. In the canonical model, ν_μ, ν_τ and their antiparticles have a temperature $T \simeq 8$ MeV, $\bar{\nu}_e$ has $T \simeq 5$ MeV, and ν_e has $T \simeq 3.5$ MeV. The temperatures differ from each other because $\bar{\nu}_e$ and ν_e have charged-current opacities (in addition to the neutral-current opacities common to all flavors), and because the proto-neutron star has more neutrons than protons. It is generally assumed that each of the six types of neutrino and antineutrino carries away about 1/6 of the total binding energy, though this has an uncertainty of at least 50%.

Elastically-scattered protons will have kinetic energies of a few MeV. Obviously, these very nonrelativistic protons will be completely invisible in any Čerenkov detector like Super-Kamiokande. However, such small energy de-

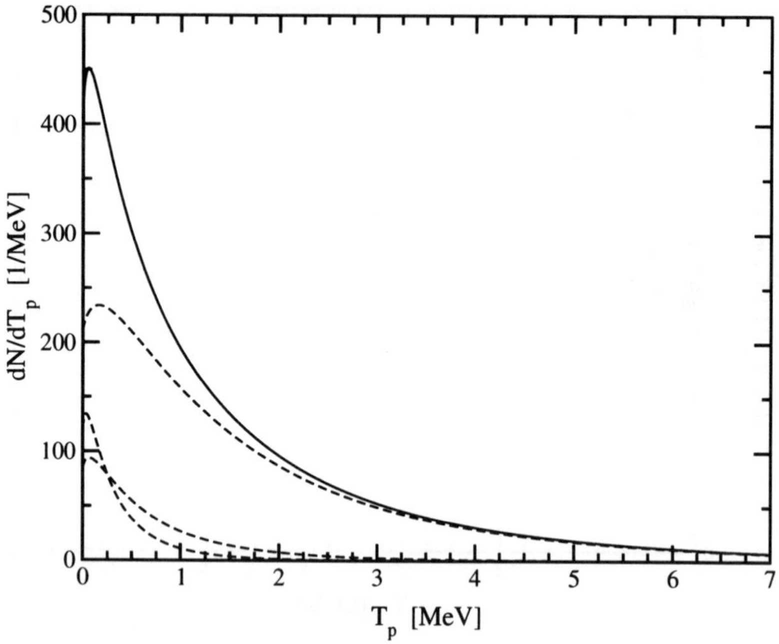

Figure 2. The true proton spectrum in KamLAND, for a standard supernova at 10 kpc. In order of increasing maximum kinetic energy, the contributions from ν_e, $\bar{\nu}_e$, and the sum of ν_μ, ν_τ, $\bar{\nu}_\mu$, and $\bar{\nu}_\tau$ are shown with dashed lines. The solid line is the sum spectrum for all flavors. Taking the detector properties into account substantially modifies these results, as shown below.

positions can be readily detected in scintillator detectors such as KamLAND and Borexino. We first consider the true proton spectrum, and then we consider how this spectrum would appear in a realistic detector. The true proton spectrum is given by

$$\frac{dN}{dT_p}(T_p) = C \int_{(E_\nu)_{min}}^{\infty} dE_\nu \, f(E_\nu) \, \frac{d\sigma}{dT_p}(E_\nu, T_p) , \tag{5}$$

where $f(E_\nu)$ is a normalized Fermi-Dirac spectrum and the differential cross section is given above. For a given T_p, the minimum required neutrino energy is

$$(E_\nu)_{min} = \frac{T_p + \sqrt{T_p(T_p + 2M_p)}}{2} \simeq \sqrt{\frac{M_p T_p}{2}} . \tag{6}$$

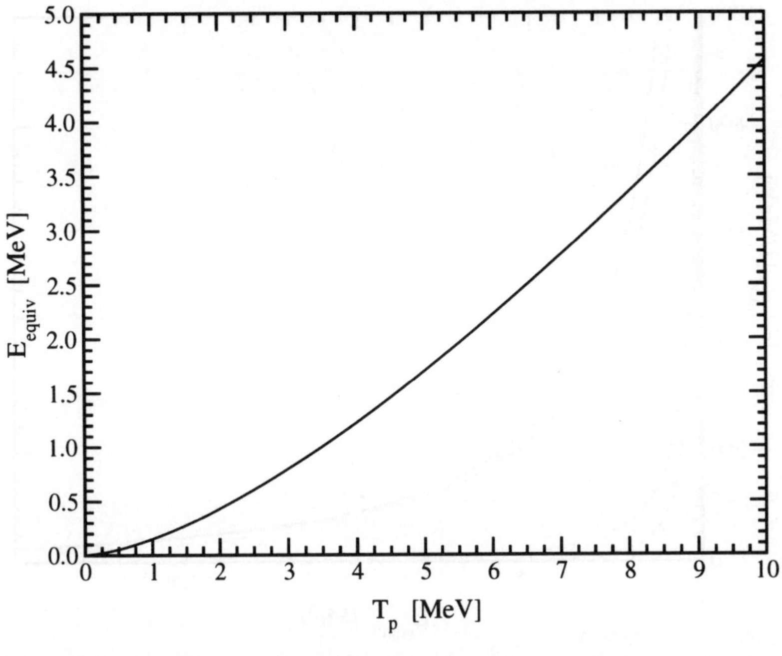

Figure 3. The quenched energy deposit (equivalent electron energy) as a function of the proton kinetic energy. The KamLAND detector properties are assumed.

For highly ionizing particles like low-energy protons, the light output is reduced or "quenched" relative to the light output for an electron depositing the same amount of energy. The observable light output E_{equiv} (i.e., equivalent to an electron of energy E_{equiv}) is given by Birk's Law:

$$\frac{dE_{equiv}}{dx} = \frac{dE/dx}{1 + k_B(dE/dx)} \tag{7}$$

where k_B is a constant of the scintillation material, and dE/dx is the energy deposition rate, now in MeV/cm (of opposite sign to the loss rate). We assume $k_B \simeq 0.015$ cm/MeV for KamLAND. For small dE/dx, the measured light output of a proton is equivalent to an electron of the same energy. But for $dE/dx \sim 100$ MeV/cm, the two terms in the denominator are comparable, and the light output is reduced. At still higher dE/dx, then dE_{equiv}/dx tends to a constant. Birk's Law can thus reflects a saturation effect: once dE/dx is large, making it larger does not increase the light output. Effectively, if all

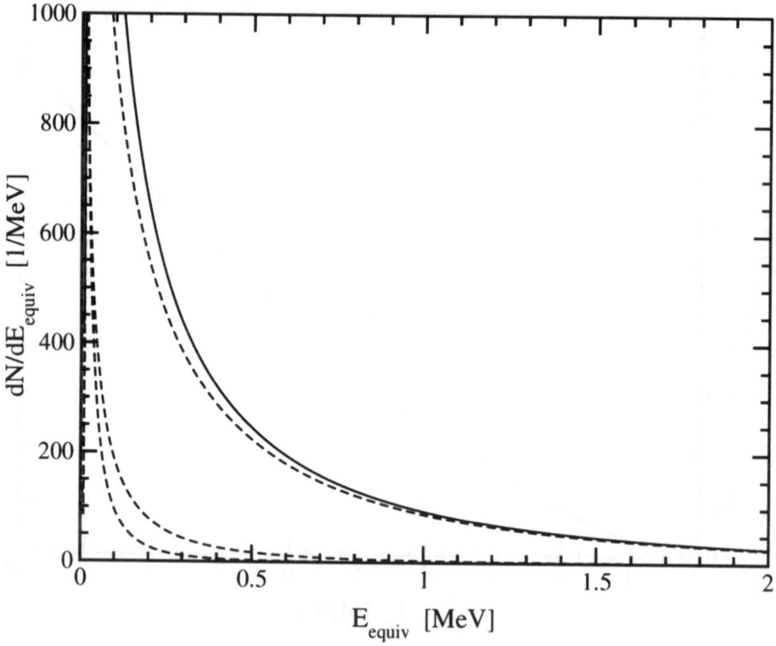

Figure 4. The struck proton spectrum for the different flavors, with quenching effects taken into account. In order of increasing maximum kinetic energy, the contributions from ν_e, $\bar{\nu}_e$, and the sum of ν_μ, ν_τ, $\bar{\nu}_\mu$, and $\bar{\nu}_\tau$ are shown with dashed lines. The solid line is the sum spectrum for all flavors. The anticipated KamLAND threshold is 0.2 MeV electron equivalent energy.

scintillation molecules are already excited, any further energy deposition is not converted to visible scintillation light.

The measured proton spectrum can be used to separately determine the total flux of ν_μ, ν_τ, $\bar{\nu}_\mu$, and $\bar{\nu}_\tau$ neutrinos *and* their time-averaged temperature. The total number of detected events is proportional to the portion of the total binding energy carried away by these four flavors, and we denote this by $E^{tot}_{4\nu_x}$. We do not have to assume that it is equal to $4(E_B/6) = 2/3E_B \simeq 2 \times 10^{53}$ ergs; it can be measured. We denote the temperature of these four flavors by T. If only the total yield were measured, as for most neutral-current reactions, there would be an unresolved degeneracy between $E^{tot}_{4\nu_x}$ and T, since

$$N \sim E^{tot}_{4\nu_x} \frac{\langle\sigma\rangle}{T_\nu} \,. \tag{8}$$

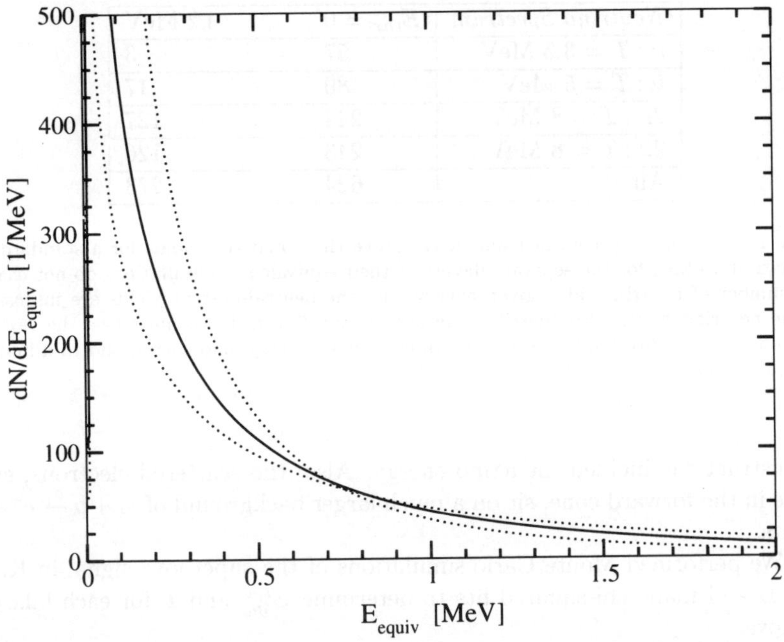

Figure 5. Example spectra with different values of $E^{tot}_{4\nu_x}$ and T, all chosen to give the *same* number of events above an electron equivalent threshold of 0.2 MeV (true proton energy 1.2 MeV). At the 0.2 MeV point, from left to right these correspond to $(E^{tot}_{4\nu_x}, T) = (4.2, 6), (2.0, 8), (1.4, 10)$, respectively, with $E^{tot}_{4\nu_x}$ in 10^{53} ergs and T in MeV.

Note that for $\sigma \sim E_\nu^n$, then $\langle\sigma\rangle \sim T^n$. For $\nu + d \to \nu + p + n$ in SNO, for example, $\sigma \sim E^2$, so $N \sim E^{tot}_{4\nu_x} T$. Thus for a given measured number of events, one would only be able to define a hyperbola in the plane of $E^{tot}_{4\nu_x}$ and T. The scaling is less simple here because of threshold effects, but the idea is the same.

Here we have crucial information on the shape of the neutrino spectrum, revealed through the proton spectrum. To remind the reader, in most neutral-current reactions there is *no* information on the neutrino energy, e.g., one only counts the numbers of thermalized neutron captures, or measures nuclear gamma rays (the energies of which depend only on nuclear level splittings). Neutrino-electron scattering is an exception. However, the distribution of electron energies for a given neutrino energy is very broad, and the electron angle can only be measured to about 25 degrees, which is too large to adequately

Neutrino Spectrum	$E_{thr} = 0$	0.2 MeV
$\nu : T = 3.5$ MeV	57	3
$\bar{\nu} : T = 5$ MeV	80	17
$2\nu : T = 8$ MeV	244	127
$2\bar{\nu} : T = 8$ MeV	243	126
All	624	273

Table 1. Numbers of events in KamLAND above the noted thresholds for a standard supernova at 10 kpc, for the separate flavors or their equivalents. Oscillations do not change the number of neutrinos at a given energy, and the neutral-current yields are insensitive to the neutrino flavor. Equipartition among the six flavors is assumed (see the text for discussion). The thresholds are in electron equivalent energy, and correspond to minimum true proton kinetic energies of 0 and 1.2 MeV.

reconstruct the incident neutrino energy. Also, the scattered electrons, even those in the forward cone, sit on a much larger background of $\bar{\nu}_e + p \rightarrow e^+ + n$ events.

We performed Monte Carlo simulations of the supernova signal in KamLAND and made chi-squared fits to determine $E^{tot}_{4\nu_x}$ and T for each fake supernova.

Three examples are shown in Fig. 6, where one can see that $E^{tot}_{4\nu_x}$ and T_ν can each be determined with roughly 10% error. These errors scale as $1/\sqrt{N}$, where N is the total number of events (i.e., if one imagines a detector of a different mass or a different assumed supernova distance). If the distance were completely uncertain, one would not be able to determine $E^{tot}_{4\nu_x}$. However, after marginalizing over the unknown $E^{tot}_{4\nu_x}$ (i.e., projecting these scatterplots onto the T axis), one would still obtain a good measurement of T.

3 Conclusions

It is important to note that the detection of recoil protons from *neutron*-proton elastic scattering at several MeV has been routinely accomplished in scintillator detectors. Since both particles are massive, the proton will typically take half of the neutron energy. This reaction provides protons in the same energy range as those struck in neutrino-proton elastic scattering with $E_\nu \sim 30$ MeV. This is an important proof of concept for all aspects of the detection of low-energy protons.

Though low-energy backgrounds will be challenging, it is also important to note that the background requirements for detecting the supernova signal are approximately 3 orders of magnitude *less* stringent than those required

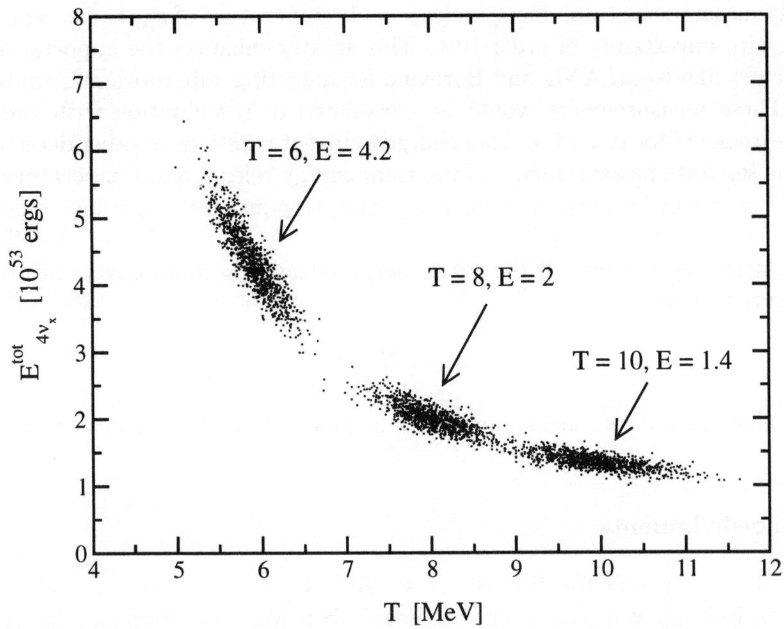

Figure 6. Scatterplot of 10^3 fitted values, in the $E^{tot}_{4\nu_x}$ and T_ν plane, for the labeled "true" values, where $E^{tot}_{4\nu_x}$ is the total portion of the binding energy carried away by the sum of ν_μ, ν_τ, $\bar{\nu}_\mu$, and $\bar{\nu}_\tau$, and T is their temperature. The values of $E^{tot}_{4\nu_x}$ and T_ν were chosen such that the numbers of events above threshold were the same. The measured shape of the proton spectrum breaks the degeneracy between these two parameters. Without that spectral information, one could not distinguish between combinations of $E^{tot}_{4\nu_x}$ and T_ν along the band in this plane that our three examples regions lie along.

for detecting solar neutrinos in the same energy range (taking quenching into account for our signal). Borexino has been designed to detect very low-energy solar neutrinos, and KamLAND hopes to do so in a later phase of their experiment.

We have shown that neutrino-proton elastic scattering, previously unrecognized as a useful detection reaction for low-energy neutrinos, in fact has a yield comparable to $\bar{\nu}_e + p \to e^+ + n$, even after taking into account the quenching of the proton scintillation light and assuming a realistic detector threshold.

In addition, the measured proton spectrum shape is closely related to the incident neutrino spectrum. We have shown explicitly that one can separately measure the total energy release and temperature of ν_μ, ν_τ, $\bar{\nu}_\mu$, and $\bar{\nu}_\tau$, each with uncertainty of order 10%. This greatly enhances the importance of detectors like KamLAND and Borexino for detecting supernova neutrinos.

These measurements would be considered in combination with similar measurements for ν_e and $\bar{\nu}_e$ from charged-current reactions in other detectors. These separate measurements of the total energy release and temperature for each flavor will be invaluable for comparing to numerical supernova models. They will also be required to make model-independent studies of the effects of neutrino oscillations. If the total energy release E_B in all flavors has been measured, then

$$E_B \simeq \frac{3}{5}\frac{GM_{NS}^2}{R_{NS}}, \tag{9}$$

thus allowing a direct measurement of the newly-formed neutron star properties.

Acknowledgments

I thank many people for discussions; see Ref. [1]. In particular, my collaborators on Ref. [1], Will Farr and Petr Vogel. JFB was supported as a Sherman Fairchild Fellow at Caltech during the initial part of this project, and as the David N. Schramm Fellow at Fermilab during the final part. Fermilab is operated by URA under DOE contract No. DE-AC02-76CH03000. JFB was additionally supported by NASA under NAG5-10842.

References

1. J.F. Beacom, W.M. Farr, and P. Vogel, in a paper to appear on www.arxiv.org in mid-May 2002.
2. Since space is limited, all references can be found in Ref. [1].

List of Participants for NOON2001

Koh ABE
University of Tokyo

abe@icepp.s.u-tokyo.ac.jp

Kazuaki ANRAKU
JSF Faculty of Engineering,
Kanagawa Univ.
anraku@phuz.b6.kanagawa-u.ac.jp

Mayumi AOKI
Theory Group
KEK
mayumi.aoki@kek.jp

Jiro ARAFUNE
NIAD

arafune@niad.ac.jp

Masako BANDO
Aichi University

bando@aichi-u.ac.jp

Alexander BARABASH
Institute of Theoretical and Experimental Physics
barabash@vxitep.itep.ru

Stephen BARR
Bartol Research Institute
University of Delaware
smbarr@bartol.udel.edu

John BEACOM
Theoretical Astrophysics,
MS-209 Fermilab
beacom@fnal.gov

Rita BERNABEI
Dip. di Fisica Universita' di Roma
"Tor Vergata"
bernabei@roma2.infn.it

Mario CAMPANELLI
DPNC,
University of Geneva
Mario.Campanelli@cern.ch

Vitaly CHEPEL
LIP-Coimbra Department of Physics,
University of Coimbra
vitaly@lipc.fis.uc.pt

Tadayoshi DOKE
Advanced Research Institute for Science and Engineering, Waseda University
tdoke@mn.waseda.ac.jp

Masaaki FUJII
University of Tokyo

fujii@hep-th.phys.s.u-tokyo.ac.jp

Yoshiyuki FUKUDA
Kamioka Observatory ICRR,
University of Tokyo
fukuda@icrr.u-tokyo.ac.jp

Yoshihito GANDO
Reserch Center for Neutrino Science
Graduate School of Science Tohoku Univ.
gando@awa.tohoku.ac.jp

Jesse GOLDMAN
Research Center for Neutrino Science
Tohoku University
goldman@awa.tohoku.ac.jp

Concha GONZALEZ-GARCIA
CERN

concepcion.gonzalez-garcia@cern.ch

Gene GUILLIAN
University of Maryland
/ Super-Kamiokande
guillian@suketto.icrr.u-tokyo.ac.jp

Jim HILL
ICRR, University of Tokyo and
State Univ. of NY at Stony Brook
jimhill@neutrino.kek.jp

Junji HISANO
KEK

junji.hisano@kek.jp

Morihiro HONDA
ICRR
University of Tokyo
mhonda@icrr.u-tokyo.ac.jp

Takanobu ISHII
IPNS
KEK
takanobu.ishii@kek.jp

Masaki ISHITSUKA
Kamioka Observatory ICRR,
University of Tokyo
ishi@suketto.icrr.u-tokyo.ac.jp

Takaaki KAJITA
ICRR,
University of Tokyo
kajita@icrr.u-tokyo.ac.jp

Mitsuru KAKIZAKI
Tohoku University

kakizaki@tuhep.phys.tohoku.ac.jp

Jun KAMEDA
Kamioka Observatory ICRR,
University of Tokyo
kameda@suketto.icrr.u-tokyo.ac.jp

Satoru KANEKO
Department of physics,
Niigata University
kaneko@muse.hep.sc.niigata-u.ac.jp

Jiro KAWADA
Nagoya University

kawad@flab.phys.nagoya-u.ac.jp

Yoshio KOIDE
Dept of Physics,
University of Shizuoka
koide@u-shizuoka-ken.ac.jp

Akira KONAKA
TRIUMF

konaka@triumf.ca

Yusuke KOSHIO
Kamioka Observatory ICRR,
University of Tokyo
koshio@icrr.u-tokyo.ac.jp

C.S. LIM
Dept. of Physics
Kobe University
lim@phys.sci.kobe-u.ac.jp

Paolo LIPARI
INFN
Unversita di Roma
paolo.lipari@roma1.infn.it

Nobuhiro MAEKAWA
Kyoto University

maekawa@gauge.scphys.kyoto-u.ac.jp

Antonio MARRONE
Dipartimento di Fisica
Universita' di Bari
antonio.marrone@ba.infn.it

Mark MESSIER
Harvard University

messier@huhepl.harvard.edu

Shoichi MIDORIKAWA
Faculty of Engineering,
Aomori University
midori@aomori-u.ac.jp

Satoshi MIHARA
ICEPP,
University of Tokyo
mihara@icepp.s.u-tokyo.ac.jp

Makoto MIURA
Kamioka Observatory ICRR,
University of Tokyo
miuram@suketto.icr.u-tokyo.ac.jp

Kohei MIZUTANI
Department of Physics
Saitama University
mizutani@cr.phy.saitama-u.ac.jp

Ben MORGAN
U.K. Dark Matter Collaboration Dept. of
Physics and Astronomy, Univ. of Sheffield
b.morgan@sheffield.ac.uk

Toshinori MORI
ICEPP,
University of Tokyo
mori@icepp.s.u-tokyo.ac.jp

Shigetaka MORIYAMA
Kamioka Observatory ICRR,
University of Tokyo
moriyama@icrr.u-tokyo.ac.jp

Takuya MOROZUMI
Hiroshima University

morozumi@theo.phys.sci.hiroshima-u.ac.jp

Tohru MOTOBAYASHI
Dept. Phys.
Rikkyo University
motobaya@rikkyo.ne.jp

Masayuki NAKAHATA
Kamioka observatory ICRR,
University of Tokyo
nakahata@suketto.icrr.u-tokyo.ac.jp

Kenzo NAKAMURA
KEK

kenzo.nakamura@kek.jp

Shogo NAKAMURA
Faculty of Engineering
Yokohama National University
snakam@ynu.ac.jp

Toshio NAMBA
Kamioka Observatory ICRR,
University of Tokyo
namba@suketto.icrr.u-tokyo.ac.jp

Kouji NARITA
Nagoya University

narita@flab.phys.nagoya-u.ac.jp

Kyoshi NISHIJIMA
Department of Physics
Tokai University
kyoshi@tkikam.sp.u-tokai.ac.jp

Jun NISHIMURA
University of Tokyo

nisimura@icrr.u-tokyo.ac.jp

Kazunori NITTA
Kamioka Observatory
Osaka University
nitta@suketto.icrr.u-tokyo.ac.jp

Takayoshi OHSHIMA
Nagoya University

ohshima@hepl.phys.nagoya-u.ac.jp

Atsushi OKADA
ICRR
University of Tokyo
okada@icrr.u-tokyo.ac.jp

Naotoshi OKAMURA
Department of Physics
Virginia Tech
nokamura@vt.edu

Wataru OOTANI
ICEPP,
University of Tokyo
wataru@icepp.s.u-tokyo.ac.jp

Takato OOYABU
ICRR
University of Tokyo
ooyabu@suketto.icrr.u-tokyo.ac.jp

Scott OSER
Sudbury Neutrino Observatory
University of Pennsylvania
oser@hep.upenn.edu

Heinrich PAES
Institut fuer Theoretische
Physik University of Wuerzburg
paes@physik.uni-wuerzburg.de

Serguey PETCOV
Elementary Particle Physics
Sector SISSA/INFN
petcov@he.sissa.it

Tomoyuki SANUKI
Tokyo University

sanuki@phys.s.u-tokyo.ac.jp

Joe SATO
Research Center for Higher Education
Kyushu University
joe@rc.kyushu-u.ac.jp

Toru SATO
Department of Physics Graduate School of Science, Osaka University
tsato@phys.sci.osaka-u.ac.jp

Stefan SCHOENERT
Max-Planck-Institute for Nuclear Physics

Stefan.Schoenert@mpi-hd.mpg.de

Qaisar SHAFI
Bartol Research Institute

shafi@bxsvr0.bartol.udel.edu

Noriyuki SHIMOYAMA
University of Niigata

simoyama@muse.hep.sc.niigata-u.ac.jp

Michael SMY
Department of Physics and Astronomy
University of California, Irvine
smy@solar1.ps.uci.edu

Neil SPOONER
U.K. Dark Matter Collaboration Dept. of Physics and Astronomy, Univ. of Sheffield
n.spooner@sheffield.ac.uk

Satoshi SUZUKI
Advanced Research Institute for Science and Engineering, Waseda University
satoshis@mu.waseda.ac.jp

Yoichiro SUZUKI
Kamioka Observatory ICRR,
University of Tokyo
suzuki@suketto.icrr.u-tokyo.ac.jp

Eiichi TAKASUGI
Dept. of Physics,
Osaka University
takasugi@het.phys.sci.osaka-u.ac.jp

Yasuo TAKEUCHI
Kamioka Observatory ICRR,
University of Tokyo
takeuchi@icrr.u-tokyo.ac.jp

Morimitsu TANIMOTO
Physics Department,
Niigata University
tanimoto@muse.hep.sc.niigata-u.ac.jp

Zurab TAVARTKILADZE
Institute for theoretical Physics
University of Heidelberg
Z.Tavartkiladze@ThPhys.Uni-Heidelberg.DE

Masahiro TOKIWA
Okayama university

tokiwa@cr2000.phys.okayama-u.ac.jp

Shoji TORII
Institute of Physics,
Kanagawa University
torii@phu2.b6.kanagawa-u.ac.jp

Toshiyuki TOSHITO
Nagoya University

toshi@flab.phys.nagoya-u.ac.jp

Mark VAGINS
University of California,
Irvine
mvagins@uci.edu

Wei WANG
Boston University

ww@bu.edu

Satoru WATANABE
Department of Astronomy,
University of Tokyo
watanabe@astron.s.u-tokyo.ac.jp

Taizan WATARI
University of Tokyo

watari@hep-th.phys.s.u-tokyo.ac.jp

Masahiro YAMAGUCHI
Tohoku University

yama@tuhep.phys.tohoku.ac.jp

Katsuji YAMAMOTO
Dept.of Nuclear Engineering
Kyoto University
yamamoto@nucleng.kyoto-u.ac.jp

Masaki YAMASHITA
Advanced Research Institute for Science and
Engineering, Waseda University
60015273@mse.waseda.ac.jp

Osamu YASUDA
Department of Physics
Tokyo Metropolitan University
yasuda@phys.metro-u.ac.jp

Masaki YASUE
Department of Physics
Tokai University
yasue@keyaki.cc.u-tokai.ac.jp

Kenji YOSHIDA
Faculty of Engineering
Kanagawa University
yoshida@kit.ie.kanagawa-u.ac.jp

Tetsuya YOSHIDA
IPNS,
KEK
tetsuya.yoshida@kek.jp

Koji YOSHIMURA
IPNS,
KEK
koji.yoshimura@kek.jp

Yuri ZDESENKO
Lepton Physics Department Institute for Nuclear Research National Academy of Sciences of Ukraine
zdesenko@kinr.kiev.ua

The 3rd Workshop on

Neutrino Oscillations and Their Origin (NOON2001)

ICRR, University of Tokyo, Kashiwa, Japan, December 5–8, 2001

SCIENTIFIC PROGRAMME

December 5 (Wednesday)

Solar neutrinos

9:00– 9:30	Super–Kamiokande	**Y. Koshio** (ICRR, U. Tokyo)
9:30–10:00	SNO	**S. Oser** (Penn)
10:00–10:30	Recent status of astrophysical S17	**T. Motobayashi** (Rikkyo)
	Coffee break	
11:00–11:20	Weak current in Deuterium and SNO experiment	**T. Sato** (Osaka)
11:20–11:50	Solar model based on Helioseismology	**S. Watanabe** (Tokyo)
	Lunch	

Solar neutrino phenomenology and future

14:00–14:30	Two flavor solar neutrino global analysis I	**M. Smy** (UCI)
14:30–15:00	Two flavor solar neutrino global analysis II	**M. C. Gonzalez–Garcia** (CERN)
15:00–15:15	Comment	**A. Marrone** (Bari)
15:15–15:45	The status of Resonant Spin Flavor Precession	**C. S. Lim** (Kobe)
	Coffee break	
16:20–16:40	KamLAND	**J. Goldman** (Tohoku)
16:40–17:10	Future Low energy solar neutrino measurements	**S. Schoenert** (MPI)
17:10–17:30	Value and sensitivity of the future low energy solar neutrino experiment	**M. Nakahata** (ICRR, U. Tokyo)

December 6 (Thursday)

Atmospheric neutrinos

9:00– 9:30	Present status of the atmospheric neutrino studies	**M. Messier** (Harvard)
9:30–10:00	Primary cosmic ray measurements	**K. Abe** (Tokyo)
	Coffee break	
10:30–11:00	Primary proton spectrum of 0.4–30TeV deconvolved from atmospheric gamma-ray spectrum at balloon ltitude	**K. Yoshida** (Kanagawa)
11:00–11:30	Systematic uncertainty in calculation of the atmospheric neutrino flux	**P. Lipari** (Rome)
	Lunch	

Oscillation phenomenology I

13:30–14:00	Oscillation-enhanced search for new interaction with neutrinos	**J. Sato** (Kyushu)
14:00–14:25	Three & four flavor analysis of solar and atmospheric neutrinos	**M. C. Gonzalez-Garcia** (CERN)
14:25–14:50	Three flavor analysis of solar and atmospheric neutrinos	**A. Marrone** (Bari)
	Coffee break	
15:20–15:50	Implications of the recent results of the solar neutrino experiments	**S. T. Petcov** (SISSA)

Absolute neutrino mass

15:50–16:20	Double beta decay and neutrino oscillation	**H. V. Klapdor-Kleingrothaus** (MPI)
16:20–16:50	Absolute neutrino mass determination	**H. Paes** (Wuerzburg)

December 7 (Friday)

Accelerator neutrinos, CPV

8:30– 9:00	K2K	**J. Hill** (ICRR & SUNY, StonyBrook)
9:00– 9:30	MINOS	**M. Messier** (Harvard)
9:30–10:00	Future CERN-based accelerator experiments	**M. Campanelli** (CERN)

Coffee break

10:30–11:00	JHF-neutrino experiment	**T. Kajita** (ICRR, U. Tokyo)
11:00–11:30	Prospects of future long baseline neutrino oscillation experiments using KEK-JAERI HIPA	**M. Aoki** (KEK)
11:30–12:00	Neutrino Factory; physics potentials and present status	**O. Yasuda** (Tokyo Metro.)

Lunch

Models and GUTs

13:50–14:30	Neutrino oscillations and extra dimensions	**Q. Shafi** (Bartol)
14:30–15:00	Semi-simple unification	**T. Watari** (Tokyo)
15:00–15:30	Leptogenesis via LHu flat direction	**M. Fujii** (Tokyo)

Coffee break

16:00–16:30	Predictive texture of realistic neutrino mixings	**Z. Tavartkilads** (WAI)
16:30–17:00	The GUT? neutrino bi-large mixing and proton decay	**N. Maekawa** (Kyoto)
17:00–17:30	Neutrino bimaximal texture and mu into e gamma	**N. Shimoyama** (Niigata)

Conference Dinner

December 8 (Saturday)

Lepton flavor violation

8:30– 9:00	tau -> mu gamma: status and prospects	**T. Ohshima** (Nagoya)
9:00– 9:30	mu -> e conversion: status and prospects	**K. Yoshimura** (KEK)
9:30–10:00	mu -> e gamma: status and prospects	**W. Ootani** (ICEPP, U. Tokyo)

Coffee break

10:30–11:00	Lepton physics in super symmetric models	**J. Hisano** (KEK)

Oscillation phenomenology II

11:00–11:30	Classification of neutrino-mixing models and bimaximal mixing	**S. Barr** (Bartol)

Supernova neutrinos

11:30–12:00	Supernova Neutrinos	**J. Beacom** (Fermilab)